Extra Class

FCC License Preparation for Element 4 Extra Class Theory

by
Gordon West
WB6NOA

with Technical Editor
Eric P. Nichols
KL7AJ

Ninth Edition

ARRL The National Association for Amateur Radio®
225 Main Street,
Newington, CT 06111-1400 USA
www.arrl.org

Cartoons by:
Jim Massara, N2EST, *based on the original Elmer by* Carson Haring, AC0BU

Thanks to the following for their assistance with this book: Suzy, N6GLF; Bob, KO6LU; George, W9DUP; Chip, K7JA; Jim, AG6IF; Jim, K3WNG; Joel, N4ALI; "RK", N1RKH; Mans, N6ALE; Andrew, N5ASE; Andrew, KJ5AOJ; Herb, KN4NTW; Bill, KC9TPR; Marv, KD9GJV; Gary, AA6GJ; Stephen, KE8TEY.

Photograph Credit:
All photographs that do not have a source identification are either courtesy of the authors or Chip Margelli, K7JA.

Editorial Team:
Acquisitions Editor
Mark Derks, KC1RVQ

Book Editor
Makenzie Ozycz

Printed in the USA

ISBN: 978-1-62595-195-3
First edition 1994
Ninth Edition and first ARRL edition 2024
28 27 26 25 2 3 4 5 6 7 8 9

About the Author

Gordon West, WB6NOA

Gordon West has been a ham radio operator for more than 60 years, holding the top Extra Class license, call sign WB6NOA. He also holds the highest FCC commercial operator license, the First Class General Radiotelephone Certificate with Radar Endorsement.

Gordon teaches evening ham radio classes and offers weekend ham radio licensing seminars on a monthly schedule. These seminars cover entry-level and upgrade licenses in ham radio. He has served on the faculty of Coastline College and Orange Coast College.

Gordon is a regular contributor to amateur radio, marine, and general two-way radio magazines. He is a fellow of the Radio Club of America, and has been a life member of the American Radio Relay for 50 years. The ARRL presented Gordon with its "Instructor of the Year" award. The Dayton Amateur Radio Association named Gordon their "2006 Amateur of the Year" for his efforts in recruiting and training new amateurs, in addition to his lifelong involvement in ham radio. Through his Gordon West Radio School, he has trained eight out of ten newly-licensed hams with his classes, books and audio courses over the past 50 years.

TWO GREAT HAMS!
TWO GREAT EDUCATORS!

We remember a pair of League members who helped shape the BEST in training — as an Extra Class operator, you too can become a training legend!

Bob Heil, K9EID, is most famous for his teaching about microphones and receiver parametric equalizers. Bob, Joe Walsh, and I enjoyed years on our video podcast, HAM NATION. See Bob's work in the Rock and Roll Hall of Fame! At 13 years old, he was building his own gear from scratch, and was front cover of QST Magazine with an early "home brew".

Chip Margelli, K7JA, loved "field operating", and training new hams how to set up portable, for total wireless, to work the world. His love of Morse code put him on the Jay Leno Tonight show, for the world to see how his simple dots and dashes could beat out text messaging. He won with his legendary smile!

Countless hams have devoted their time and talent to help all of us learn from these experts! Let's follow these leaders, on the air, with fun operating skills!

Eric P. Nichols, KL7AJ
Technical Editor

Technical Editor Eric P. Nichols, KL7AJ Eric has been a licensed amateur operator since 1972, holding the top Extra Class license, as well as the FCC GROL commercial license. He has spent his career in various aspects of radio research, broadcast communications, and teaching.

A prolific author, he has written many articles for ***QST, QEX,*** and a number of other ham radio magazines, as well as professional journals. In 2010 and again in 2014 he was the recipient of the William Orr, W6SAI, Technical Writing Award conferred by the American Radio Relay League for articles he authored for ***QST*** magazine. He is the author of ***Radio Science for the Radio Amateur***, and ***The Opus of Amateur Radio Knowledge and Lore***. Along with Gordon West, WB6NOA, he is Technical Editor of the ***GROL+RADAR*** study manual for the FCC commercial radio licenses.

Eric lives in Fairbanks, AK. He has operated just about every mode available to the radio amateur, but always returns to CW on the lower HF bands as his "default" mode. He enjoys restoring vintage "boat anchor" radio equipment as well as designing cutting edge radio instrumentation with Arduino, his latest obsession. Being in interior Alaska, he gets a lot of requests from new hams (and some older ones) for a first-time Alaskan radio contact, and is always willing to try a "sked" on any HF band or mode!

Table of Contents

QUESTION POOL NOMENCLATURE
The latest nomenclature changes recommended by the Volunteer Examiner Coordinator's Question Pool Committee (QPC) for the Element 4 question pool effective July 1, 2024 and valid to June 30, 2028 have been incorporated in this book.

FCC RULES, REGULATIONS AND POLICIES
The NCVEC QPC releases revised question pools on a regular cycle, and deletions as necessary. The FCC releases changes to FCC rules, regulations and policies as they are implemented. This book includes the most recent information released by the FCC when this copy was printed.

Preface

Amateur Extra Class is the highest-level Federal Communications Commission license you can hold in the amateur radio service. Achieving the Extra Class license opens up exclusive sub-bands for voice operation on worldwide frequencies, along with other exclusive Extra Class sub-bands for Morse code and data. Your Extra Class license gives you full operating privileges across the complete spectrum of every ham radio band!

On February 23, 2007, all Morse code testing for amateur radio licenses was completely eliminated by the FCC. From that time until now with the elimination of the code "stress test" a strange phenomenon has occurred. More hams than ever are learning the code and practicing CW over the worldwide airwaves. We encourage you to learn the code. It's fun!

Achieving an Extra Class license allows you to administer all amateur radio exams. As an Extra Class licensee, you may be accredited as a Volunteer Examiner to prepare and administer ham exams as a member of a 3-person VE team. You also could qualify to become a contact Volunteer Examiner to help coordinate ham exam teams in your area. This is a great way to help our hobby and service grow!

The new Element 4 Extra Class question pool included in this book is valid for written examinations from July 1, 2024 through June 30, 2028. The total number of questions in the new pool is about the same as in years past – with 50 of these multiple choice questions appearing on the Extra Class exam. No additional math complexity has been added to this fresh new pool, but there are many updated questions on newer operating modes and techniques. Our book explains everything about the Extra Class exam:

- ***Chapter 1*** reviews the operating privileges you'll earn when you pass the Extra Class exam.
- ***Chapter 2*** is a brief review of the history of ham radio licensing and explains the exam requirements for all of the current FCC Amateur Radio licenses.
- ***Chapter 3*** provides information on preparing for and taking the Element 4 exam.
- ***Chapter 4*** contains all 603 Element 4 questions and answers, along with our description of the correct answer, followed by the correct answer key.
- ***Chapter 5*** tells you what to expect on examination day and reviews the latest FCC e-filing requirements.

Eric and I welcome Advanced and General Class operators to Extra Class exam preparation! Get set for new privileges and increased VE opportunities. We hope to hear you on our top bands very soon!

Gordon West, WB6NOA   *Eric P. Nichols, KL7AJ*

1

Extra Class Privileges

The Extra Class amateur license permits you to use all segments of the worldwide bands designated for amateur communications for long-distance voice, data, and video communications and expands your General or Advanced Class operating privileges. Of course, you keep all of the privileges you enjoyed as a General or Advanced Class operator.

When you pass your Extra Class examination, you will gain a 25-kHz "window" of exclusive CW privileges at the bottom end of each of the 80, 40, 20, and 15 meter bands. You also gain a total of 150 kHz of additional "windows" of exclusive Extra Class phone privileges on 75, 20, and 15 meters. You will gain valuable DX operating privileges on the Extra Class bands.

Extra Class has always been recognized as the "top" amateur radio license because of the unlimited skywave privileges it provides on each of the worldwide ham bands. Earning Extra Class privileges requires passing written exams for Element 2, Technician Class, and Element 3, General Class, prior to taking your Element 4 exam.

Never in the history of amateur radio has it been easier to achieve an Extra Class license. There is no longer a Morse code test required for the Extra Class license. In February, 2007, the Federal Communications Commission ended code test requirements for all amateur radio licenses.

Just because the code test has been eliminated, it doesn't mean earning your Extra Class license will be a snap. For most of us, it requires time dedicated to developing our operating skills and means spending some serious time diligently studying the electronic theory and operating techniques covered in the questions. Becoming an Extra Class means you've reached the top of our hobby and service.

The Extra Class license allows you to give something back to the amateur service as a volunteer who conducts examinations as part of a 3-person accredited Volunteer Examiner team. As an accredited volunteer, you are qualified to prepare and administer all levels of amateur radio exams. You can help recruit and organize other hams to serve on your VE team. This is a great opportunity to help our hobby grow and prosper!

As an Extra Class operator, you will be eligible for the short call signs that are reserved exclusively for those amateurs who have made it to the very highest level of license in the amateur service. You may wish to take advantage of the Vanity Call Sign program to select specific call letters of your choice.

Best of all, when you achieve your Extra Class license, you do not have to prepare for anymore license upgrade tests. You're the top!

Now, let's take a look at privileges of the Extra Class license.

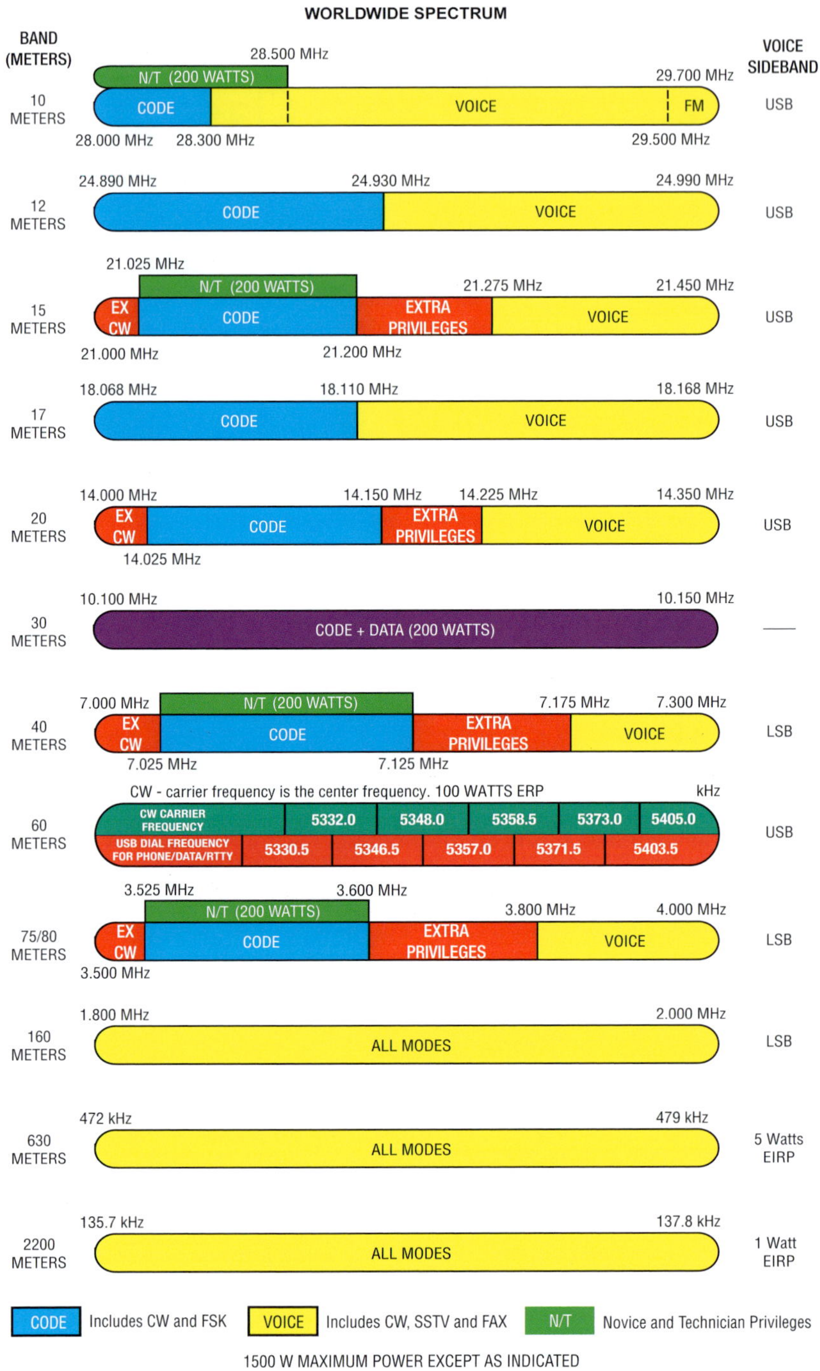

Figure 1-1. Extra Class Privileges

EXTRA CLASS LICENSE PRIVILEGES

Figure 1-1 graphically illustrates your Extra Class code, data, and voice privileges on low frequency (LF) (30 kHz – 300 kHz), medium frequency (MF) (300 kHz – 3 MHz), and high frequency (HF) (3 MHz – 30 MHz) bands. Code and data privileges are in the designated areas on the left side of each band edge. Voice privileges are on the right side of each band, other than 30 meters, which is reserved exclusively for CW and data.

Figure 1-1 includes areas designated in red "EX" on the 15, 20, 40, and 80 meter bands where Extra Class licensees have operating privileges. In other words, *as an Extra Class operator, you get everything!*

2200 AND 630 METERS

For home brew experimenters, we now have two new bands below the AM broadcast band! You will need to build your own gear!

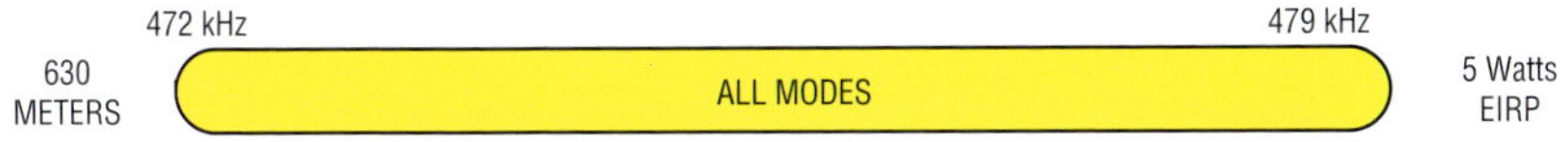

160 METERS: 1.8 MHz - 2.0 MHz.

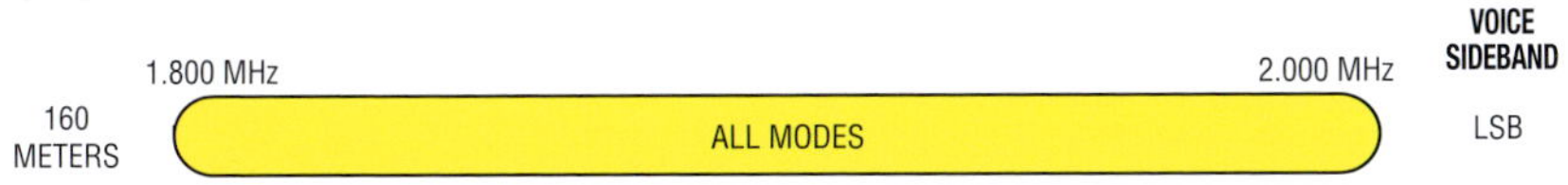

Extra Class privileges let you operate voice and code from one end to the other of the 160 meter band. These privileges are shared with General and Advanced Class operators. The 160 meter band is great for long-distance nighttime communications. At night, you work the world on this band.

75/80 METERS: 3.5 MHz - 4.0 MHz.

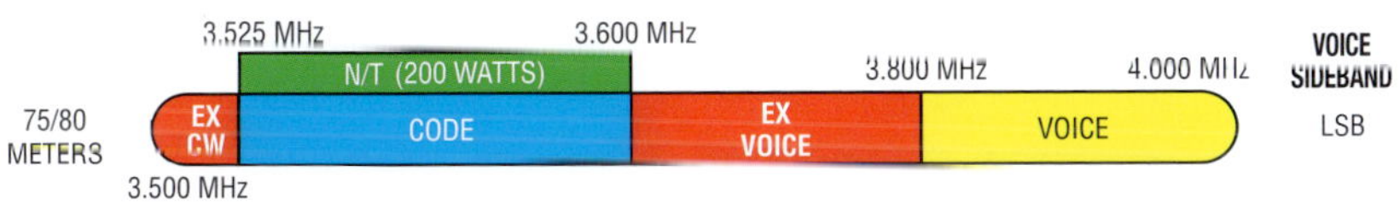

Extra Class privileges on 75/80 meters give you exclusive voice privileges that were recently expanded to include 3.600 to 3.800 MHz. This part of the band is not shared with any other ham license class and gives the Extra Class operator a huge increase for nighttime DX voice contacts; some well over 6000 miles away. Extra Class operators also have exclusive CW privileges at the bottom of the 75/80 meter band.

60 METERS: 5 SPECIFIC CHANNELS

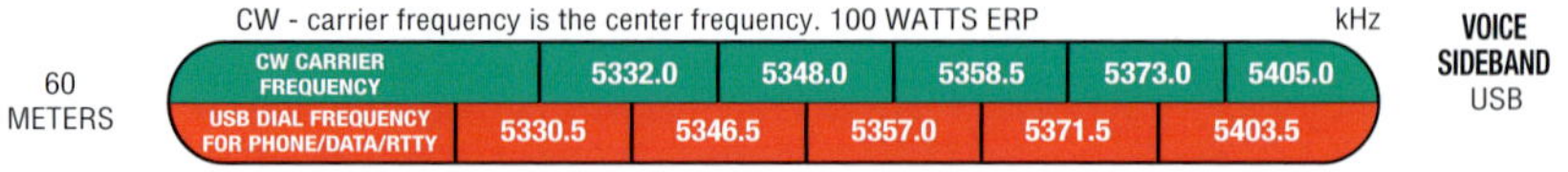

In 2003, a portion of the 60 meter band was allocated by the FCC for amateur radio use. Unlike other ham bands where we have a range of frequencies, 60 meter privileges are assigned to 5 specific channels. Recently, the middle channel on 60 meters was changed from 5366.5 kHz on your dial to 5357.0 kHz, upper sideband. Effective radiated power is limited to 100 watts when using a half-wave dipole antenna. If you use a different type of antenna, you must keep a log and adjust your power to stay within the FCC's rules. On SSB voice, your signal cannot occupy more than 2.8 kHz bandwidth. Narrow bandwidth data, such as PSK-31 and PACTOR III are allowed, as is CW, on any of the channels. Your transmitted signal must end up exactly on the channel CENTER, which is 1.5 kHz ABOVE the carrier frequencies listed on our chart. See page 26 for the questions and answer explanations related to 60 meters, and Eric's Elmer sidebar explaining these operating rules.

40 METERS: 7 MHz - 7.3 MHz.

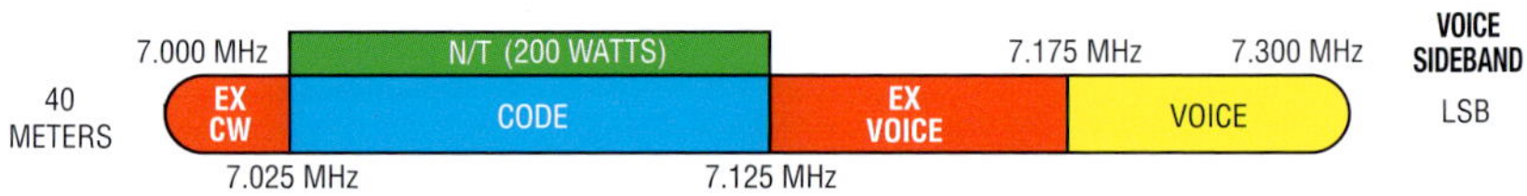

Extra Class operators have an exclusive 25 kHz of CW operating privileges at the bottom of the of the 40 meter band; 7.000 to 7.025 MHz. Voice privileges extend from to 7.125 to 7.300 MHz giving more "elbow room" on this band. During daylight hours, 40 meters is a great band for base station and mobile contacts up to 500 miles away. At night, 40 meters skips all over the country, and many times all over the world! Beware, though, since 40 meters is shared with worldwide AM shortwave broadcast stations, at night you should be prepared to dodge megawatt carriers playing everything from rock and roll to political broadcasts.

30 METERS: 10.1 MHz - 10.15 MHz.

Only code, data, and RTTY are permitted on this band. Privileges on 30 meters are shared by General, Advanced and Extra Class amateur operators. Thirty meters is located just above the 10 MHz WWV time broadcasts on shortwave radio so voice transmissions are not allowed on this band by any class of amateur operator. Power is limited to 200 watts PEP for any of the operating modes.

20 METERS: 14 MHz - 14.350 MHz.

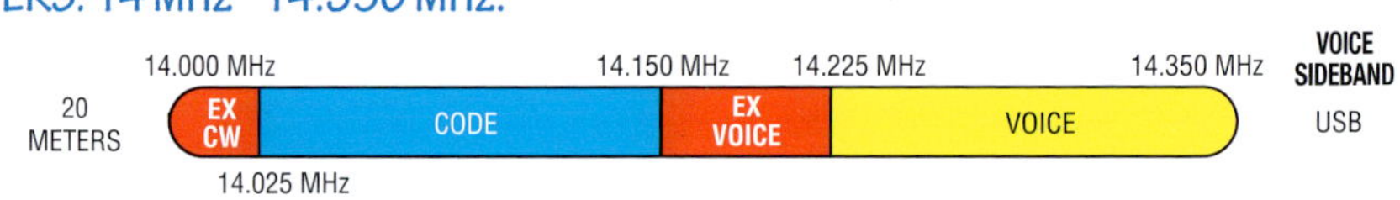

20 meters is where the real DX activity takes place! In addition to the parts of the band shared with the General and Advanced class licensees, Extra Class operators have two band segments of exclusive frequency privileges on this band. For CW, you gain the bottom 25 kHz of the band (14.000 – 14.025 MHz). For voice, you have exclusive use of 14.150 to 14.175 MHz. Happy DXing!

Almost 24 hours a day, you should be able to work stations in excess of 5,000 miles with a modest antenna set up. If you are a mariner, most maritime mobile operations take place on the 20 meter band. If you are into recreational vehicles (RVs), there are nets all over the country especially for you. The "where it's at" band is 20 meters when you want to work the world from your car, boat, RV, or home shack!

17 METERS: 18.068 MHz - 18.168 MHz.

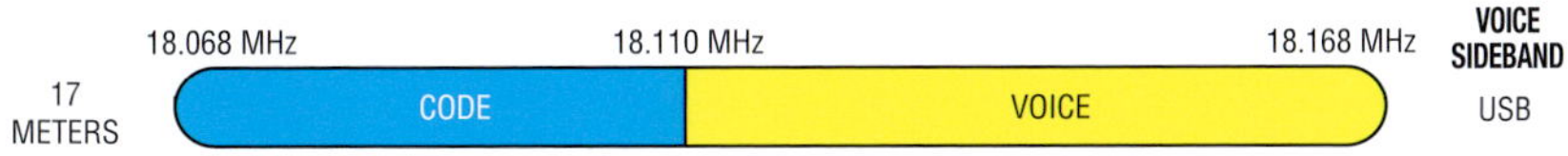

All emission types are authorized on this Amateur Radio band. There is plenty of elbow room here with lots of foreign DX coming in day and night. The 17 meter band is shared with General and Advanced class operators giving you plenty of opportunity for fun operations and lots of contacts locally as well as from foreign lands.

15 METERS: 21.0 MHz - 21.450 MHz.

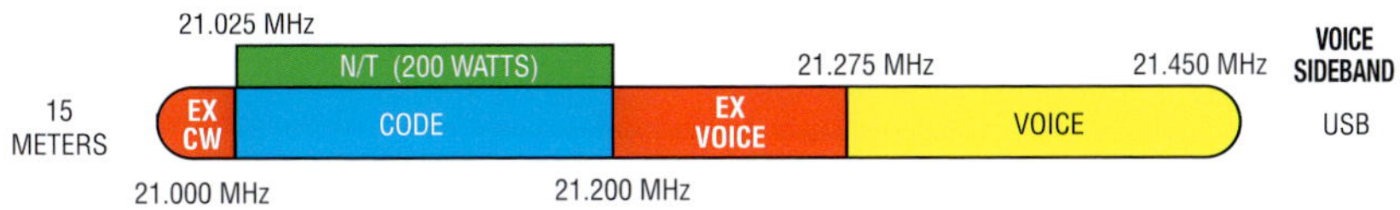

Along with privileges shared with General and Advanced class amateur operators on 15 meters, Extra Class operators have two exclusive operating segments on this band: the bottom 25 kHz from 21.00 to 21.025 MHz and another 25 kHz segment from 21.200 to 21.225 MHz.

The 15-meter band is loved by hams throughout the world because of the extremely low noise and strong signal levels at these frequencies. Band conditions on 15 meters usually favor daytime and evening contacts in the direction of the Sun. Skywave coverage on 15 meters fades as the Sun sets. This is a popular band for mobile operators because antennas are smaller.

12 METERS: 24.890 MHz - 24.990 MHz.

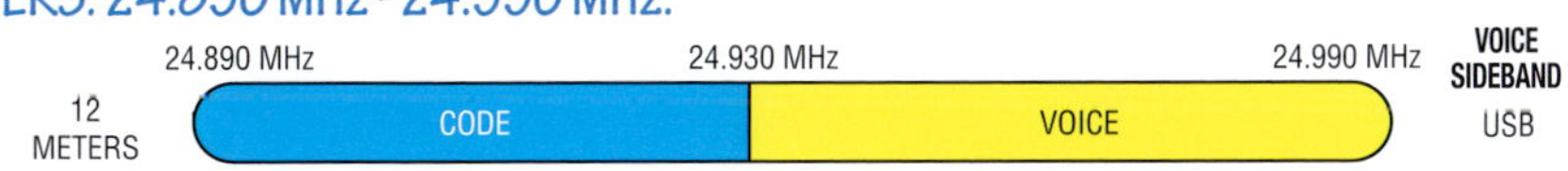

Code, data, and RTTY privileges on the 12 meter band extend from 24.890 to 24.930 MHz. Voice privileges are from 24.930 to 24.990 MHz. Extra Class operators share this band with General and Advanced class amateur operators. Although this is a very narrow band, expect excellent daytime range throughout the world. At night, range is limited to groundwave coverage because the ionosphere is not receiving sunlight to produce skywave coverage on this band.

10 METERS: 28.0 MHz - 29.7 MHz.

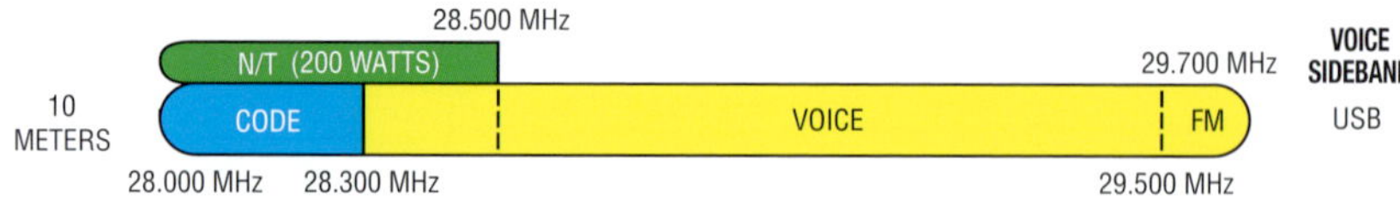

There is plenty of excitement shared by all classes of amateur operators on the 10 meter band. Worldwide and continental band openings may pop up anytime there are extra amounts of solar activity. During the summertime, Sporadic-E ionospheric conditions make 1,500-mile band openings exciting for all operators on "10." The Extra Class operator has the entire 10 meter band from top to bottom (28.000 to 29.700 MHz). There is plenty of CW activity at 28.1 MHz, propagation beacons at 28.2 MHz, phone excitement between 28.3 to 29.0 MHz with all sorts of elbow room. And wait until you try the full fidelity of frequency modulation (FM) on 29.600 MHz! There are even FM repeaters from 29.5 to 29.7 MHz at the top of 10 meters.

VOICE BAND EXPANSION

In late 2006, the FCC issued rules that provide expanded voice privileges and removed the 200 watt power limitation for CW on the worldwide bands for General, Advanced and Extra Class operators. Novice and Technician licensees were given additional CW privileges on the 80, 40, 15, and 10 meter bands. This Report and Order granted ham radio operators more "elbow room" on frequencies that FCC viewed as "grossly underused" and "wasted" spectrum within the CW portion of these bands. Here's what we got:

On 75 meters: General class operators may operate voice from 3800 to 4000 kHz, an increase of 50 kHz. Advanced Class licensees are able to operate voice from 3700 to 4000 kHz, a 75 kHz increase. Extra Class operators gained 150 kHz of bandwidth for voice operations from 3600 to 4000 kHz.

On 40 meters: Advanced and Extra Class operators are able to use voice from 7125 to 7300 kHz, an increase of 25 kHz. General Class may now use voice from 7175 to 7300 kHz, a 50 kHz increase.

On 15 meters: Generals gained an additional 25 kHz of voice from 21.275 to 21.450 MHz. Novice and Technician operators gained CW privileges in an expanded portion in the lower half of the 10, 15, 40, and 80 meter bands. And Technician Class gained voice communications on 10 meters from 28.300 to 28.500 MHz, giving these brand new ham operators a taste of skywave communications on this worldwide band. As a new Extra, spend some time in this part of "10" to encourage these new hams!

On 60 meters: While the Report did not change the number of frequencies available to all amateur operators on 60 meters, one of the 60 meter frequencies changed. Dial frequency 5366.5 kHz was replaced by dial frequency 5357.0 kHz. Other changes in the Report for this band included an increase of power to 100 watts ERP and digital modes plus CW operation on all 5 channels.

6 Meters and Up: Your Extra Class license allows unlimited band privileges using all emission types on the frequencies detailed in the following chart. These privileges are shared with all amateur operators.

Meters	Frequency
6 meters	50-54 MHz
2 meters	144-148 MHz
1.25 meters	222-225 MHz (219-220 MHz Digital)
70 centimeters	420-450 MHz
35 centimeters	902-928 MHz
23 centimeters	1240-1300 MHz

Microwave Bands: The Extra Class license allows for unlimited privileges using all emission types on the following microwave bands:

Frequency	Frequency
2300-2310 MHz	47.0-47.2 GHz
2390-2450 MHz	75.5-81.0 GHz
3.3-3.45 GHz	119.98-120.02 GHz
5.65-5.925 GHz	142-149 GHz
10.0-10.5 GHz (popular X band)	241-250 GHz
24.0-24.25 GHz	All above 300 GHz

Look for plenty of Extra Class excitement on the 10 GHz band. This may soon become the new frontier for those amateur operators establishing records on microwave frequencies.

BECOME A VOLUNTEER EXAMINER

Earning the Extra Class license provides you with the opportunity to give back to amateur radio! As an Extra Class licensee you may be accredited as a Volunteer Examiner to prepare and administer ham exams as a member of a three-person VE team. You also could qualify to become a contact Volunteer Examiner to help coordinate ham exam teams in your area. This is a great way to help our hobby and service grow and prosper! To learn more, visit **www.arrl.org/volunteer-examiners**.

BE AN EXTRA CLASS ELMER — TEACH A CLASS

When you pass your Extra Class exam, you will have arrived at the TOP! We encourage you to share your enthusiasm for our amateur radio service by becoming an instructor. The ARRL National Ham Instructor program is designed to help you spread the word – and training – to radio enthusiasts and encourage them to earn their ham radio licenses.

Ham Instructors may find kids and scouting a great place to introduce ham radio and develop amateur radio training classes. Boaters, flyers, and RVers are *all* interested in what amateur radio can do to make their traveling safer and more enjoyable. You might become an Elmer to help them locate club or local weekend classes in your area where they can earn their entry-level Technician Class license.

The radio industry supports Ham Instructors with free band plan charts, frequency guides, and other training materials. Extra Class Ham Instructors help the amateur radio service continually grow. If you are interested in becoming an ARRL Volunteer Examiner and serving the Amateur Radio community, it's easy and free! Visit **www.arrl.org/become-an-arrl-ve** to find ARRL's Volunteer Examiner Manual and *Volunteer Examiner application*. Join the outreach program to bring more kids and radio enthusiasts into our hobby and public service.

Join the ARRL National Instructor program and help our hobby grow!

SUMMARY

Whether you are upgrading to Extra Class from a grandfathered Advanced Class license or a General Class license, you will gain exciting new privileges on the bands for worldwide communications. This is where the rare DX happens!

But your biggest privilege as an Extra Class operator is the ability for you to work closely with a Volunteer Examination Coordinator and serve as a VE team leader. Now that General and Advanced Class operators may assist Extra Class operators in administering license exams, your role as a Volunteer Examiner will be an important one for the testing teams – and the growth of our hobby – throughout the country.

Join the ARRL National Instructor Program Today!

Help our hobby and service grow! Sign up today as an ARRL National Instructor and join the ranks of teachers, scout leaders, Elmers, and hams using Gordo's classroom techniques to successfully train new hams. ARRL provides FREE support to registered instructors, including:

- ✔ Instructor Guides for all 3 license classes
- ✔ Pre-study homework assignments for your students
- ✔ Downloadable Power Point presentations for Tech, General, and Extra
- ✔ Postings of your ham radio class dates
- ✔ Discounted Gordon West training materials

Learn more and register as a Gordon West / ARRL National Ham Instructor by visiting **http://www.arrl.org/Gordon-West**.

2

A Little Ham History!

Ham radio has changed a lot in the 100-plus years since radio's inception. In the past 25 years, we have seen some monumental changes! So, before we get started preparing for the exam, we're going to give you a little refresher lesson about our hobby, its history, and an overview of how you'll progress through the amateur ranks to the top amateur ticket – the Extra Class license.

In this chapter you'll learn all of the licensing requirements under the FCC rules that became effective April 15, 2000. And you'll learn about the six classes of license that were in effect *prior* to those rules changes. That way, when you run into a Novice or Advanced Class operator on the air you'll have some understanding of their skill level, experience, and frequency privileges.

WHAT IS THE AMATEUR SERVICE?

There are more than 750,000 Americans who are licensed amateur radio operators in the U.S. today. The Federal Communications Commission, the Federal agency responsible for licensing amateur operators, defines our radio service this way:

"The amateur service is for qualified persons of all ages who are interested in radio technique solely with a personal aim and without pecuniary interest."

Ham radio is first and foremost a fun hobby! In addition, it is a service. And note the word "qualified" in the FCC's definition – that's the reason you're studying for an exam; by passing the exam you prove that you're qualified to get on the air.

Millions of operators around the world exchange ham radio greetings and messages by voice, radioteletype, telegraphy, facsimile, and television worldwide. Japan, alone, has more than a million hams! It is very commonplace for U.S. amateurs to communicate with Russian amateurs. China is just getting started with its amateur service. Being a ham operator is a great way to promote international good will.

The benefits of ham radio are countless! Ham operators are probably known best for their contributions during times of disaster. In recent years, many recreational sailors in the Caribbean who were attacked by modern-day pirates had their lives saved by hams directing rescue efforts. Following the 9/11 terrorist attacks on the World Trade Center and Pentagon, literally thousands of local hams assisted with emergency communications. After Katrina blew through New Orleans, ham radio operators were the first to report the levee breach and to warn that flood waters were rushing into the city.

The ham community knows no geographic, political or social barrier. If you study and make the effort, you are going to earn your upgrade to Extra Class with its expanded privileges on the HF worldwide bands. Follow the suggestions in our book and your chances of passing the written exam are excellent. Learning what you need to know to be an Extra Class operator will be easy and fun!

A BRIEF HISTORY OF AMATEUR RADIO LICENSING

Before government licensing of radio stations and amateur operators was instituted in 1912, hams could operate on any wavelength they chose and could even select their own call letters. The Radio Act of 1912 mandated the first Federal licensing of all radio stations and assigned amateurs to the short wavelengths of less than 200 meters. These "new" requirements didn't deter them, and within a few years there were thousands of licensed ham operators in the United States.

Since electromagnetic signals do not respect national boundaries, radio is international in scope. National governments enact and enforce radio laws within a framework of international agreements that are overseen by the International Telecommunications Union (ITU). The ITU is a worldwide United Nations agency headquartered in Geneva, Switzerland. It divides the radio spectrum into a number of frequency bands, with each band reserved for a particular use. Amateur radio is fortunate to have many bands allocated to it all across the radio spectrum.

In the U.S., the Federal Communications Commission is the government agency responsible for the regulation of wire and radio communications. The FCC allocates frequency bands in accordance with the ITU plan to the various services – including the amateur service – and regulates stations and operators.

By international agreement in 1927, the alphabet was apportioned among various nations for basic call sign use. The prefix letters K, N and W were assigned solely to the United States. The letter A is shared by the United States and other countries.

In the early years of amateur radio licensing in the U.S., the classes of licenses were designated by the letters "A," "B," and "C." The highest license class with the most privileges was "A." In 1951, the FCC dropped the letter designations and gave the license classes names – names that are familiar to us today. Class A became Advanced; Class B became General; and Class C became Conditional. At the same time, the FCC also created the Extra Class and Technician Class licenses. They also added a new Novice Class – a one-year, non-renewable license for beginners that required a 5-wpm Morse code speed proficiency test and a 20-question written examination on elementary theory and regulations, with both tests taken in the presence of one licensed ham.

The General exam required 13-wpm code speed, and Extra required 20-wpm. Each of the five written exams was progressively more comprehensive and formed what came to be known as the *Incentive Licensing System*.

In the '70s, the Technician Class license became very popular because of the number of repeater stations appearing on the air that extended the range of VHF and UHF mobile and handheld radios. It also was very fashionable to be able to patch your mobile radio into the telephone system, which allowed hams to make telephone calls from their automobiles long before the advent of cell phones.

In 1979, the international amateur service regulations were changed to permit all countries to waive the manual Morse code proficiency requirement for "...stations making use exclusively of frequencies above 30 MHz." This set the stage for creation of the Technician "no-code" license in 1991, when the 5-wpm Morse code requirement was eliminated. New "no-code" Technician licensees were permitted to operate on all amateur bands above 30 MHz. Applicants for the "no-code" Technician license had to pass the 35-question Novice and 30-question Technician

Class written examinations, but no Morse code test.

"No-code" Technician Class amateurs who passed a 5-wpm code test were awarded a Technician-Plus license. Besides their 30 MHz and higher no-code frequency privileges, Tech-Plus licensees gained the Novice CW privileges and a sliver of the 10 meter voice spectrum.

With these changes in 1991, there were six amateur service license classes – Novice, Technician, Technician-Plus, General, Advanced, and Extra – along with five written exams and three Morse code tests used to qualify hams for their various licenses.

Then in 1984 the FCC made another important change to amateur radio rules, when the length of term for amateur radio licenses was increased from five to ten years.

The Amateur Service is Restructured

As you can see, through the years ham radio licensing and privileges underwent a myriad of changes. In 1998, the FCC began working with the ham radio community to restructure our amateur radio service to make it more effective. The objective of this restructuring was to streamline the license process, eliminate unnecessary and duplicated rules, and reduce the emphasis on the Morse code test. The result of this review was a complete restructuring of the U.S. amateur service, which became effective April 15, 2000. The restructuring included a change in the Morse code requirement for General and Extra classes to 5-wpm and the reduction from six to three amateur radio license classes. Since 2000 there have been only 3 amateur radio license classes:

- Technician Class, the HF/VHF/UHF entry level license.
- General Class, the HF entry level license, which required a 5-wpm code test.
- Amateur Extra Class, a technically oriented senior license, based on-5wpm code credit.

The next major change to FCC Amateur Radio rules occurred on February 23, 2007, when the FCC completely eliminated the Morse code examination for all amateur radio licenses! The FCC action was based on 6,200 written comments, with most supporting the elimination of all code tests. At the same time, the FCC also "upgraded" No-code Technician Class operators, awarding them use of the Novice and Technician-Plus high frequency privileges on 80, 40, 15, and 10 meters.

In its public notice announcing elimination of the Morse code test requirement, the FCC Commissioners wrote: "We believe that because the international requirement for telegraphy proficiency has been eliminated, we should treat Morse code telegraphy no differently than other amateur service communication techniques. This change eliminates an unnecessary regulatory burden that may discourage current amateur radio operators from advancing their skills and participating more fully in the benefits of amateur radio."

Eliminating the Morse code test in no way diminishes the enthusiasm many ham operators have for the technique of sending dits and dahs over the air. In fact, we believe we have more hams than ever before learning the code now that they can do it on high frequency where practicing code with other hams is fun!

Self-Testing in the Amateur Service

Prior to 1984, all amateur radio exams were administered by FCC personnel at FCC Field Offices around the country. In 1984, the FCC adopted a two-tier system beneath it called the Volunteer Examiner Coordinator (VEC) system to handle amateur radio license exams. The VEC System was formed after Congress passed laws that allowed the FCC to accept the services of Volunteer Examiners (VEs) to prepare and administer amateur service license examinations. Testing activity of the VEs is managed by the VECs. The VEC acts as the administrative liaison between the VEs, who administer the various ham examinations, and the FCC, which grants the license.

A team of three VEs, who must be accredited by a VEC, is required to conduct amateur radio examinations. General Class amateurs may serve as examiners for the Technician class. Advanced Class and Extra Class amateurs may administer exams for Technician and General class. The Extra Class exam may only be administered by VEs who hold an Extra Class license.

In 1986, the FCC turned over responsibility for maintenance of the examination question pools for amateur radio licenses to the National Conference of Volunteer Examiner Coordinators (NCVEC). The Conference appoints a Question Pool Committee (QPC) to develop and revise the various question pools according to a schedule. As a rule, each of the three question pools is changed once every four years. The FCC requires at least ten times as many questions in each of the pools as may appear on an examination. Both the Technician Class Element 2 and General Class Element 3 written examinations contain 35 multiple-choice questions. The Extra Class Element 4 written examination has 50 questions.

OPERATOR LICENSE CLASSES AND EXAM REQUIREMENTS

Anyone is eligible to become a U.S. licensed amateur operator (including foreign nationals, if they are not a representative of a foreign government). There is no age limitation – if you can pass the examinations, you can become a ham!

Today, there are three amateur operator licenses issued by the FCC – Technician, General, and Extra. Each license requires progressively higher levels of learning and proficiency, and each gives you additional operating privileges. This is known as incentive licensing – a method of strengthening the amateur service by offering more radio spectrum privileges in exchange for more operating and electronic knowledge.

There is no waiting time required to upgrade from one amateur license class to another. Similarly, there is no required waiting time to retake a failed exam. You can even take all three examinations at one sitting if you're really brave! *Table 2-1* details the amateur service license structure and required examinations.

Table 2-1: Current Amateur License Classes and Exam Requirements

License Class	Exam Element	Type of Examination
Technician Class	2	35-question, multiple-choice written examination. Minimum passing score is 26 questions answered correctly (74%).
General Class	3	35-question, multiple-choice written examination. Minimum passing score is 26 questions answered correctly (74%).
Extra Class	4	50-question, multiple-choice written examination. Minimum passing score is 37 questions answered correctly (74%).

ABOUT THE WRITTEN EXAMS

What is the focus of each of the written examinations, and how does it relate to gaining expanded amateur radio privileges as you move up the ladder toward your Extra Class license? *Table 2-2* summarizes the subjects covered in each written examination element.

Table 2-2. Question Element Subjects

Exam Element	License Class	Subjects
Element 2	Technician	Elementary operating procedures, radio regulations, and a smattering of beginning electronics. Emphasis is on VHF and UHF operating.
Element 3	General	HF (high-frequency) operating privileges, amateur practices, radio regulations, and a little more electronics. Emphasis is on HF bands.
Element 4	Extra	A technical examination covering specialized operating procedures, more radio regulations, formulas and heavy math. Also covers the specifics on amateur testing procedures.

No Jumping Allowed

You may not by-pass a required examination as you upgrade from Technician to General to Extra. For example, to obtain a General Class license, you must first take and pass the Element 2 written examination for the Technician Class license then take and pass the Element 3 written examination. To obtain an Extra Class license, you must first pass the Element 2 (Technician) and Element 3 (General) written examinations and then successfully pass the Element 4 (Extra) written examination.

TAKING THE ELEMENT 4 EXAM

Here's a summary of what you can expect when you go to the session to take the Element 4 written exam for your Extra Class license. Detailed information about how to find an exam session, what to expect at the session, what to bring to the session, and more is included in Chapter 4.

Examination Administration

All amateur radio examinations are administered by a team of at least three Volunteer Examiners (VEs) who have been accredited by a Volunteer Examiner Coordinator (VEC). The VEs are licensed hams who volunteer their time to help our hobby grow.

How to Find an Exam Session

Examination sessions are organized under the auspices of an approved VEC. A list of VECs is located in the Appendix on page 255. The ARRL-VEC is the largest VEC in the country. ARRL tests in all 50 states. Their 3-member, accredited examination teams are just about everywhere. So when you call the VEC, you can be assured they probably have an examination team only a few miles from where you are reading this book right now!

Taking the Exam

The Element 4 exam is a multiple-choice format. The VEs will give you a test paper that contains the 50 questions and multiple choice answers and an answer sheet for you to complete. Take your time! Make sure you read each question carefully and select the correct answer. Once you're finished, double check your work before handing in your exam papers.

Since 2020, online examinations have really taken off. They now account for about 40% of all examinations given in the US! When taking an exam online, you will likely connect with the VEs through a VOIP system like Zoom or Microsoft Teams. The VEs will give you access to a web page containing the test questions, where you can simply click on the answer you choose. The VEs will monitor your location and computer usage. Generally, you're not allowed to navigate to pages other than the testing page during the exam. The VECs that offer online testing also offer in-person tests, so be sure to verify the type of test you're signing up for when you register.

The VEs will score your exam immediately. You'll know before you leave the exam site whether you've passed. Chances are very good that, if you've studied hard, you'll get that passing grade!

GETTING/KEEPING YOUR CALL SIGN

Once the VE team scores your exam and you've passed, the process of upgrading your official FCC Amateur Radio License begins – usually that same day.

At the exam site, you will complete NCVEC Form 605, which is your application to the FCC for your license. If you pass the exam, the VE team will send on the required paperwork to their VEC. The VEC reviews the paperwork then files your application with the FCC. This filing is done electronically and your license upgrade will be granted and posted on the FCC's website within a few days.

Should you keep your old call sign or ask for a new one? If you check the box on the Form 605 asking for a new call sign, the FCC's computer will assign one to you. But you might not like what you get. So it is best to stick with your old call sign, and if you want a new, specific, call file an application for a Vanity Call Sign. See Chapter 4 for more details.

As soon as you pass your exam and have the CSCE, you are permitted to go on the air with your Extra Class HF privileges – even before your license upgrade appears on the FCC's Universal License System database. See Chapter 4 for more details on this process.

Vanity Call Signs

Once you've upgraded to Extra Class, you may want to get a personalized call sign that has special meaning to you. If so, you can apply for a Vanity Call Sign with a different letter combination. This process is explained in more detail In Chapter 4.

HOW MANY CLASSES OF LICENSES?

Once you've passed your Element 4 exam and go on the air as a new Extra Class operator, you'll be talking to more hams throughout the U.S. and around the world. Here's a summary of the new and "grandfathered" licenses that your fellow amateurs may hold and a recap of the level of expertise they have demonstrated in order to gain their licenses.

Current License Classes

Following the FCC's restructuring of Amateur Radio licensing that took effect April 15, 2000, there are just three license classes – Technician, General, and Extra. Persons who hold licenses issued prior to April 15, 2000 may continue to hold onto their license class and renew it every 10 years for as long as they wish.

Older License Classes Still on the Air

Once you get on the air with your new Extra Class privileges, every now and then you might meet a Novice operator while yakking on 10 meters, or sending CW on 15, 40, or 80 meters. And then there are the Advanced Class operators who may continue to hold onto their license class designation until they finally decide to move up to Extra Class. When you meet an operator holding an Advanced Class license, you know without a doubt that he had to successfully complete a 13 word per minute Morse code test along with the written exam to earn that license.

When you look at the Frequency Charts in this book, you'll see that we have updated them to reflect the expanded privileges on many high frequency bands for Extra Class operators. We also list the sub-band privileges for Extra Class, Advanced Class, General Class, Technician Class and even the Novice.

"Paper Upgrade" to General

The FCC Rules (Part 97.505) provides credit for General Class Element 3, without examination, to those applicants who can prove they held an unexpired Technician Class license granted before March 21, 1987. No other class of license receives this element credit.

To see if a candidate qualifies for any of these opportunities to regain their former license privileges, or for a "paper upgrade" to General Class, contact the ARRL VEC at 860-594-0200.

Element Credit for Prior Licenses

Recent rule changes by the FCC make it possible for hams whose licenses have expired to regain their privileges simply by taking and passing the Element 2 Technician Class written exam. If you were licensed as a General, passing the Tech exam will earn you a new General Class license. If you were Advanced, passing the Tech exam will also get you a new General Class license. And if you were an Extra, passing the Tech exam will earn you a new Extra Class license!

This rule applies to those with expired General, Advanced, and Extra Class licenses that are beyond the two-year grace period. In order to take advantage of this, you must show proof of holding your expired license. If you do not have a copy of your old license, one resource to help verify your former status is to call the ARRL VEC at 1-860-594-0123.

Paperless Licenses

Effective February 17, 2015 the FCC no longer routinely issues/mails a paper license document to amateur radio licensees and other WTB licensed services. The Commission maintains that the official Amateur Radio license authorization is the electronic record that exists in its Universal Licensing System database. Licensees may access their current, official authorization, and print out an unofficial "reference copy" of their license, using the FCC ULS License Search feature at: **https://wireless2.fcc.gov/UlsApp/UlsSearch/searchLicense.jsp**.

You can request a paper license by logging into the ULS license record using your FRN and CORES password and selecting to receive a paper copy by mail. If you do not know your CORES password, you can reset it on the FCC website: **https://apps.fcc.gov/cores/userLogin.do**

> **NOTE:** The renewal period on your new Extra Class license is NOT 10 years. It remains the SAME renewal date that is now on your General or Advanced Class license. Make sure your address is up-to-date, too! An upgrade to your license class does NOT change the original grant date of your previous license class.

IT'S EASY!

The primary pre-requisite for passing any amateur radio operator license exam is the "will" to do it. If you follow our suggestions in this book, your chances of passing the Extra Class exam are excellent.

The 2007 FCC rule changes that eliminated the Morse code test requirement for Extra Class make it even easier for you to earn your Extra Class license and enjoy voice privileges on the high frequency bands that were expanded in those same changes.

When you become an Extra Class operator, GET ON THE AIR! Operating is an important step to becoming a good ham. So GET ON THE AIR and enjoy your new Extra Class worldwide HF privileges. Remember, you are now at the top of the hobby!

> *Want to find a test site fast?*
> Call 1-860-594-0200, or visit **www.arrl.org/exam**.

3

Getting Ready for the Exam

The complete new Element 4 Question Pool of 603 question and answers that will be used to make-up your 50-question Extra Class written theory examination is included in this chapter. The questions and answers are presented *exactly as they will appear on your exam.* Each question has a correct answer and three distracters (the incorrect answers - those that distract you from the truth!). A score or 74%, or no more than 13 wrong answers, will earn you a passing grade.

THIS EXTRA CLASS QUESTION POOL IS VALID FROM JULY 1, 2024, THROUGH JUNE 30, 2028.

The Volunteer Examiners who make up the exam are not permitted to reword the questions, nor are they permitted to change any of the answers (but they can change the A B C D *order* of the answer choices). *Every question in this book, letter for letter, word for word, number for number, and every right and wrong answer, will be exactly the same on your upcoming 50-question Extra Class Element 4 exam.*

This new ninth edition study guide is designed to educate you about Extra Class operating techniques, rules and regulations, propagation, electrical theory, practical circuits, antennas, and more. We do this by including an explanation of the correct answer for each question. We also offer some memory tricks to help you remember the correct answer. Key words that help you associate the question and correct answer are highlighted in ***blue*** in the explanations. For those pesky distracters, we provide hints on how to identify wrong answers that may look very much like the correct one.

This new Extra Class Element 4 question pool was carefully developed by the Question Pool Committee (QPC) of the National Conference of Volunteer Examiner Coordinators. They began by carefully reviewing the most current Element 4 question pool, deleting obsolete or duplicate questions, modifying some questions, and adding new ones. The previous pool contained 621 questions. The new pool contains 603. All of this effort results in a pool that is up-to-date with questions on newer amateur radio technologies and current FCC rules and regulations.

This new 2024-2028 question pool was developed by the QPC, and many additional volunteers (including Gordo and Eric) recommended potential new questions to the Question Pool Committee. We all want to thank the members of the Question Pool Committee for their dedication and the hundreds of hours spent updating the Extra Class pool. Thanks, Team, for a fresh, new Extra Class question pool!

WHAT THE EXAM CONTAINS

Your exam will be comprised of 50 questions taken from the current Element 4 question pool. The pool is divided into 10 subelements. *Table 3-1* shows you the subelement topics, the total number of questions in each, and the number of questions from each that will appear on your exam. Carefully study *Table 3-1* to gain an understanding of how your exam will be constructed.

You will find that each subelement is divided into topic groups. There will be one question on your exam taken from each topic group. If there is a specific group of questions that are giving you a hard time, study them later. Only one question will be taken from any one group. If you just can't get a good understanding of that one group, remember that will cause you to miss only one question.

Table 3-1. Question Distribution for the Extra Class Element 4 Exam

Subelement Number	Subelement Topic	No. of Questions in Pool	No. of Questions on Exam
E1	***Commission's Rules*** (FCC rules for the Amateur Radio services)	68	6
E2	***Operating Procedures*** (Amateur station operating procedures)	61	5
E3	***Radio Wave Propagation*** (Radio wave propagation characteristics)	39	3
E4	***Amateur Practices***	64	5
E5	***Electrical Principles*** (Electrical principles as applied to amateur station equipment)	49	4
E6	***Circuit Components*** (Amateur station equipment circuit components)	69	6
E7	***Practical Circuits*** (Practical circuits employed in amateur station equipment)	99	8
E8	***Signals and Emissions*** (Signals and emissions transmitted by amateur stations)	48	4
E9	***Antennas and Feed Lines*** (Amateur station antennas and feed lines)	94	8
E0	***Safety***	12	1
Total		603	50

QUESTION POOL VALID THROUGH JUNE 30, 2028

The Element 4 question pool in this book is valid from July 1, 2024, until June 30, 2028. During this period, the questions and answers will not be changed. *Every question in this book, letter for letter, word for word, number for number, and every right and wrong answer, will be exactly the same on your 50-question Extra Class Element 4 exam.* If FCC rules change or advancements in technology cause any question to become obsolete, that question will be deleted from the question pool and no longer appear on an exam. There will be no surprises in the exam room between what you study here and the questions and answers you see on your test.

The new 2024-2028 Extra Class Element 4 pool covers many new subject areas designed to keep the Extra Class questions fresh:

- *Ham Rules and Regulations on band edge SSB operation*
- *Satellite operation and modes*
- *ATV and DTV technicalities*
- *Contest operating*
- *Contest log files*
- *PSK, FT8, and Q65*
- *Antenna takeoff patterns*
- *Antenna analyzer versus SWR Bridge*
- *NPN transistors*
- *Impedance calculations*
- *Transmission link budgets*
- *dBm noise floor calculations*
- *Roofing filter topics*
- *Receiver blocking dynamic range*
- *Solar flares and CMEs*
- *Intermodulation interference*
- *Third order intercept levels*
- *Noise blankers*
- *RF attenuators*
- *Schottky barrier diodes*
- *Emitter amplifier bias*
- *Third order intermodulation distortion*
- *Parametric amplifiers*
- *High voltage power supply filters*
- *Frequency counters*
- *Antenna modeling*
- *Safety practices, RF radiation hazards, hazardous materials*
- *Mesh Networks*
- *Smith Charts*

QUESTION CODING

Each question in the Element 4 Extra Class pool is assigned a coded number using a system of letters and numbers. *Figure 3-1* explains the coding for question E5D09. Understanding how the questions are numbered in the pool will help you know when you encounter a question on your exam that is from a section that was giving you trouble. Not to panic… there will be only one question from that subelement group on the exam.

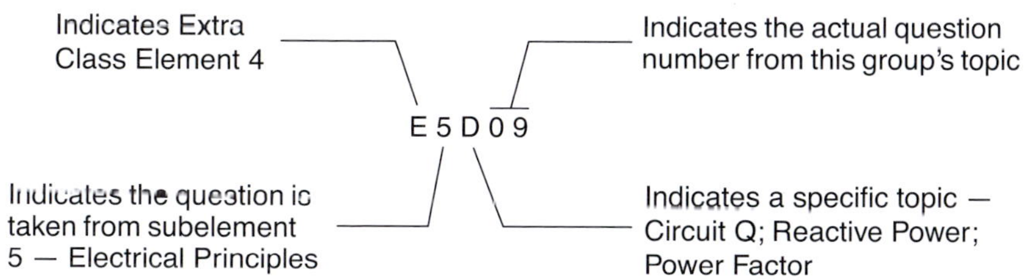

Figure 3-1. Examination Question Coding

STUDY HINTS

You're probably saying: "603 questions and answers! This is an overwhelming number to learn for the Extra Class exam!" But good news – almost every question is asked at least twice with just slightly different wording. Many of the math questions are based on a single formula. Once you know it, you can easily answer the 4 or 5 questions based on that formula.

You should learn the formulas, understand how to use them, and make sure you know the correct units to use. Many times you may have to convert kilohertz (kHz) to megahertz (MHz), or vice versa. Understand the formulas and learn how to apply them to arrive at a correct answer. In many cases, we will give you handy shortcuts to help you through the longer formulas. These shortcuts and a little humor now and then should keep you on track.

Work your book! Put a check mark beside questions you have down cold. Circle those that may need a little bit more study. Highlight those that appear to be brain-busters and figure out ways to remember the correct answers. Remember, do not memorize the answer key (A, B, C, or D) since the order of the answers can be changed for your exam.

If you study the questions and answers for a half hour in the morning and a half hour in the evening, it should take you approximately 30 days to prepare for the Extra Class exam. If you are an Advanced Class operator, it may take you only 15 days to prepare for your upgrade. One week before the exam, highlight the correct answer to each question and spend that week looking at the question and only the highlighted correct answer. The day before the exam, study the ***BLUE KEY WORDS*** in each explanation.

QUESTIONS REARRANGED FOR SMARTER LEARNING

The first thing you'll notice when you look at how the Element 4 question pool is presented in the *Extra Class* book is that the entire question pool is arranged into logical topic areas for easier learning. This arrangement is the same as Gordo uses when he teaches his 4-day Extra Class seminars. Rather than jumping back and forth between subelement topic areas to study similar-topic questions, he and Eric take you LOGICALLY through the entire question pool in 16 topic areas. Here is a list of the topic areas and the page numbers on which each begins in this book:

A numerical cross-reference of all of the Element 4 questions is found in the Appendix, on page 255. If you need to look up a specific question by its number, we tell you precisely the page that it is on. This logical arrangement of questions is an exclusive training method found in all Gordon West ham radio test preparation books.

Every single question in the Element 4 pool has been placed within the proper topic area for logical learning as a group. Gordo's classroom students love this approach to learning, and you will, too!

ADDITIONAL STUDY MATERIALS

Passing your Extra Class exam will require plenty of study. There are additional resources that will help you learn – and understand – the material presented in this book.

Gordo has been offering audio training materials for over 40 years. Remember tape cassettes, VCR's, and then CDs? Time for another media upgrade, so stand by for a fresh modern way to tune into his license preparation audio courses!

Download his free Technician, General, and Extra "On the Air Sounds" of radio calls at **http://www.arrl.org/Gordon-West**.

If you like learning at your PC, study software is available from the ARRL. When you take a practice exam, the software shows you the correct answer, scores your work, and tells you where you need to focus your study efforts.

When you're ready to learn more about the electronic principles introduced in the questions of the Element 4 pool, ARRL offers books on every aspect of ham radio. You've probably already used the Gordon West Tech and General class prep books, but those are handy references, so hold on to your copies! Every ham should have a copy of the *ARRL Handbook for Radio Communications*. It's the gold standard for ham radio electronics and operating, and its companion the *ARRL Antenna Book* provides information on antennas to suit any operating activity or style. For those interested in computers and electronics, we offer volumes on RF design and Arduino projects. You can also find books that cover operating activities and science, like HF digital, portable operating, Parks on the Air®, propagation, and a variety of other activities. Visit **www.arrl.org/shop** to view our current titles.

4

Element 4 Question Pool

Let's get started reviewing the questions and answers. Read over the question and see if you already know the answer. Next, spot the likely answer and then read the explanation of the correct answer. Finally, at the end of the explanation, we give you the correct A, B, C, or D answer. Use a small card to cover up the explanations and then slowly reveal the 4 possible answers, the explanation, and then the correct answer. Put a check mark beside the question number if you answered it correctly. Put an X if you need more study with a specific question.

And one more time, be assured that every question in this book, letter for letter, word for word, number for number, and every right and wrong answer, will be exactly the same on your upcoming 50-question Extra Class Element 4 exam.

Work the book! In Gordo's classes, he can usually tell who will pass the exam the first time by looking at their marked up textbooks.

Ready to get started? Here we go with your first look into the world of the amateur Extra Class license.

Rules & Regs

E1A01 Why is it not legal to transmit a 3 kHz bandwidth USB signal with a carrier frequency of 14.348 MHz?

A. USB is not used on 20-meter phone
B. The lower 1 kHz of the signal is outside the 20-meter band
C. 14.348 MHz is outside the 20-meter band
D. The upper 1 kHz of the signal is outside the 20-meter band

FCC rules clearly state that ALL of an amateur radio emission must fall within a designated band or sub-band. You must consider your entire transmitted bandwidth when "crowding" a bad edge. It's always safest to not even try to crowd a band edge, but if you ***must***, just be sure none of your signal "slops over" the boundary. If you're operating USB, you need to be careful of the upper band edge; if you're operating LSB, you need to be conscious of the lower band edge. **ANSWER D.**

E1A03 What is the highest legal carrier frequency on the 20-meter band for transmitting a 2.8 kHz wide USB data signal?

A. 14.0708 MHz
B. 14.1002 MHz
C. 14.1472 MHz
D. 14.3490 MHz

Recent FCC rulemaking has removed High Frequency baud rate limitations and implemented a 2.8 kHz bandwidth maximum for digital emissions, enabling amateur radio operators with more modern data efficiency, especially with emergency data communications. The top of the 20-meter band for data is 14.150 MHz, and no data emission may extend beyond this band limit. This means a USB data signal carrier frequency must be placed ***no higher than 14.147.2 MHz, 2.8 kHz below the top of the USB data band.*** **ANSWER C.**

TOO CLOSE! Stay at least 2.8 KHz below the top of the digital band edge.

E1A02 When using a transceiver that displays the carrier frequency of phone signals, which of the following displayed frequencies represents the lowest frequency at which a properly adjusted LSB emission will be totally within the band?

A. The exact lower band edge
B. 300 Hz above the lower band edge
C. 1 kHz above the lower band edge
D. 3 kHz above the lower band edge

This question has us on lower sideband, maybe 40 meters, so we need to ***stay 3 kHz ABOVE the lower band edge*** to stay within the band. It would be considered poor operating practice to intentionally hug a band edge and switch to the non-conforming sideband. [97.301, 97.305] **ANSWER D.**

E1A04 May an Extra class operator answer the CQ of a station on 3.601 MHz LSB phone?

A. Yes, the entire signal will be inside the SSB allocation for Extra class operators
B. Yes, the displayed frequency is within the 75-meter phone band segment
C. No, the sideband components will extend beyond the edge of the phone band segment
D. No, US stations are not permitted to use phone emissions below 3.610 MHz

Lower sideband is used on 75 meters. The lowest frequency near the voice band edge that we would select as our dial carrier frequency is 3.603 MHz. ***U.S. operators on 3.601 MHz*** would have ***sideband components extending illegally below the band edge*** of the phone segment. [97.301, 97.305] **ANSWER C.**

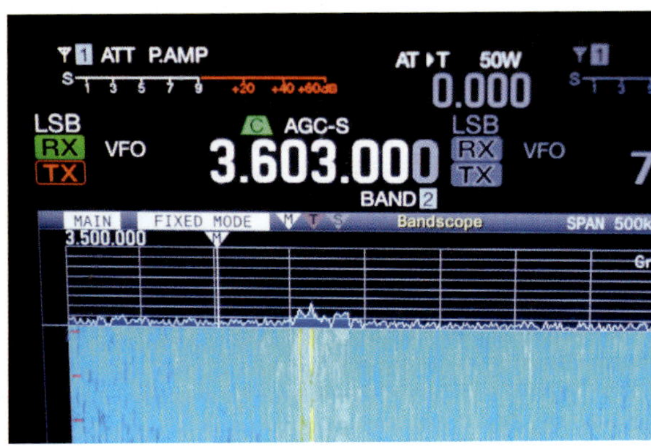

Stay 3 kHz above the lower voice limit on 75m.

E1C10 What is the maximum mean power level for a spurious emission below 30 MHz with respect to the fundamental emission?

A. - 43 dB
B. - 53 dB
C. - 63 dB
D. - 73 dB

Down on high frequency (below 30 MHz), the FCC requires any spurious emission to be at least 43 dB below the mean power output of that amplifier. Amplifiers contain both harmonic and band pass filters to reduce ***spurious emissions down at least 43 dB***. [97.307] **ANSWER A.**

E1C01 What is the maximum bandwidth for a data emission on 60 meters?

A. 60 Hz
B. 170 Hz
C. 1.5 kHz
D. 2.8 kHz

FCC Rule 97.303 states that amateur radio operators must ensure that ***emissions from their 60 meter stations do not occupy more than 2.8 kHz bandwidth***, centered on the channel center frequency. PSK31 audio is on channel center frequencies, generally 1500 Hz above the suppressed carrier frequency. However, CW 150HA1A must be centered on channel, where channel sharing and cooperation are of paramount importance. [97.303] **ANSWER D.**

CW - carrier frequency is the center frequency. 100 WATTS ERP

60 METERS						VOICE SIDEBAND USB
CW CARRIER FREQUENCY	5332.0	5348.0	5358.5	5373.0	5405.0	
USB DIAL FREQUENCY FOR PHONE/DATA/RTTY	5330.5	5346.5	5357.0	5371.5	5403.5	

E1A06 What is the required transmit frequency of a CW signal for channelized 60 meter operation?

A. At the lowest frequency of the channel
B. At the center frequency of the channel
C. At the highest frequency of the channel
D. On any frequency where the signal's sidebands are within the channel

The CW designator 150HA1A precludes tone modulated CW where different tones could help sort out different CW stations transmitting on the same channel. The new rules read: ***"For CW emissions, the carrier frequency is set to the CENTER frequency."*** This means that if all ham sets sending CW have zero error on the center channel display, all the CW signals will come out with the same exact tone! Band planning may help solve the problem of multiple CW signals sharing a specific channel sounding the same! [97.303(h)(1)] **ANSWER B.**

E1C07 What notifications must be given before transmitting on the 630- or 2200-meter bands?

A. A special endorsement must be requested from the FCC
B. An environmental impact statement must be filed with the Department of the Interior
C. Operators must inform the FAA of their intent to operate, giving their call sign and distance to the nearest runway
D. Operators must inform the Utilities Technology Council (UTC) of their call sign and coordinates of the station

Some utilities, such as those employing power line monitoring, use frequencies near our two newest amateur radio bands on 630 and 2200 meters. While the potential for interference is minimal, if you anticipate operating on these bands you need to ***inform the local Utilities Telecom Council*** of your call sign and coordinates. Unless there is some issue, you will not receive any response from the UTC and are free to operate (with legal power, of course!) after a wait of 30 days. [97.303(g)] **ANSWER D.**

THREE SIMPLE RULES FOR 60 METERS

When operating 60 meters, it is crucial that you transmit on the correct frequency within the authorized channel. Compliance with the rules when using Single Sideband phone or sound card digital modes is very straightforward. CW operation is slightly different. 60 Meter channels are assigned by the following table, where the frequency represents the VFO display (suppressed carrier) frequency of your transceiver, which is also the low frequency boundary of the channel:

Channel 1: 5330.5 kHz
Channel 2: 5346.5 kHz
Channel 3: 5357.0 kHz
Channel 4: 5371.5 kHz
Channel 5: 5403.5 kHz

If you heed the following three rules, you will be in compliance with the FCC regulations:

(1) ***Phone Operation:*** *Set your transmitter to Upper Sideband (USB) and your VFO to display the frequency of the chosen channel from the table. Be sure your transmitter is set to limit the bandwidth to 2.8 kHz. Check your operator's manual if you are unsure about how to do this. Most modern rigs default to this setting, but READ THE MANUAL!*

(2) ***Sound Card Digital Modes:*** *Set your transmitter to USB, select the channel frequency on your VFO, and click on 1.5 kHz on your waterfall or spectrum display. This will place your transmitted signal properly in the CENTER of the channel.*

(3) ***CW operation:*** *Place your transmitted signal in the CENTER of the channel, 1.5 kHz above the assigned carrier frequency shown in the table. Depending on your particular transceiver, this MAY OR MAY NOT correspond to your VFO display frequency! READ YOUR OPERATOR'S MANUAL TO BE SURE.*

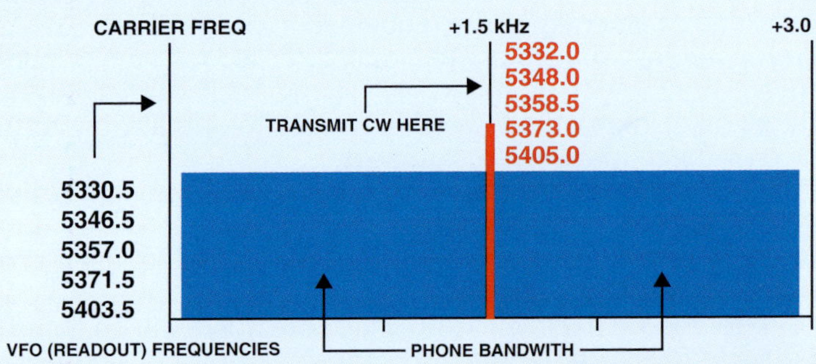

The diagram above shows the position of your transmitted signal within a 60 meter channel.

One final note: If you're using **non-sound card digital modes**, *such as "old school" RTTY with direct FSK modulation, you must assure that your transmitted signal is exactly centered in the channel, using any reliable means at your disposal.*

CW DXer Chip Margelli, K7JA, can now run code on the 60 meter band!

E1C03 How long must an operator wait after filing a notification with the Utilities Technology Council (UTC) before operating on the 2200-meter or 630-meter band?

A. Operators must not operate until approval is received
B. Operators may operate after 30 days, providing they have not been told that their station is within 1 km of PLC systems using those frequencies
C. Operators may not operate until a test signal has been transmitted in coordination with the local power company
D. Operations may commence immediately, and may continue unless interference is reported by the UTC

In nearly all cases, informing your local Utilities Telecom Commission of your intention of operating on 630 or 2200 meters will result in no response, which is good news! But you must ***wait 30 days*** to give the UTC time to respond. If you are within 1 km of a utility using those frequencies, you will be notified of this fact, but this does not necessarily mean you are prohibited from operating, but specific restrictions in your particular case may apply. [97.303(g)] **ANSWER B.**

E1C12 In what portion of the 630-meter band are phone emissions permitted?

A. None
B. Only the top 3 kHz
C. Only the bottom 3 kHz
D. The entire band

Although most activity on our two lowest frequency amateur bands is either slow speed CW (QRSs) or narrow-band digital modes, because of the extremely weak signals involved, there is nothing legally to prevent the use of phone modes like AM or SSB, and a number of hams are experimenting with phone modes on 630 meters, and to a lesser degree on 2200 meters. Like 160 meters, ***there are no FCC mandated sub-bands on 630 or 2200 meters***, but gentlemen's agreements may eventually be adopted, and should be followed. [97.305(c)] **ANSWER D.**

E1A09 Except in some parts of Alaska, what is the maximum power permitted on the 630-meter band?

A. 50 watts PEP (peak envelope power)
B. 100 watts PEP (peak envelope power)
C. 1 watt EIRP (equivalent isotropic radiated power)
D. 5 watts EIRP (equivalent isotropic radiated power)

Operation on 630 meters is similar to operation on 2200 meters in that the maximum EIRP is specified. However, ***on 630 meters the EIRP limit is 5 watts***, and there is a limit of 500 watts transmitter output power. There is an additional EIRP power restriction to 1 watt if you are within 800 km of the Russian Federation, which includes some parts of western Alaska. EIRP stands for equivalent isotropic radiated power. [97.313(l)] **ANSWER D.**

E1A07 What is the maximum power permitted on the 2200-meter band?

A. 50 watts PEP (peak envelope power)
B. 100 watts PEP (peak envelope power)
C. 1 watt EIRP (equivalent isotropic radiated power)
D. 5 watts EIRP (equivalent isotropic radiated power)

Unlike most of the other amateur bands, the power limits on our 2200 meter low frequency (LF) band are specified in terms of equivalent isotropic radiated power, (EIRP), not transmitter output power. Antennas on 2200 meters are very inefficient, so it will take a great deal more than 1 watt of transmitter power to achieve anywhere near 1 watt EIRP. In fact, you are allowed the same full legal limit *transmitter* power of 1500 watts on ***2200 meters*** as long as ***your EIRP does not exceed 1 watt!*** [97.313(k)] **ANSWER C.**

E1A05 Who must be in physical control of the station apparatus of an amateur station aboard any vessel or craft that is documented or registered in the United States?

A. Only a person with an FCC Marine Radio license grant
B. Only a person named in an amateur station license grant
C. Any person holding an FCC issued amateur license or who is authorized for alien reciprocal operation
D. Any person named in an amateur station license grant or a person holding an unrestricted Radiotelephone Operator Permit

If you go offshore aboard your U.S. documented or registered boat, ***anyone*** on board ***holding an FCC issued amateur radio license*** of the correct grade can work the ham radio equipment. This also holds true for your non-U.S.-citizen friend who holds an ***FCC issued reciprocal license***. Just because you may be in international waters, non-licensed personnel (the ship's Captain, for instance) would not be permitted to operate on the ham bands, even though they may hold a Marine Radiotelephone Operator Permit. [97.5] **ANSWER C.**

E1A11 What licensing is required when operating an amateur station aboard a US-registered vessel in international waters?

A. Any amateur license with an FCC Marine or Aircraft endorsement
B. Any FCC-issued amateur license
C. Only General class or higher amateur licenses
D. An unrestricted Radiotelephone Operator Permit

This is a great question dealing with maritime mobile operation when out on the high seas. If the vessel is flying the American flag, registered in the United States, ***any ham radio transmissions MUST be from an FCC-licensed amateur operator or person holding a reciprocal permit for an alien amateur license***. Years ago, there was the false belief that once someone was in international waters anyone could transmit over ham radio without a license. Not so! Your ham license is required and is separate from the ship's marine radio license or the captain's Marine Radio Operator Permit. [97.5] **ANSWER B.**

Maritime Mobile (M/M) is found on 14.300 MHz.

E1A10 If an amateur station is installed aboard a ship or aircraft, what condition must be met before the station is operated?

A. Its operation must be approved by the master of the ship or the pilot in command of the aircraft
B. The amateur station operator must agree not to transmit when the main radio of the ship or aircraft is in use
C. The amateur station must have a power supply that is completely independent of the main ship or aircraft power supply
D. The amateur station must operate only in specific segments of the amateur service HF and VHF bands

If you are planning a long cruise, or taking a ride in a friend's airplane, make sure you have received ***approval by the captain (master of the ship) or aircraft pilot before you begin to transmit***. Look at the three incorrect answers – they sound logical, but only the answer we just gave you is correct. [97.11] **ANSWER A.**

E1B05 What is the National Radio Quiet Zone?

A. An area surrounding the FCC monitoring station in Laurel, Maryland
B. An area in New Mexico surrounding the White Sands Test Area
C. An area surrounding the National Radio Astronomy Observatory
D. An area in Florida surrounding Cape Canaveral

A national ***radio quiet zone*** covering parts of Maryland, West Virginia, and Virginia has been established near the ***National Radio Astronomy Observatory***. The quiet zone restriction applies to "constant-on" stations, like beacons and repeaters. Operations from your home or vehicle would not be affected unless you actually enter the Observatory property in Greenbank, WV. [97.3] **ANSWER C.**

E1B03 Within what distance must an amateur station protect an FCC monitoring facility from harmful interference?

A. 1 mile
B. 3 miles
C. 10 miles
D. 30 miles

The Federal Communications Commission recently automated its monitoring stations. Now there are remote-controlled, unmanned monitoring stations, and all amateur ***operators within one mile of those stations must contact the local FCC engineer in charge*** to ensure they do not cause harmful interference to the remote-controlled monitoring units. [97.13] **ANSWER A.**

E1B04 What must the control operator of a repeater operating in the 70-centimeter band do if a radiolocation system experiences interference from that repeater?

A. Reduce the repeater antenna HAAT (Height Above Average Terrain)
B. File an FAA NOTAM (Notice to Air Missions) with the repeater system's ERP, call sign, and six-character grid locator
C. Cease operation or make changes to the repeater that mitigate the interference
D. All these choices are correct

The term "radiolocation systems" refers to government and military early warning radar installations all over the country guarding us from attack. ***Both 70 centimeter (440 MHz) and 1.2 GHz repeaters must never interfere!*** **ANSWER C.**

E1F08 Which of the following types of amateur station communications are prohibited?

A. Communications transmitted for hire or material compensation, except as otherwise provided in the rules
B. Communications that have a political content, except as allowed by the Fairness Doctrine
C. Communications that have a religious content
D. Communications in a language other than English

This question deals with permissible communications from one ham to another. Political, religious, and foreign language communications are permitted, but ***communications transmitted for hire or material compensation are normally prohibited***. [97.113(c)] **ANSWER A.**

You can speak any language on the ham bands, but always give your call sign in English.

E1F07 When may an amateur station send a message to a business?

A. When the pecuniary interest of the amateur or his or her employer is less than $25
B. When the pecuniary interest of the amateur or his or her employer is less than $50
C. At no time
D. When neither the amateur nor their employer has a pecuniary interest in the communications

In 1993, the prohibited business communications rules were redefined by the FCC to allow an amateur to ***send a message to a business as long as neither the amateur nor his or her employer had a pecuniary interest in the communications***. If you run a tree-trimming business, it would not be legal to use an autopatch to call your answering machine to find out where your next customer is waiting. On the other hand, if you are a school teacher and discover that a big pine tree has fallen down across the school's driveway, it would be perfectly acceptable to place an autopatch call to a tree-trimmer and have them come out to do their job. But could the school teacher call into the school district and tell them that she is going to be late for work? Probably not legal. [97.113] **ANSWER D.**

E1C02 Which of the following apply to communications transmitted to amateur stations in foreign countries?

A. Third party traffic must be limited to that intended for the exclusive use of government and non-Government Organization (NGOs) involved in emergency relief activities
B. All transmissions must be in English
C. Communications must be limited to those incidental to the purpose of the amateur service and remarks of a personal nature
D. All these choices are correct

With your new Extra Class privileges you will likely make even more contacts in foreign countries. Remember, no business communications allowed, so keep your communications to purely ***personal remarks or communications of a technical nature***. [97.117] **ANSWER C.**

E1F04 Which of the following geographic descriptions approximately describes "Line A"?

A. A line roughly parallel to and south of the border between the US and Canada
B. A line roughly parallel to and west of the US Atlantic coastline
C. A line roughly parallel to and north of the border between the US and Mexico
D. A line roughly parallel to and east of the US Pacific coastline

The ***United States and Canada*** have adopted the ***"Line A" area as a buffer zone*** for certain types of radio stations. This keeps normally "line of sight" radio energy from interfering with the other country's communications. Amateur transmissions are restricted on the 70-cm band between 420-430 MHz if the station is located north of "Line A." This is because Canada has a different type of radio allocation in the bottom part of the 70-cm band. [97.3] **ANSWER A.**

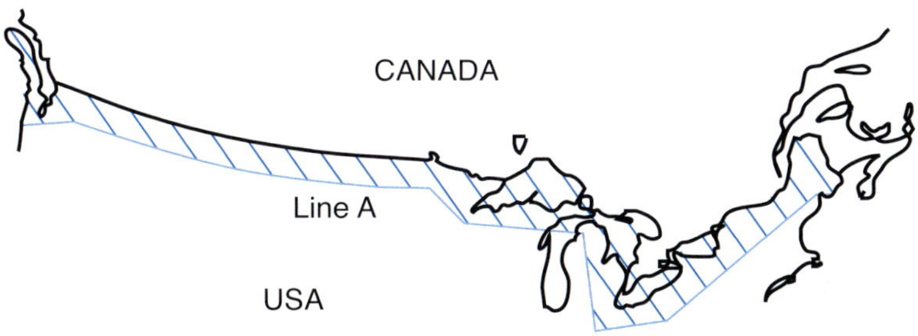

Amateur Radio Restrictions

E1F05 Amateur stations may not transmit in which of the following frequency segments if they are located in the contiguous 48 states and north of Line A?

A. 440 MHz - 450 MHz
B. 53 MHz - 54 MHz
C. 222 MHz - 223 MHz
D. 420 MHz - 430 MHz

Frequencies between 420-430 MHz are normally used for digital modes, fast-scan television, and auxiliary links. ***Within the "Line A" area*** along the Canadian border, ***U.S. hams are prohibited from operating below 430 MHz*** to keep their transmissions from interfering with the different radio allocations Canada has in this frequency range. [97.303] **ANSWER D.**

Good To Know:

Going "down under?" The Australian Communications and Media Authority (ACMA) has issued a new class of license to allow visiting amateurs to operate in Australia for up to 90 days. All you have to do is use your home call sign followed by the suffix "VK" followed by "portable" and then announce the location of your station, without doing anything more.

E1F02 What privileges are authorized in the US to persons holding an amateur service license granted by the government of Canada?

A. None, they must obtain a US license
B. Full privileges of the General class license on the 80-, 40-, 20-, 15-, and 10-meter bands
C. The operating terms and conditions of the Canadian amateur service license, not to exceed US Amateur Extra class license privileges
D. Full privileges, up to and including those of the Amateur Extra class license, on the 80-, 40-, 20-, 15-, and 10-meter bands

When operating in a foreign country, you need to be sure you comply with both your own license terms *and* those of the country in which you operate. This works both ways, as in the example of Canada. While amateur radio privileges are approximately matched in most countries, in many cases the "band borders" are not exactly the same... even those of Canada and the U.S. ***So be sure you comply with both country's rules at all times***. [97.107] **ANSWER C.**

E1C04 What is an IARP?

A. A permit that allows US amateurs to operate in certain countries of the Americas
B. The internal amateur radio practices policy of the FCC
C. An indication of increased antenna reflected power
D. A forecast of intermittent aurora radio propagation

Amateur Radio is an international hobby, and we should take advantage of every opportunity to enhance our international good will. This often involves traveling to foreign countries. ***The International Amateur Radio Permit (IARP) allows you to operate in certain countries in the Americas.*** **ANSWER A.**

E1C11 Which of the following operating arrangements allows an FCC-licensed US citizen to operate in many European countries, and amateurs from many European countries to operate in the US?

A. CEPT
B. IARP
C. ITU reciprocal license
D. All these choices are correct

The ***CEPT agreement*** (European Conference of Postal & Telecommunications Administrations – CEPT) is great news for U.S. Advanced and Extra Class amateur operators ***when visiting many countries in Europe*** – no more reciprocal paperwork! Before leaving for Europe, pack a copy of FCC Public Notice DA 16-1048 along with the original and copies of your U.S. ham license. This paperwork is necessary in case you are asked about the equipment you are bringing into that country at customs. Have fun in Europe with ham radio! [97.5] **ANSWER A.**

E1C06 Which of the following is required in order to operate in accordance with CEPT rules in foreign countries where permitted?

A. You must identify in the official language of the country in which you are operating
B. The US embassy must approve of your operation
C. You must have a copy of FCC Public Notice DA 16-1048
D. You must append "/CEPT" to your call sign

The Central European Postal and Telecommunications (CEPT) authority is the "FCC" for much of Europe. If you travel there, you may operate in participating countries by presenting proof of U.S. Citizenship, your FCC license, and a copy of ***FCC Public Notice DA 16-1048***. Note that CEPT for much of Europe is only *automatically* available to Advanced and Extra Class amateurs, which is yet another incentive to go for the Extra Class license. **ANSWER C.**

Participating CEPT countries, as of February, 2024, are:

Albania
Andora
Austria
Azerbaijan
Belgium
Bosnia and Herzegovina
Bulgaria
Croatia
Cyprus
Czech Republic
Denmark
Estonia
Finland
France
Georgia
Germany
Greece
Hungary
Iceland
Ireland
Italy
Latvia
Liechtenstein
Lithania
Luxembourg
Malta
Moldova
Monaco
Montenegro
Netherlands
North Macedonia
Norway
Poland
Portugal
Romania
San Marino
Serbia
Slovak Republic
Slovenia
Spain
Sweden
Switzerland
Turkey
Ukraine
Vatican
United Kingdom

Good To Know:

European Reciprocal Licenses Now Limited to Advanced & Extra Class Licensees

The European Conference of Postal and Telecommunications Administrations (CEPT) has revised its table of equivalence between FCC amateur licenses and the CEPT license. Effective February 4, 2008, Recommendation T/R 61-01 (as amended) now grants **full CEPT privileges only to those U.S. citizens who hold an FCC-issued Advanced or Amateur Extra Class license**. *This means that those U.S. licensees who hold an FCC-issued General or Technician license are no longer eligible for full operating privileges in countries where CEPT-reciprocal operation had previously been permitted. U.S. Novice class licensees have had no reciprocal operating privileges under the CEPT provisions.*

E1A08 If a station in a message forwarding system inadvertently forwards a message that is in violation of FCC rules, who is primarily accountable for the rules violation?

A. The control operator of the packet bulletin board station
B. The control operator of the originating station
C. The control operators of all the stations in the system
D. The control operators of all the stations in the system not authenticating the source from which they accept communications

A few years ago, dozens of amateur operators were sent "greetings" from the FCC for taking part in forwarding a packet radio message that contained materials specifically prohibited by the rules and regulations. The hams forwarding the messages claimed everything was in the automatic mode and it was not up to them to censor what was coming down the electronic data line! Hence, new rules hold the ***control operator of the station originating the message as primarily accountable*** and the control operator of the first forwarding station accountable for any violation of the rules. More specifically, everyone else "down the line" is no longer responsible for the message content, but must shut down their packet bulletin boards if informed that there is a message being automatically forwarded that violates FCC rules. [97.219] **ANSWER B.**

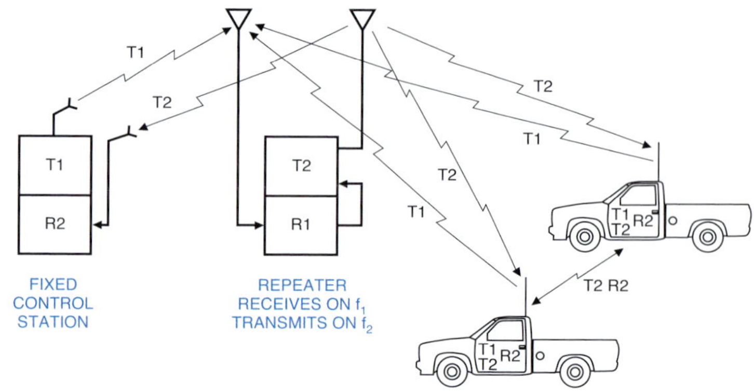

Repeater

E1C05 Under what situation may a station transmit third party communications while being automatically controlled?

A. Never
B. Only when transmitting RTTY or data emissions
C. Only when transmitting SSB or CW
D. On any mode approved by the National Telecommunication and Information Administration

Now that the FCC allows data and RTTY at greater speeds over past, restrictive, HF baud rates, we will see more third party digital messages being passed. As long as an ***automatically controlled station*** is passing this traffic from a station under ***local or remote*** control, they are good to go. **ANSWER B.**

E2C01 What indicator is required to be used by US-licensed operators when operating a station via remote control and the remote transmitter is located in the US?

A. / followed by the USPS two-letter abbreviation for the state in which the remote station is located
B. /R# where # is the district of the remote station
C. / followed by the ARRL Section of the remote station
D. No additional indicator is required

In the past, non-home-based amateur communications were much more restricted and required specific identifiers. ***Remote control designators are no longer required*** during your ID, but you are certainly free to indicate your remote control status. **ANSWER D.**

E1C08 What is the maximum permissible duration of a remotely controlled station's transmissions if its control link malfunctions?

A. 30 seconds
B. 3 minutes
C. 5 minutes
D. 10 minutes

If a distant, remotely-controlled station is undergoing remote control tone commands from an auxiliary station and the control link on microwave fails, that remotely-controlled station must ***shut down within three minutes***. [97.213] **ANSWER B.**

E2E12 How do ALE stations establish contact?

A. ALE constantly scans a list of frequencies, activating the radio when the designated call sign is received
B. ALE radios monitor an internet site for the frequency they are being paged on
C. ALE radios send a constant tone code to establish a frequency for future use
D. ALE radios activate when they hear their signal echoed by back scatter

---

ALE (Automatic Link Establishment) has been around for a while, primarily on military and commercial HF circuits, but has only recently become common on the amateur bands. The ***ALE protocol scans a list of HF frequencies*** and "pings" a remote station to find out which frequency has the best propagation. ALE can be used with any transmission mode; so is not really a "mode" in itself. **ANSWER A.**

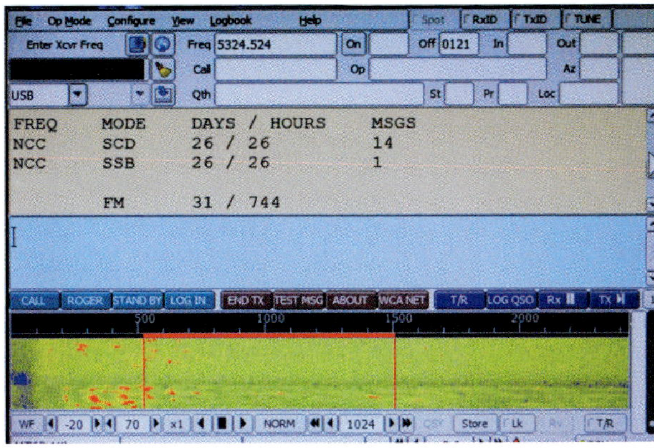

Operating digital is a great way to marry your computer to your ham rig as an Extra Class operator!

E1F10 Who may be the control operator of an auxiliary station?

A. Any licensed amateur operator
B. Only Technician, General, Advanced, or Amateur Extra class operators
C. Only General, Advanced, or Amateur Extra class operators
D. Only Amateur Extra class operators

An amateur radio auxiliary station is a transceiver under your local control which remotely controls another station, like a mountaintop repeater or a distant, high-ridge cross-band system. ***Any amateur radio operator, EXCEPT NOVICE CLASS, may set up and be the control operator of an auxiliary station***. While Novice and Advanced Class licenses are no long being issued, there are hams who still hold valid Novice and Advanced Class licenses. [97.201] **ANSWER B.**

E1D12 Which of the following amateur stations may transmit one-way communications?

A. A space station, beacon station, or telecommand station
B. A local repeater or linked repeater station
C. A message forwarding station or automatically controlled digital station
D. All these choices are correct

As a general rule, amateur radio stations are prohibited from *broadcasting,* which is the ***one-way transmission*** to unspecified recipients. Exceptions to the "non-broadcasting" rule are provided ***for space, beacon, and telecommand stations*** (as well as code practice or general bulletin stations such as W1AW). [97.207(e), 97.203(g)] **ANSWER A.**

E1D06 What is the maximum permitted transmitter output power when operating a model craft by telecommand?

A. 1 watt
B. 2 watts
C. 5 watts
D. 100 watts

Radio control of model vehicles is as old as amateur radio itself. Because radio control transmissions are generally not identified, there is a restriction of "RC" power to ***1 watt***. This is more than enough power for just about any practical remote control application. [97.215(c)] **ANSWER A.**

E1B07 To what type of regulations does PRB-1 apply?

A. Homeowners associations
B. FAA tower height limits
C. State and local zoning
D. Use of wireless devices in vehicles

PRB-1 is an FCC ruling that clarifies that the ***federal government has preemption over state and local zoning regulations***. This means that states and local zoning governments have no jurisdiction over radio transmission and equipment, per se. However, hams who voluntarily enter into agreements with homeowner associations or other covenants are still subject to some antenna restrictions. [97.15] **ANSWER C.**

E1B11 What does PRB-1 require of state and local regulations affecting amateur radio antenna size and structures?

A. No limitations may be placed on antenna size or placement
B. Reasonable accommodations of amateur radio must be made
C. Such structures must be permitted when use for emergency communications can be demonstrated
D. Such structures must be permitted if certified by a registered professional engineer

Because of federal pre-emption, ***state and local governments must make reasonable accommodations for amateur radio*** (primarily regarding antennas). What "reasonable" actually means is currently under debate in many communities and will continue to be so for the foreseeable future. PRB-1 does not negate HOAs, and covenants voluntarily entered into by radio amateurs. [97.15] **ANSWER B.**

E1F06 Under what circumstances might the FCC issue a Special Temporary Authority (STA) to an amateur station?

A. To provide for experimental amateur communications
B. To allow use of a special event call sign
C. To allow a licensee who has passed an upgrade exam to operate with upgraded privileges while waiting for posting on the FCC database
D. To provide temporary operation while awaiting normal licensing.

Ham radio operators are always on the prowl for new types of emissions and vacant frequencies where the ***FCC may allow us special temporary authority***. While obtaining the FCC's STA requires diligence and numerous filings, ham operators have been successful in obtaining permission ***to experiment with new emissions*** that are not even mentioned in the rules and to operate on unused very low frequencies (VLF) for specific propagation studies. These experimental communications may lead to future rule makings on our behalf! [1.931] **ANSWER A.**

E1B09 Which amateur stations may be operated under RACES rules?

A. Only those club stations licensed to Amateur Extra class operators
B. Any FCC-licensed amateur station except a Technician class
C. Any FCC-licensed amateur station certified by the responsible civil defense organization for the area served
D. Only stations meeting the FCC Part 97 technical standards for operation during an emergency

All amateur radio operators are eligible for RACES membership once they are properly ***certified by the civil defense organization responsible for the area served***. You must plan ahead if you want to be part of RACES. It takes official certification and training before you can help. In an emergency the best an uncertified and untrained operator can do is listen and stay out of the way. [97.407] **ANSWER C.**

E1B10 What frequencies are authorized to an amateur station operating under RACES rules?

A. All amateur service frequencies authorized to the control operator
B. Specific segments in the amateur service MF, HF, VHF, and UHF bands
C. Specific local government channels
D. All these choices are correct

Anyone with a valid amateur radio license, including Novice Class, is eligible to be certified by a civil defense organization as a RACES operator. However, ***you do not gain any out-of-band privileges as a RACES operator***. [97.407] **ANSWER A.**

The RACES logo.

E1E03 What is a Volunteer Examiner Coordinator?

A. A person who has volunteered to administer amateur operator license examinations
B. An organization paid by the volunteer examiner team to publicize and schedule examinations
C. An organization that has entered into an agreement with the FCC to coordinate, prepare, and administer amateur operator license examinations
D. The person who has entered into an agreement with the FCC to be the VE session manager

A Volunteer Examiner Coordinator is ***an organization that has an agreement with the FCC to coordinate amateur radio license exams***. Two of the biggest VECs are the American Radio Relay League and the W5YI-VEC. [97.521] **ANSWER C.**

E1E02 Who is tasked by Part 97 with maintaining the pools of questions for all US amateur license examinations?

A. The VEs
B. The FCC
C. The VECs
D. The ARRL

The National Conference of Volunteer Examiner Coordinators (NCVEC) has an internal Question Pool Committee that ensures all exam questions are valid. Each of the three amateur radio question pools is updated every four years. ***The question pools are maintained by all of the VECs***. Hams are encouraged to work with their VECs to help keep the pools stocked with fresh questions. The questions do not come from, nor are they maintained by the FCC! [97.523] **ANSWER C.**

E1E04 What is required to be accredited as a Volunteer Examiner?

A. Each General, Advanced and Amateur Extra class operator is automatically accredited as a VE when the license is granted
B. The amateur operator applying must pass a VE examination administered by the FCC Enforcement Bureau
C. The prospective VE must obtain accreditation from the FCC
D. A VEC must confirm that the VE applicant meets FCC requirements to serve as an examiner

VE accreditation is received from a Volunteer Examiner Coordinator ***(VEC)*** that ***makes sure each VE applicant meets specific FCC requirements*** and has a clean record. [97.509, 97.525] **ANSWER D.**

The accredited Extra Class Volunteer Examiner may test all levels of ham radio licensing.

E1E01 For which types of out-of-pocket expenses do the Part 97 rules state that VEs and VECs may be reimbursed?

A. Preparing, processing, administering, and coordinating an examination for an amateur radio operator license
B. Teaching an amateur operator license examination preparation course
C. No expenses are authorized for reimbursement
D. Providing amateur operator license examination preparation training materials

Volunteer examiners and VECs are allowed to recoup out-of-pocket ***expenses for preparing, processing, administering, and coordinating an examination*** for an amateur radio license. [97.527] **ANSWER A.**

E1E08 To which of the following examinees may a VE not administer an examination?

A. Employees of the VE
B. Friends of the VE
C. Relatives of the VE as listed in the FCC rules
D. All these choices are correct

As a volunteer examiner, ***you may not test your own close relatives, such as kids, grandkids, brothers and sisters, and parents***, as indicated in the FCC Rules. [97.509] **ANSWER C.**

Working with kids as an Elmer or VE is both fun and a great way to help ham radio grow

E1E06 Who is responsible for the proper conduct and necessary supervision during an amateur operator license examination session?

A. The VEC coordinating the session
B. The designated monitoring VE
C. Each administering VE
D. Only the VE session manager

Each one of the administering Volunteer Examiners must continuously supervise everyone in the room taking the exams. The VEs are the final authority for any issue that might arise during the exam session. [97.509] **ANSWER C.**

E1E10 What must the administering VEs do after the administration of a successful examination for an amateur operator license?

A. They must collect and send the documents directly to the FCC
B. They must collect and submit the documents to the coordinating VEC for grading
C. They must submit the application document to the coordinating VEC according to the coordinating VEC instructions
D. They must return the documents to the applicant for submission to the FCC according to the FCC instructions

If you are the head VE for an exam session, called the Administering VE, ***you must submit the application documents to the coordinating VEC***. Follow the instructions of the VEC for how to submit the paperwork. [97.509(h)] **ANSWER C.**

E1E05 What must the VE team do with the application form if the examinee does not pass the exam?

A. Maintain the application form with the VEC's records
B. Return the application document to the examinee
C. Send the application form to the FCC and inform the FCC of the grade
D. Destroy the application form

As an accredited volunteer examiner, you need to know the number of questions an applicant can miss while still achieving a ***minimum passing grade of 74%***. Technician Class and General Class cannot miss more than 9 questions, and pass with 26 correct out of 35. Extra Class cannot miss more than 13 questions, and pass with 37 correct out of 50. [97.509(j)] **ANSWER B.**

E1E11 What must the VE team do if an examinee scores a passing grade on all examination elements needed for an upgrade or new license?

A. Photocopy all examination documents and forward them to the FCC for processing
B. Three VEs must certify that the examinee is qualified for the license grant and that they have complied with the administering VE requirements
C. Issue the examinee the new or upgrade license
D. All these choices are correct

When an examinee upgrades to a new, higher license, the 3-member VE team must ***certify the exam paperwork and sign the CSCE*** (Certificate of Successful Completion of Examination). The applicant holds onto the CSCE as proof of their successful upgrade until the license is issued by the FCC. [97.509(m)] **ANSWER B.**

The Effect of the Ionosphere on Radio Waves

To help you with the questions on radio wave propagation, here is a brief explanation on the effect the ionosphere has on radio waves.

The ionosphere is the electrified atmosphere from 40 miles to 400 miles above the Earth. You can sometimes see it as "northern lights." It is charged-up daily by the Sun, and does some miraculous things to radio waves that strike it. Some radio waves are absorbed during daylight hours by the ionosphere's D layer. Others are bounced back to Earth. Yet others penetrate the ionosphere and never come back again. The wavelength of the radio waves determines whether the waves will be absorbed, refracted, or will penetrate. Here's a quick way to memorize what the different layers do during day and nighttime hours:

The D layer is about 40 miles up. The D layer is a Daylight layer; it almost disappears at night. D for Daylight. The D layer absorbs radio waves between 1 MHz to 7 MHz. These are long wavelengths. All others pass through.

The E layer is also a daylight layer, and it is very Eccentric. E for Eccentric. Patches of E layer ionization may cause some surprising reflections of signals on both high frequency as well as very-high frequency. The E layer height is usually 70 miles.

The F1 layer is one of the layers farthest away. The F layer gives us those Far away signals. F for Far away. The F1 layer is present during daylight hours, and is up around 150 miles. The F2 layer is also present during daylight hours, and it gives us the Furthest range. The F2 layer is 250 miles high, and it's the best for the Farthest range on medium and short waves. The F2 layer is strongest in the summer months. During winter months, both the F1 and F2 layers may become unpredictable, but always strong enough to support exciting skywaves! At nighttime, the F1 and F2 layers combine to become just the F layer at 180 miles. This F layer at nighttime will usually bend radio waves between 1 MHz and 15 MHz back to earth. At night, the D and E layers disappear.

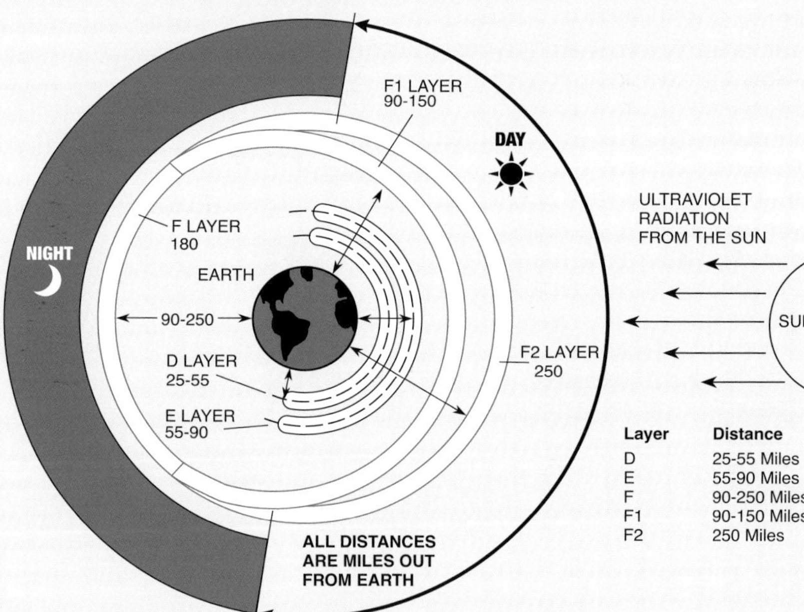

Layer	Distance
D	25-55 Miles
E	55-90 Miles
F	90-250 Miles
F1	90-150 Miles
F2	250 Miles

Ionosphere Layers

Source: *Antennas — Selection and Installation*, © 1986, Master Publishing, Inc., Arlington, Texas

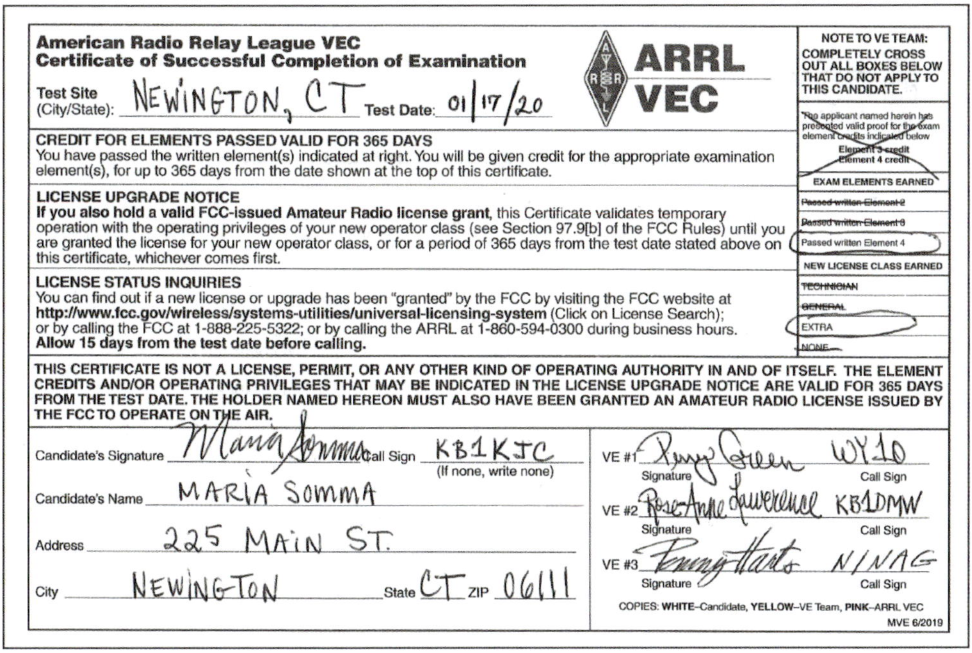

American Radio Relay League VEC
Certificate of Successful Completion of Examination

ARRL VEC

Test Site (City/State): NEWINGTON, CT Test Date: 01/17/20

NOTE TO VE TEAM: COMPLETELY CROSS OUT ALL BOXES BELOW THAT DO NOT APPLY TO THIS CANDIDATE.

The applicant named herein has presented valid proof for the exam element credits indicated below
Element 3 credit
Element 4 credit

CREDIT FOR ELEMENTS PASSED VALID FOR 365 DAYS
You have passed the written element(s) indicated at right. You will be given credit for the appropriate examination element(s), for up to 365 days from the date shown at the top of this certificate.

EXAM ELEMENTS EARNED
~~Passed written Element 2~~
~~Passed written Element 3~~
Passed written Element 4

LICENSE UPGRADE NOTICE
If you also hold a valid FCC-issued Amateur Radio license grant, this Certificate validates temporary operation with the operating privileges of your new operator class (see Section 97.9[b] of the FCC Rules) until you are granted the license for your new operator class, or for a period of 365 days from the test date stated above on this certificate, whichever comes first.

NEW LICENSE CLASS EARNED
~~TECHNICIAN~~
~~GENERAL~~
EXTRA
~~NONE~~

LICENSE STATUS INQUIRIES
You can find out if a new license or upgrade has been "granted" by the FCC by visiting the FCC website at **http://www.fcc.gov/wireless/systems-utilities/universal-licensing-system** (Click on License Search); or by calling the FCC at 1-888-225-5322; or by calling the ARRL at 1-860-594-0300 during business hours. **Allow 15 days from the test date before calling.**

THIS CERTIFICATE IS NOT A LICENSE, PERMIT, OR ANY OTHER KIND OF OPERATING AUTHORITY IN AND OF ITSELF. THE ELEMENT CREDITS AND/OR OPERATING PRIVILEGES THAT MAY BE INDICATED IN THE LICENSE UPGRADE NOTICE ARE VALID FOR 365 DAYS FROM THE TEST DATE. THE HOLDER NAMED HEREON MUST ALSO HAVE BEEN GRANTED AN AMATEUR RADIO LICENSE ISSUED BY THE FCC TO OPERATE ON THE AIR.

Candidate's Signature Maria Somma Call Sign KB1KJC (If none, write none)
Candidate's Name MARIA SOMMA
Address 225 MAIN ST.
City NEWINGTON State CT ZIP 06111

VE #1 Signature Call Sign WY1O
VE #2 Signature Rose-Anne Lawrence Call Sign KB1DMW
VE #3 Signature Call Sign N1NAG

COPIES: WHITE–Candidate, YELLOW–VE Team, PINK–ARRL VEC
MVE 6/2019

Sample CSCE.

E1E07 What should a VE do if a candidate fails to comply with the examiner's instructions during an amateur operator license examination?

A. Warn the candidate that continued failure to comply will result in termination of the examination
B. Immediately terminate the candidate's examination
C. Allow the candidate to complete the examination, but invalidate the results
D. Immediately terminate everyone's examination and close the session

If an examinee fails to comply with your instructions, you ***immediately terminate that candidate's examination***. [97.509, 97.511] **ANSWER B.**

E1E09 What may be the penalty for a VE who fraudulently administers or certifies an examination?

A. Revocation of the VE's amateur station license grant and the suspension of the VE's amateur operator license grant
B. A fine of up to $1,000 per occurrence
C. A sentence of up to one year in prison
D. All these choices are correct

Any VE who is caught administering a fraudulent exam may have his or her ***station license revoked and the VE's amateur radio license grant suspended***. [97.509] **ANSWER A.**

Skywaves & Contesting

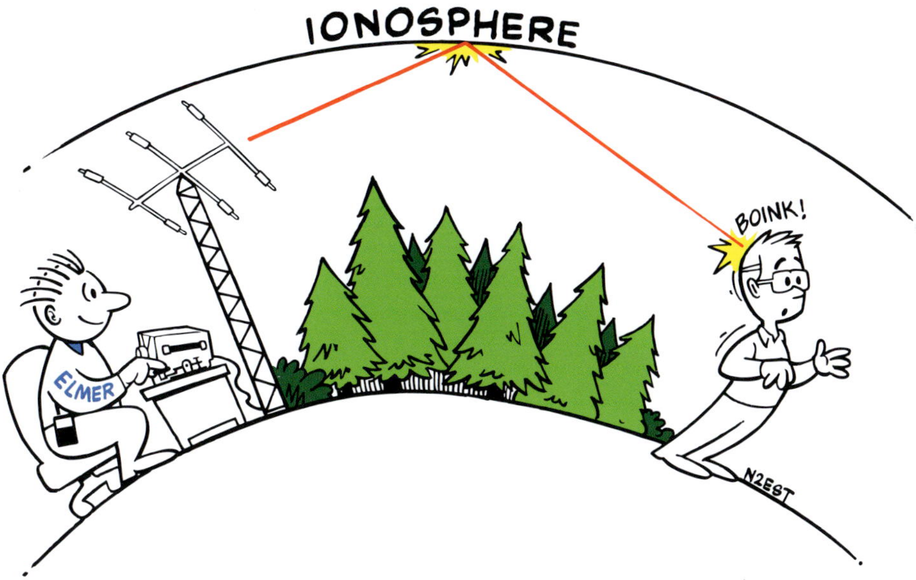

E3B04 What are "extraordinary" and "ordinary" waves?

A. Extraordinary waves exhibit rare long-skip propagation, compared to ordinary waves, which travel shorter distances
B. Independently propagating, elliptically polarized waves created in the ionosphere
C. Long-path and short-path waves
D. Refracted rays and reflected waves

Any time you launch an ***HF signal*** into the ionosphere, it ***splits into two elliptically polarized "characteristic waves,"*** one being clockwise (O mode or "ordinary"), and one being counterclockwise (X mode or "extraordinary"). These two waves follow very different paths, especially near the polar regions where the Earth's magnetic field is strong. **ANSWER B.**

E3A14 What are circularly polarized electromagnetic waves?

A. Waves with an electric field bent into a circular shape
B. Waves with rotating electric and magnetic fields
C. Waves that circle Earth
D. Waves produced by a loop antenna

Some satellite enthusiasts will use circular polarization to improve their transmission and reception through an orbiting satellite. With ***circular polarization, the electric and magnetic fields rotates***. It could be right-hand circular or left-hand circular, depending on the antenna feed system. **ANSWER B.**

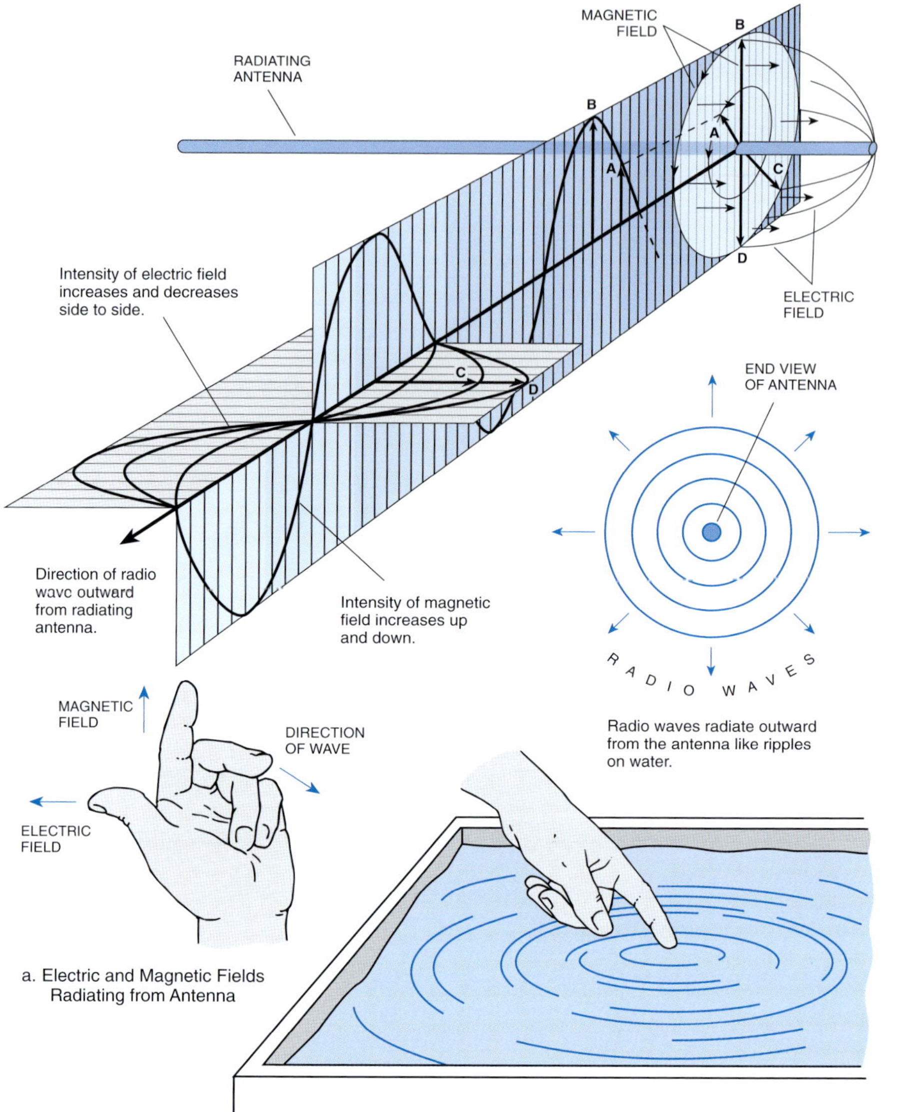

a. Electric and Magnetic Fields Radiating from Antenna

b. Radiation Pattern of Radio Waves Around Antenna

Radio Waves

Source: Basic Electronics © 1994, 2000 Master Publishing, Inc., Arlington, Texas

---

E3C11 What does VOACAP software model?

A. AC voltage and impedance
B. VHF radio propagation
C. HF propagation
D. AC current and impedance

The Voice of America Coverage Analysis Program, or ***VOACAP***, is one of the oldest and most familiar ***HF radio propagation*** tools. Other programs such as IONCAP were derived from VOACAP. If you're an avid contester or casual DXer, you can benefit from learning how to use VOACAP and its kin. **ANSWER C.**

E3C01 What is the cause of short-term radio blackouts?

A. Coronal mass ejections
B. Sunspots on the solar equator
C. North-oriented interplanetary magnetic field
D. Solar flares

As we ride the peak of Solar Cycle 25, the sun regularly erupts with solar flares when the magnetic fields leap out and arc to re-connect back. These powerful flares emit UV and hard X-rays traveling at the speed of light. If these flares are strong enough, and facing the earth, there is a near instant increase in disturbed ionization to our ionospheric regions, to the point where High Frequency refracted signals quickly disappear on your radio dial! This is a short-term radio blackout, lasting up to an hour at most, caused by this solar flare and the near instant extra radiation it emits. This radiation could also be a danger to satellites and our Space Station. The incorrect answer, coronal mass ejection, will not create a radio blackout, as these CME's are slower moving particles that take about three days to raise the noise floor by a couple of S units on HF. ***It is the solar flare that may create a radio blackout.*** **ANSWER D.** **http://www.spaceweather.com**

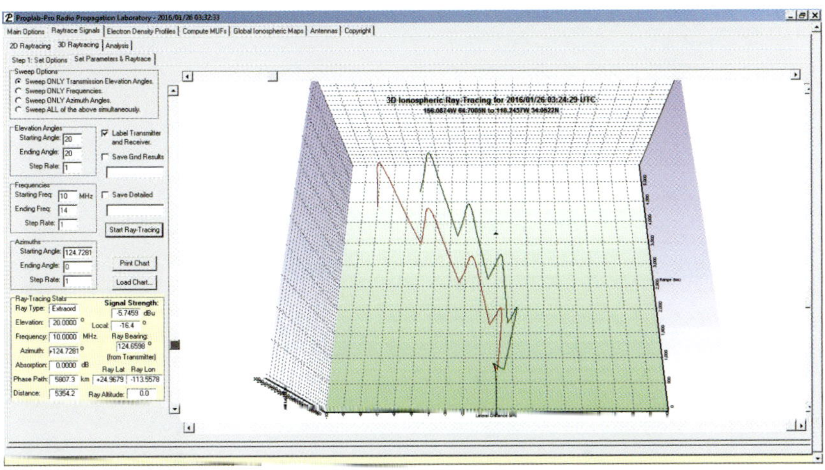

Ray Tracing can give you a 3D view of the path your signal travels around the world.

E3B06 On which of the following amateur bands is long-path propagation most frequent?

A. 160 meters and 80 meters
B. 40 meters and 20 meters
C. 10 meters and 6 meters
D. 6 meters and 2 meters

The term "long path" describes working a station in the opposite direction from the one that is closest to your location. The amateur radio ***20 and 40 meter bands are the best ones for long-path*** DX with a relatively modest directional antenna system. When you pass your upcoming Extra Class exam, you will gain an exclusive 20 meter CW window from 14.000 to 14.025 MHz and an exclusive Extra Class window from 14.150 to 14.175 MHz for CW, voice and data. These are great for working rare 20 meter DX on a band shared only with other Extra Class operators in the United States! **ANSWER B.**

E3B05 Which of the following paths is most likely to support long-distance propagation on 160 meters?

A. A path entirely in sunlight
B. Paths at high latitudes
C. A direct north-south path
D. A path entirely in darkness

Long-range, ***long-path*** propagation can be found on the worldwide bands from ***10 meters down to 160 meters***. Expect almost daily long-path propagation on 20 meters, which is best in the late evening or early morning hours. **ANSWER D.**

Good To Know:

Once you earn your Extra Class license, you will stay up for long hours working rare DX. Just as the Sun is rising here, it is setting over Europe or Asia. Working stations in the twilight at both ends of the circuit may lead to some extraordinary DX contacts that can last up to 30 minutes. This is called grayline propagation.

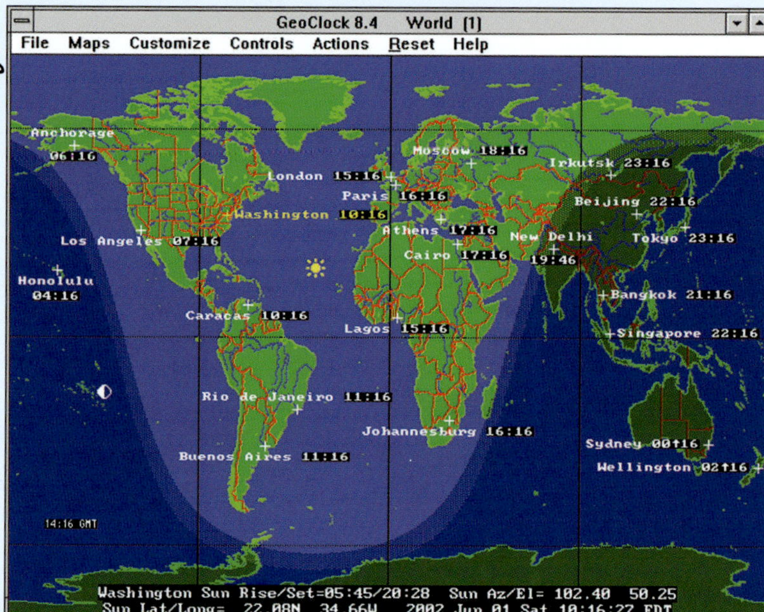

Gray line conditions follow the Sun's terminator. It's a 10 minute opportunity at great DX for stations at each end of the gray line terminator.

E3B08 How does the maximum range of ground-wave propagation change when the signal frequency is increased?

A. It stays the same
B. It increases
C. It decreases
D. It peaks at roughly 8 MHz

Once you earn your Extra Class license, you will stay up for long hours working rare DX. Just as the Sun is rising here, it is setting over Europe or Asia. Working stations in the twilight at both ends of the circuit may lead to some extraordinary DX contacts that can last up to 30 minutes. This is called ***grayline propagation***. **ANSWER C.**

E3A06 What should be done to continue a long-distance contact when the MUF for that path decreases due to darkness?

A. Switch to a higher frequency HF band
B. Switch to a lower frequency HF band
C. Change to an antenna with a higher takeoff angle
D. Change to an antenna with greater beam width

We are at the peak of Solar Cycle 25, with all High Frequency bands teaming with activity day and night! After your local sunset, you may find that 10 meter and 15 meter DX is disappearing, but drop down to a lower band, like 20 and 40 meters, and continue to work the world! **ANSWER B.**

E3B10 What is the effect of chordal-hop propagation?

A. The signal experiences less loss compared to multi-hop propagation, which uses Earth as a reflector
B. The MUF for chordal-hop propagation is much lower than for normal skip propagation
C. Atmospheric noise is reduced in the direction of chordal-hop propagation
D. Signals travel faster along ionospheric chords

Chordal hop signals can be unusually strong, experiencing ***less loss*** compared to skip propagation. There is even some tantalizing evidence that *amplification* can take place along a chordal hop path. (This process can be duplicated in the plasma physics laboratory, while it is unknown if all the proper conditions indeed exist "in the wild." This is wide open for further amateur experimentation.) Chordal hop signals always occur along the "dark path" opposite the Sun. **ANSWER A.**

E3B12 What is chordal-hop propagation?

A. Propagation away from the great circle bearing between stations
B. Successive ionospheric refractions without an intermediate reflection from the ground
C. Propagation across the geomagnetic equator
D. Signals reflected back toward the transmitting station

Chordal hop propagation can result in some surprisingly loud signals, especially on 80 meters. Since there are ***no ground reflections*** in between sky reflections, the losses are much less than "normal" multi-hop propagation. One additional indication that chordal hop is in play is that there's a huge dead "skip zone" in the middle of the path, which is along the dark side of the Earth. For further information and some great diagrams showing chordal hop propagation, check out this website: **http://www.antennex.com/ prop/prop0806/prop0806.pdf**
ANSWER B.

Chip's home brew 10 meter beam gets a shot at chordal hop propagation on the 10 meter band for loud DX signals!

E3B01 Where is transequatorial propagation (TEP) most likely to occur?

A. Between points separated by 2,000 miles to 3,000 miles over a path perpendicular to the geomagnetic equator

B. Between points located 1,500 miles to 2,000 miles apart on the geomagnetic equator

C. Between points located at each other's antipode

D. Through the region where the terminator crosses the geographic equator

One unusual and often exciting propagation mode is transequatorial (TE) propagation, which travels over a path of 2,000 to 3,000 miles perpendicular to the geomagnetic equator. There is still a great deal of mystery about TE, and it is a wide-open area for experimentation by the VHF and UHF enthusiast. There are a limited number of locations where TE can be experienced in the US. (Alaska is certainly not one of them!) **ANSWER A.**

E3B02 What is the approximate maximum range for signals using transequatorial propagation?

A. 1,000 miles
B. 2,500 miles
C. 5,000 miles
D. 7,500 miles

The maximum total distance for a TE (transequatorial) contact is about ***5,000 miles***. **ANSWER C.**

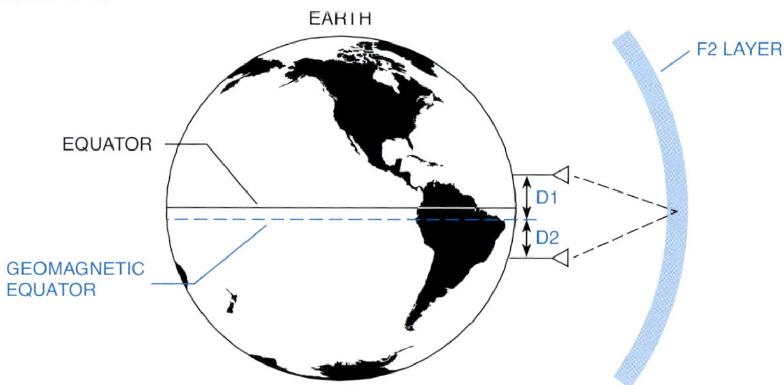

For transequatorial propagation to occur, stations at D1 and D2 must be equidistant from the geomagnetic equator. Popular theory is that reflections occur in and off the F2 layer during peak sunspot activity.

Transequatorial Propagation

E3B03 At what time of day is transequatorial propagation most likely to occur?

A. Morning
B. Noon
C. Afternoon or early evening
D. Late at night

The ***best time of day for TE*** propagation is ***mid-afternoon or early evening***. By this time of day, the Sun's ultraviolet radiation has had time to "warm up" the possible path. **ANSWER C.**

E3C12 Which of the following is indicated by a sudden rise in radio background noise across a large portion of the HF spectrum?

A. A temperature inversion has occurred

B. A coronal mass ejection impact or a solar flare has occurred

C. Transequatorial propagation on 6 meters is likely

D. Long-path propagation on the higher HF bands is likely

While having a steady bath of UV radiation is good for the ionosphere, being blasted with hard X-rays and charged particles from a ***solar flare*** is not. During such an event, radio absorption in the D layer can be almost total. In addition, gigantic blobs of fast-moving charged particles tend to wrinkle the ionosphere to the point that it's no longer a reflective or refractive layer. An ***increase in noise*** also results as ionized particles, primarily in the D layer, recombine and emit vast quantities of random radio waves (and even visible light). **ANSWER B.**

E3C02 What is indicated by a rising A-index or K-index?

A. Increasing disruption of the geomagnetic field
B. Decreasing disruption of the geomagnetic field
C. Higher levels of solar UV radiation
D. An increase in the critical frequency

A and K indices are provided by data from terrestrial *magnetometers.* They are sensitive indicators of solar activity, but need to be interpreted carefully to be truly useful. Learn to read the A and K indices; as they rise they are telling you there is an ***increasing disturbance of the geomagnetic field***. **ANSWER A.**

E3C03 Which of the following signal paths is most likely to experience high levels of absorption when the A-index or K-index is elevated?

A. Transequatorial
B. Through the auroral oval
C. Sporadic-E
D. NVIS

Since the Earth's magnetic field lines are concentrated near the magnetic poles (not the geographic poles), magnetic disruptions are also mostly concentrated in these regions. The ionosphere near the auroral oval polar regions is a magnetized plasma which has some interesting properties. Not only is it absorptive to radio waves, but it is *directionally* absorptive. Don't let anyone tell you there's no such thing as one- way propagation. It is the *norm* in ***polar regions!*** **ANSWER B.**

E3C04 What does the value of B_z (B sub z) represent?

A. Geomagnetic field stability
B. Critical frequency for vertical transmissions
C. North-south strength of the interplanetary magnetic field.
D. Duration of long-delayed echoes

The Sun and Earth have an ***Interplanetary magnetic field*** which contributes to the character of the Earth's ionosphere. Just like a familiar bar magnet, the magnets embedded in the Sun and the Earth have poles that either attract or repel depending on their orientation. The greatest effect occurs when the Sun's north pole magnetically connects with the Earth's south pole. The magnetic field orientation is defined relative to the interplanetary, which is the plane of the orbit of the Earth around the Sun. The Z axis of B (the magnetic field) is the north/south axis of the interplanetary field, perpendicular to the ecliptic, which is roughly the same direction as the Earth's north/south axis. The X and Y axes have minimal effect on the Sun/Earth interactions and are generally ignored. **ANSWER C.**

E3C05 What orientation of B_z (B sub z) increases the likelihood that charged particles from the Sun will cause disturbed conditions?

A. Southward
B. Northward
C. Eastward
D. Westward

It is standard practice to designate the type of field with a subscript denoting the AXIS. In this case B designates a magnetic field, the interplanetary magnetic field, and Z designates a Z axis, or UP/DOWN orientation relative to the ecliptic, the plane of the Earth's orbit around the Sun. Just as two bar magnets interact most strongly when the opposite poles are near each other, ***the Earth has the greatest interaction with a magnetic field when that field is oriented north-to-south (or southward).*** Now the real interesting part is that, unlike the Earth's magnetic field, which is fairly fixed, an interplanetary solar field is carried by solar wind, which means it's as variable as the weather! Interplanetary magnetism can range from almost nothing to huge, and it can change very suddenly. Any magnetic field that is not parallel to the Earth's field causes very little interaction. **ANSWER A.**

Good To Know:

WWV AND WWVH SPACE WEATHER BROADCASTS

The National Oceanic and Atmospheric Administration (NOAA) uses WWV and WWVH to broadcast geophysical alert messages that provide information about solar terrestrial conditions. Geophysical alerts are broadcast from WWV at 18 minutes after the hour and from WWVH at 45 minutes after the hour. The messages are less than 45 seconds in length and are updated every 3 hours (typically at 0000, 0300, 0600, 0900, 1200, 1500, 1800, and 2100 UTC). More frequent updates are made when necessary.

The geophysical alerts provide information about the current conditions for long distance HF radio communications. The alerts use a standardized format and terminology that requires some explanation. Before looking at a sample message, let's define some of the terminology:

Solar flux is a measurement of the intensity of solar radio emissions with a wavelength of 10.7 cm (a frequency of about 2800 MHz). The daily solar flux measurement is recorded at 2000 UTC by the Dominion Radio Astrophysical Observatory of the Canadian National Research Council located at Penticton, British Columbia. The value broadcast is in solar flux units that range from a theoretical minimum of about 50 to numbers larger than 300. During the early part of the 11-year sunspot cycle, the flux numbers are low; but they rise and fall as the cycle proceeds. The numbers will remain high for extended periods around sunspot maximum.

The A and K indices are a measurement of the behavior of the magnetic field in and around the Earth. The K index uses a scale from 0 to 9 to measure the change in the horizontal component of the geomagnetic field. A new K index is determined and added to the broadcast every 3 hours based on magnetometer measurements made at the Table Mountain Observatory, north of Boulder, Colorado, or an alternate middle latitude observatory. The A index is a daily value on a scale from 0 to 400 to express the range of disturbance of the geomagnetic field. It is obtained by converting and averaging the eight, 3-hour K index

When we get news of a flare on the Sun like this, expect some unsettled band conditions about 40 hours later.

E3C07 Which of the following indicates the greatest solar flare intensity?

A. Class A
B. Class Z
C. Class M
D. Class X

Think of X class as being Xtreme. (As we've noted before, physicists can't spell). In any case, ***X class solar flares are extremely disruptive*** as they emit large quantities of charged particles and, as you might suspect, a lot of X-rays; which can certainly create a lot of ionization, but not the type that you want. They penetrate the atmosphere so far that they excite the absorptive D layer far beyond what is normal. The nice thing is that solar flares don't last forever. Visit Spaceweatherwoman.com to learn more about space weather coming from the Sun. **ANSWER D.**

values of a given day. An estimate of the A index is first announced at 2100 UTC, based on 7 measurements and 1 estimated value. At 0000 UTC, the announced A index consists entirely of known measurements, and the word "estimated" is dropped from the announcement.

Space Weather describes the conditions in space that affect Earth and its technological systems. Space weather is a consequence of the behavior of the Sun, the nature of Earth's magnetic field and atmosphere, and our location in the solar system.

Space Weather storms observed and expected are characterized using the NOAA Space Weather scales. The abbreviated table below shows the levels of activity that are included in the announcements and the associated terminology. The descriptor used to identify observed or expected conditions is the maximum level reached or predicted. The NOAA Space Weather Scales are further described at the Space Environment Center's web site.

Space Weather Storm Levels

The NOAA Space Weather Scales were introduced as a way to communicate to the general public the current and future space weather conditions and their possible effects on people and systems. The scales describe the environmental disturbances for three event types: geomagnetic storms G1 thru G5), solar radiation storms (S1 thru S5), and radio blackouts (R1 thru R5). The scales have numbered levels, analogous to hurricanes, tornadoes, and earthquakes that convey severity. In addition, the scales for radio blackouts also show the peak X-ray levels and flux generated by these solar events.

Read the full explanation of these scales and learn more about space weather at → **swpc.noaa.gov/noaa-scales-explanation**.
To hear current space weather levels dial 303-497-3235.

STORM LEVELS	G Scale	S Scale	R Scale	Peak X-Ray Levels & Flux	K Index
Minor	G1	S1	R1	M1 (10^{-5})	K5 or less
Moderate	G2	S2	R2	M5 (5×10^{-5})	K6
Strong	G3	S3	R3	X1 (10^{-4})	K7
Severe	G4	S4	R4	X10 (10^{-3})	K8
Extreme	G5	S5	R5	X20 (2 x 10-3)	K9

E3C09 What type of data is reported by amateur radio propagation reporting networks?

A. Solar flux
B. Electric field intensity
C. Magnetic declination
D. Digital-mode and CW signals

The amateur radio reverse beacon network (RBN), the WSPR net, and several other networks allow you to receive your signal from a distant location to determine propagation characteristics, as well as your own transmitted power. These networks use CW or digital modes. You can see a map of many, if not most RBN monitor stations, here: **https://www.reversebeacon.net/main.php?rows=10&max_age=10,hours&hide=distance_km>** **ANSWER D.**

E3C08 Which of the following is the space-weather term for an extreme geomagnetic storm?

A. B9
B. X5
C. M9
D. G5

Radio propagation on Earth is closely connected to what's happening on the Sun. Many of the parameters that affect propagation are defined by the type of space weather activity, designated by a letter, and the intensity of that activity specified by a number. A G5 report tells us there's a geomagnetic disturbance, with an intensity of five on a scale of one to five. For a complete listing of all the space weather parameters, check out NOAA's website: **https://www.swpc.noaa.gov/content/space-weather-glossary#x** Also, take a look at Appendix A of the glossary: **https://www.swpc.noaa.gov/noaa-scales-explanation** **ANSWER D.**

E3C10 What does the 304A solar parameter measure?

A. The ratio of X-ray flux to radio flux, correlated to sunspot number
B. UV emissions at 304 angstroms, correlated to the solar flux index
C. The solar wind velocity at an angle of 304 degrees from the solar equator, correlated to geomagnetic storms
D. The solar emission at 304 GHz, correlated to X-ray flare levels

Ultraviolet emissions are a good indication of sunspot activity. UV is the primary ingredient necessary for the normal ionosphere to form. ***The 304A solar parameter measures UV emissions at 304 angstroms.*** Ultraviolet actually encompasses a large chunk of the electromagnetic spectrum, as can be seen here: **http://csep10.phys.utk.edu/astr162/lect/light/spectrum.html** Because of its wide range of wavelengths, ultraviolet is usually subdivided into two or more regions. **ANSWER B.**

E3B11 At what time of day is sporadic-E propagation most likely to occur?

A. Between midnight and sunrise
B. Between sunset and midnight
C. Between sunset and sunrise
D. Between sunrise and sunset

While Sporadic-E appears to be a seasonal phenomenon, it doesn't seem to care much about the time of day. This further adds to the mystery and excitement of the mode. You can count on the sun for exciting VHF, short-lived, sporadic-E region refraction, from sunup to sundown. It typically lasts for only a few minutes on 2 meters, so keep your contact short to let others work this DX! **ANSWER D.**

E3B09 At what time of year is sporadic-E propagation most likely to occur?

A. Around the solstices, especially the summer solstice
B. Around the solstices, especially the winter solstice
C. Around the equinoxes, especially the spring equinox
D. Around the equinoxes, especially the fall equinox

Sporadic E is, as the name suggests, ***sporadic.*** It can actually occur any time of the year and is often a surprise when it does. However, it is most likely to occur during the ***winter solstice, and especially during the summer solstice.*** The reasons for this are not yet fully understood, but the phenomenon is well documented. **ANSWER A.**

Here is a link dedicated to Sporadic E "sightings" and some possible predictions: **http://www.uksmg.org/content/sporade.htm**

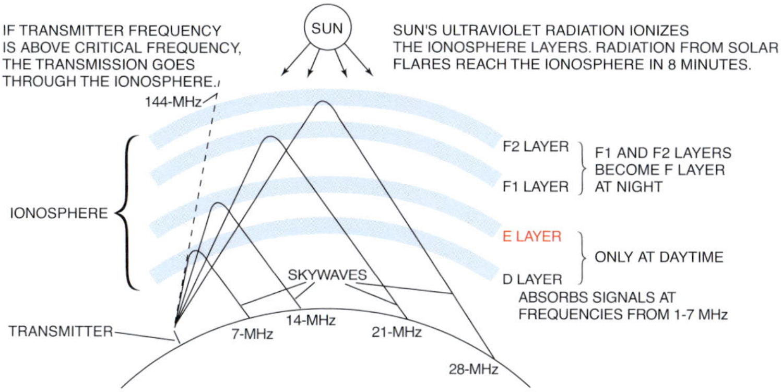

Ionosphere and Its Layers

E3A08 When a meteor strikes the Earth's atmosphere, a linear ionized region is formed at what region of the ionosphere?

A. The E region
B. The F1 region
C. The F2 region
D. The D region

It is within the ***E-region*** of the ionosphere that 6- and 2-meter excitement like Sporadic-E and ***meteor scatter contacts*** take place. Think mEtEor! **ANSWER A.**

E3A09 Which of the following frequency ranges is most suited for meteor-scatter communications?

A. 1.8 MHz - 1.9 MHz
B. 10 MHz - 14 MHz
C. 28 MHz - 148 MHz
D. 220 MHz - 450 MHz

During a recent meteor shower, Gordo was able to maintain 30-second meteor-scatter QSOs on the 10 meter band; 10-second QSOs on the 6 meter band, and 3-second QSOs on the 2 meter band. The higher you go in frequency, the shorter the contact time. ***28-148 MHz are the best*** areas to hear the effects of ***meteor scatter***. **ANSWER C.**

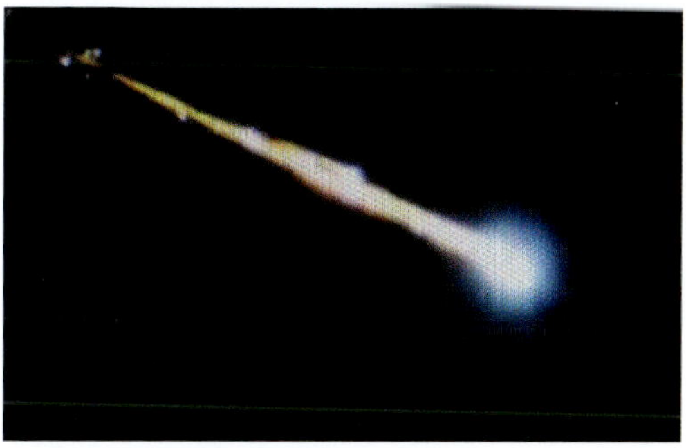

Making meteor contacts is a great way to work CW during a 3-second "burn."

E2D01 Which of the following digital modes is designed for meteor scatter communications?

A. WSPR
B. MSK144
C. Hellschreiber
D. APRS

If you are into computers, ham radio digital signaling is just the thing for you! There are many software programs that work with your computer's sound card and there are excellent add-on radio signal controllers that let the more sophisticated transceivers both transmit and receive ham radio digital calls. ***MSK 144*** is the fastest and most common digital code ***for meteor-scatter contacts***, with strong error correction, increased sensitivity over earlier modes, and can lead to an exciting "catch" of other hams bouncing signals off of meteor trails. **ANSWER B.**

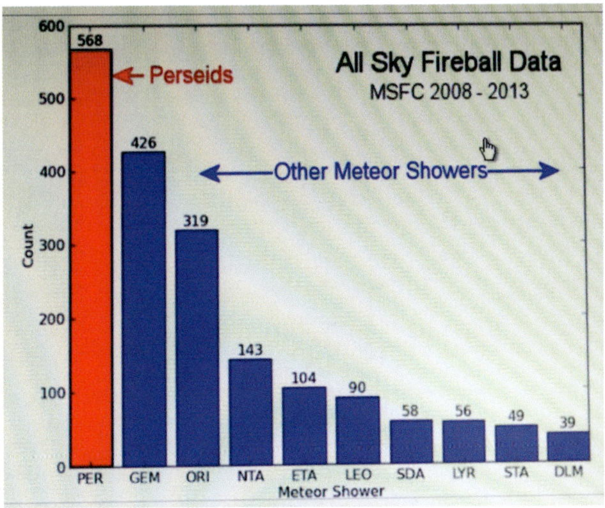

Perseids in August, Geminids in December, and Orionids in October are favorite ham radio meteor shower events.

E2D02 What information replaces signal-to-noise ratio when using the FT8 or FT4 modes in a VHF contest?

A. RST report
B. State abbreviation
C. Serial number
D. Grid square

Formatted digital modes, such as FT8 and FT4 (as opposed to "free-lance" modes such as PSK-31), have severe limitations on the amount of information that can be exchanged. Since VHF and UHF operators usually like to exchange grid square data (which is less important for HF operation), some information must be sacrificed in the data block to accommodate a station's grid square information. It is standard practice on VHF and UHF to insert grid square data into the signal-to-noise ratio field when using FT4 or FT6. **ANSWER D.**

E3A12 What is most likely to result in auroral propagation?

A. Meteor showers
B. Quiet geomagnetic conditions
C. Severe geomagnetic storms
D. Extreme low-pressure areas in polar regions

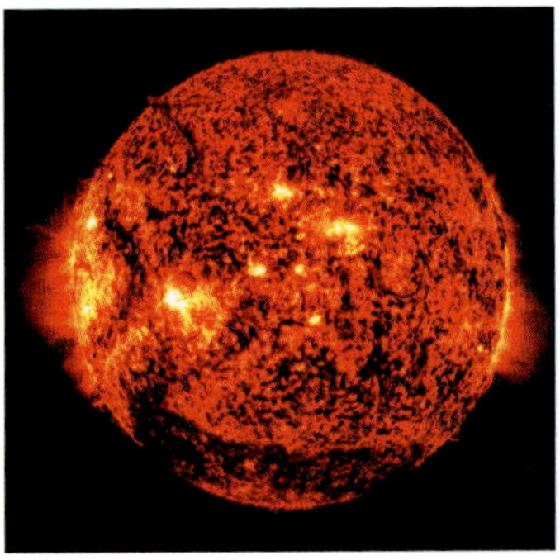

Solar flares and sunspots affect radiowave propagation.
Photo courtesy of N.A.S.A.

The same processes that create beautiful auroras, i.e. severe geomagnetic storms, also usually wreak havoc with radio propagation. In interior Alaska, when there is an active aurora, the best thing one can do is go outside and look at the aurora, because there certainly won't be any ham radio fun! In lower latitudes, active auroras CAN result in some unusual propagation, especially at VHF and UHF. **ANSWER C.**

E3A13 Which of these emission modes is best for auroral propagation?

A. CW
B. SSB
C. FM
D. RTTY

Bouncing a signal off of an aurora is difficult. CW always provides the best opportunity for someone hearing raspy sounds coming into their receiver. On many ***auroral contacts on CW***, we don't even hear a tone only short and long rushes of super-imposed noise. Play Gordo's audio Extra Class course and hear for yourself. **ANSWER A.**

This aurora photo is from Chip and Janet Margelli, K7JA and KL7MF, during their week of radio fun in Coldfoot, Alaska.

E3C06 How does the VHF/UHF radio horizon compare to the geographic horizon?

A. It is approximately 15 percent farther
B. It is approximately 20 percent nearer
C. It is approximately 50 percent farther
D. They are approximately the same

Radio waves bend slightly over the horizon because of the difference in the air's refractive index at higher altitudes. You can calculate your approximate VHF range to the horizon with the formula: D (miles) = 1.415 x $\sqrt{2H}$ (in feet). Determine H (the height in feet) of the station you are wishing to communicate with, calculate its distance to the horizon, and add the two answers together for your rock-solid communications range. Depending on local weather conditions, a ***15%*** to 30% ***range enhancement over the geometric horizon*** will usually take place at VHF and UHF radio frequencies. **ANSWER A.**

E3A05 How are the component fields of an electromagnetic wave oriented?

A. They are parallel
B. They are tangential
C. They are at right angles
D. They are 90 degrees out of phase

The Electric and Magnetic components of an electromagnetic wave are perpendicular, or orthogonal, to one another. The direction of wave propagation is orthogonal to the magnetic and electric fields. All three components are known as mutually orthogonal. **ANSWER C.**

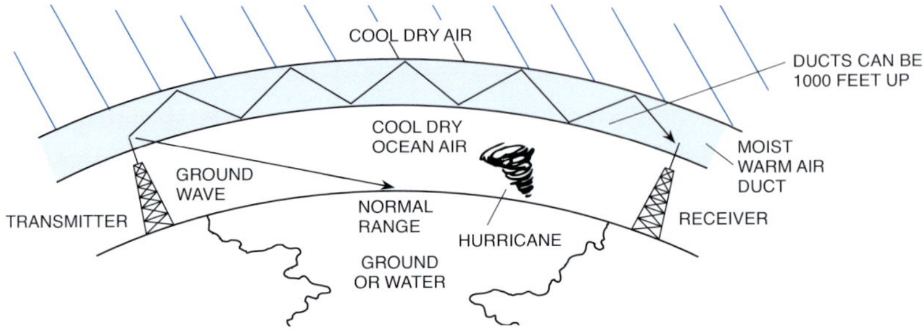

Tropospheric Ducting

E3A07 Atmospheric ducts capable of propagating microwave signals often form over what geographic feature?

A. Mountain ranges
B. Stratocumulus clouds
C. Large bodies of water
D. Nimbus clouds

Ducting works best when the two "walls" forming the duct are flat and parallel. This is most likely to happen over a large flat surface such as a ***body of water***, but can also occur over a large expanse of desert. **ANSWER C.**

VHF and UHF signals will travel freely within the red and orange areas of troposphere ducting. See Hawaii at bottom left? We work it tropo every summer for days on end on 2 meters through 5.6 GHz!

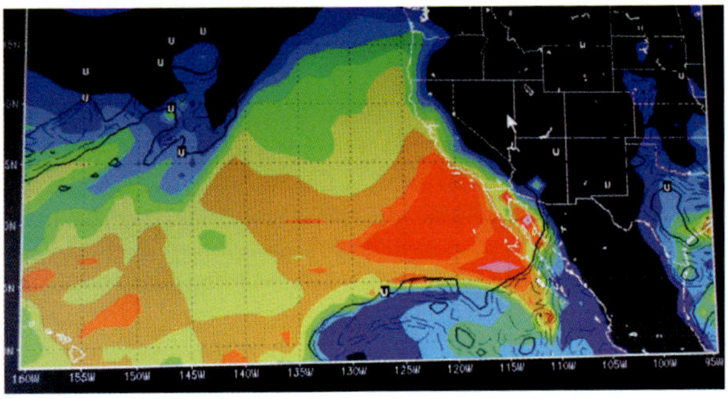

Hepburn Report: www.dxinfocentre

E3A10 What determines the speed of electromagnetic waves through a medium?

A. Resistance and reactance
B. Evanescence
C. Birefringence
D. The index of refraction

Here is a visual of a tropo duct—a mirage as images are refracted from above to the roadway below.

In free space, electromagnetic waves of any frequency travel at the speed of light. (In "normal" air, the speed of propagation is very nearly the same as in a vacuum.) Radio waves will travel more slowly through most solid and liquid substances. The index of refraction of the substance determines the velocity of propagation through the substance. For the simplest case, the index of refraction is inversely proportional to the velocity factor. **ANSWER D.**

E3A11 What is a typical range for tropospheric duct propagation of microwave signals?

A. 10 miles to 50 miles
B. 100 miles to 300 miles
C. 1,200 miles
D. 2,500 miles

Because the troposphere is so low compared to ionospheric regions, it stands to reason that "skip distances" will be a lot less for tropospheric modes than, say F2 ionospheric reflections. ***Tropo propagation is typically on the order of 100 to 300 miles***, but can occasionally be much farther because of multiple hops, or "mixed mode" propagation. **ANSWER B.**

E3A04 In what direction does an electromagnetic wave travel?

A. It depends on the phase angle of the magnetic field
B. It travels parallel to the electric and magnetic fields
C. It depends on the phase angle of the electric field
D. It travels at the right angle to the electric and magnetic fields

In an electromagnetic wave, the electric and magnetic fields are perpendicular to each other, but in phase, the direction of propagation is at right angles to either. The direction of propagation can be determined by the ***RIGHT HAND RULE***. **ANSWER D.**

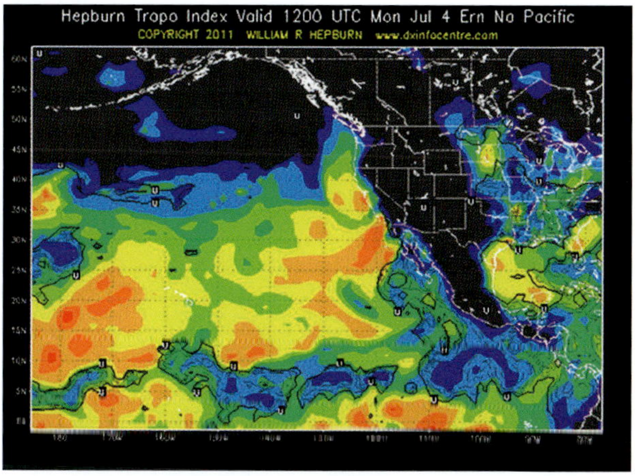

Great tropo conditions seen on the West Coast and in the Gulf between Texas and Florida!

E3B13 What type of polarization is best for ground-wave propagation?

A. Vertical
B. Horizontal
C. Circular
D. Elliptical

We use ***vertical polarization for ground-wave propagation***. Knife-edge bending is a good example of vertically polarized ground waves that drag the bottom half of the wave over a mountain top, resulting in the ground-wave signal getting into valleys where normally there would be no reception. On High Frequency bands, vertical ground wave propagation range increases on the longer wavelength, lower frequency, bands. **ANSWER A.**

E2C11 How should you generally identify your station when attempting to contact a DX station during a contest or in a pileup?

A. Send your full call sign once or twice
B. Send only the last two letters of your call sign until you make contact
C. Send your full call sign and grid square
D. Send the call sign of the DX station three times, the words "this is," then your call sign three times

Always make sure to ***send your full call sign*** when operating DX, not just the last letter or two in your call sign, which is technically illegal. Rules also require that you momentarily listen to your transmit frequency to ensure the frequency is not in use. **ANSWER A.**

E2C10 Why do DX stations often transmit and receive on different frequencies?

A. Because the DX station may be transmitting on a frequency that is prohibited to some responding stations
B. To separate the calling stations from the DX station
C. To improve operating efficiency by reducing interference
D. All these choices are correct

As a new Extra Class operator, learn from listening during your first contest in the Extra Class portion of the band. If it is an international contest, you will hear rare DX stations calling for a contact "listening up 10." They are not listening on their own transmit frequency but are listening 10 kHz higher for a response. An example might be a foreign DX station operating just below your Extra Class band limits on 20 meters, at 14.145 MHz, listening "up 10." This means you would transmit back to them on 14.155 MHz. To do this, you tune VFO-A to the receive frequency of 14.145 MHz and put VFO-B on your transmit frequency of 14.155 MHz. Now engage SPLIT VFO OPERATION and make sure you are transmitting on VFO-B, while receiving on VFO-A. You must be in SPLIT for this to occur. If you transmit back to the foreign station outside of your Extra Class band edge you are illegally out of band and the foreign station can't hear you because they are listening "up 10 kHz." ***Split operation reduces interference, separates calling stations from the DX station, and lets the DX station stay in the clear out of our U.S. phone band.*** **ANSWER D.**

Veteran DXers operate split to work rare DX foreign stations.

E2C07 What is the Cabrillo format?

A. A standard for submission of electronic contest logs
B. A method of exchanging information during a contest QSO
C. The most common set of contest rules
D. A digital protocol specifically designed for rapid contest exchanges

The various computer logging programs (such as CT by K1EA, TR Log, N1MM, etc.) have built-in utilities that ***automatically generate a Cabrillo-formatted log*** after the contest is over. It's one of those features where you click on a box and the program automatically ker-chunks out a Cabrillo file for submission to the contest sponsor. **ANSWER A.**

E2C02 Which of the following file formats is used for exchanging amateur radio log data?

A. NEC
B. ARLD
C. ADIF
D. OCF

Most radio amateurs have converted from paper logs to electronic logging (or sometimes, a combination of both). For contesters and DX operators alike, computer logging has greatly improved the convenience and accuracy of amateur radio log keeping. There are several different file formats for submitting various files, such as Cabrillo, ADIF, and even raw text. The ***Amateur Data Interchange Format (ADIF)*** is the preferred format for submission to most contest organizers, but always check the official rules of a contest for which file format is required. There are several on-line file converting programs to allow you to reformat your log data into the required format, such as ADIF. **ANSWER C.**

E2C06 During a VHF/UHF contest, in which band segment would you expect to find the highest level of SSB or CW activity?

A. At the top of each band, usually in a segment reserved for contests
B. In the middle of each band, usually on the national calling frequency
C. In the weak signal segment of the band, with most of the activity near the calling frequency
D. In the middle of the band, usually 25 kHz above the national calling frequency

During a VHF/UHF contest, ***most of the weak signal single sideband activity is found on each side of the calling frequency***. **ANSWER C.**

These two frequencies are where to hang out looking for a random DX call. Once you make contact, move at least 5 kHz from these calling channels.

E2C03 From which of the following bands is amateur radio contesting generally excluded?

A. 30 meters
B. 6 meters
C. 70 centimeters
D. 33 centimeters

The 30 meter, 10 MHz, band allows CW and data only with a maximum power of 200 watts output. ***There is no contesting allowed on the 30 meter band.*** **ANSWER A.**

E2C05 What is the function of a DX QSL Manager?

A. Allocate frequencies for DXpeditions
B. Handle the receiving and sending of confirmation for a DX station
C. Run a net to allow many stations to contact a rare DX station
D. Communicate to a DXpedition about propagation, band openings, pileup conditions, etc.

It's fun going out on DXpedition. Sometimes, on a big scale DXpedition, minding the paper QSL cards and the electronic QSL requests is best handled by a ***DX QSL manager***. They carefully ***check the logs to insure all of the contact information is valid, and then send out the coveted QSL card confirmation***. It is a huge job, so if you know a DX QSL manager, give them a big pat on the back! **ANSWER B**.

E2C08 Which of the following contacts may be confirmed through the Logbook of The World (LoTW)?

A. Special event contacts between stations in the US
B. Contacts between a US station and a non-US station
C. Contacts for Worked All States credit
D. All these choices are correct

Old time hams (like me) love to show off our QSL cards, confirming contacts all over the world. But there is a better way for near-instant confirmation of a contact, and that is the League's ***Logbook of The World (LoTW)***. This is used to computer confirm special event station contacts, US to US contacts, US contacts to foreign stations, worked all states, you name it! LoTW will save postage and wall space to confirm ALL these contacts.**ANSWER D.**

Outer Space Comms

E1D11 Which amateur stations are eligible to operate as Earth stations?

A. Any amateur licensee who has successfully completed the AMSAT space communications course
B. Only those of General, Advanced or Amateur Extra class operators
C. Only those of Amateur Extra class operators
D. Any amateur station, subject to the privileges of the class of operator license held by the control operator

An amateur Earth station is located down here on planet Earth. Operating as an ***Earth station*** for space communications ***is available to all amateurs***. Be sure to use only those frequency privileges allowed by your license class. [97.209] **ANSWER D.**

E1D03 What is a space telecommand station?

A. An amateur station located on the surface of the Earth for communication with other Earth stations by means of Earth satellites
B. An amateur station that transmits communications to initiate, modify, or terminate functions of a space station
C. An amateur station located in a satellite or a balloon more than 50 kilometers above the surface of the Earth
D. An amateur station that receives telemetry from a satellite or balloon more than 50 kilometers above the surface of the Earth

Telecommand means ***initiating, modifying, or terminating a function of another station***, such as a space station. [97.3(a) (45)] **ANSWER B.**

E1D10 Which amateur stations are eligible to be telecommand stations of space stations, subject to the privileges of the class of operator license held by the control operator of the station?

A. Any amateur station approved by AMSAT
B. Any amateur station so designated by the space station licensee
C. Any amateur station so designated by the ITU
D. All these choices are correct

Any amateur station so designated by the space station licensee may be a telecommand station. [97.211] **ANSWER B.**

E1D01 What is the definition of telemetry?

A. One-way transmission of measurements at a distance from the measuring instrument
B. Two-way transmissions in excess of 1000 feet
C. Two-way transmissions of data
D. One-way transmission that initiates, modifies, or terminates the functions of a device at a distance

We hear the word "telemetry" a lot in satellite service. Satellite telemetry may give ground controllers vital information on satellite battery power, power loads, and satellite solar power charging. ***Telemetry is a one-way transmission of measurements at a distance*** from the measuring instrument. [97.3] **ANSWER A.**

E1D02 Which of the following may transmit encrypted messages?

A. Telecommand signals to terrestrial repeaters
B. Telecommand signals from a space telecommand station
C. Auxiliary relay links carrying repeater audio
D. Mesh network backbone nodes

Encrypted Amateur Radio communications are strictly forbidden except for one special case. ***Telecommands from a space station may be encrypted*** to prevent unauthorized access to the control functions of the spacecraft. [97.211(b)] **ANSWER B.**

E1D05 What must be posted at the location of a station being operated by telecommand on or within 50 kilometers of the Earth's surface?

A. A photocopy of the station license
B. A label with the name, address, and telephone number of the station licensee
C. A label with the name, address, and telephone number of the control operator
D. All these choices are correct

The FCC rules tell us that any amateur radio station may be a space station and any amateur radio station may be a telecommand station. This is great news for any experimentally minded radio amateur or radio club. If you operate a telecommand station, the station must have posted a photocopy of the station license, and a label with the name, address, and telephone number of the station licensee and control operator (both of which may be the same person). ***So, all of these choices are correct***. An amateur radio space station is not generally required to have these documents, as there is often no way to put such documents on a space-restricted satellite or other spacecraft. (Not to mention, it would be a bit difficult for the FCC to read them!) [97.213(d)] **ANSWER D.**

E2A06 What are Keplerian elements?

A. Parameters that define the orbit of a satellite
B. Phase reversing elements in a Yagi antenna
C. High-emission heater filaments used in magnetron tubes
D. Encrypting codes used for spread spectrum modulation

Keplerian elements, or "keps" (named after Renaissance astronomer Johannes Kepler) are ***standardized sets of data to help you find a particular satellite in the sky*** as well as telling you in which direction it's moving. Keps are regularly broadcast (or e-mailed) from W1AW and other sources. Many computer programs are capable of accepting the keps and automatically steering your antenna, but a lot of folks still work with them manually to help find their favorite "bird." **ANSWER A.**

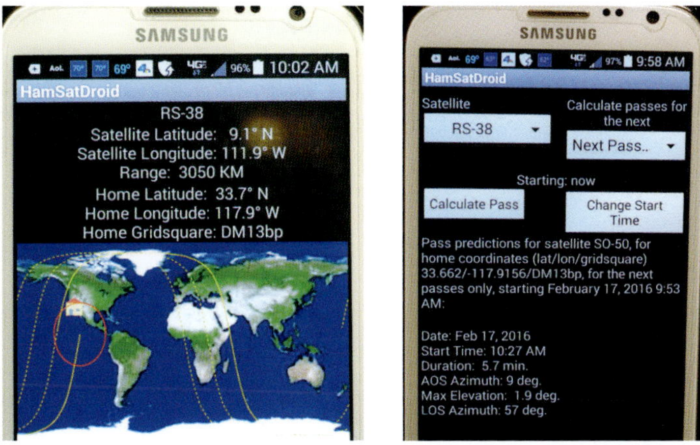

You can use smart phone apps to access computer programs and websites that can show you where and when an amateur satellite or the Space Station will be in range of your ham station.

Good to Know:

Where's the amateur satellite?

AMSAT is the North American distributor of SatPC32, a tracking program designed for ham satellite applications. It has a host of features that help you locate and track amateur satellites, set your antenna rotator for automatic control, and provides visual mapping of satellite locations as they orbit the Earth. You can download a demo version from www.dk1tb.de/indexeng.htm, or order online at www.amsat.org or by calling 1-888-322-6728. BTW – the author of SatPC32, DK1TB, donated the program to AMSAT and all proceeds support AMSAT.

E2A01 What is the direction of an ascending pass for an amateur satellite?

A. From west to east
B. From east to west
C. From south to north
D. From north to south

Some satellites travel over the Equator from north to south (descending) while others pass over the Equator ***from south to north (ascending)***. Direction, height, and speed all affect where a satellite is at a specific time. It's easy to remember if you think of the North Pole as being at the top of the earth. **ANSWER C.**

E2A10 What type of satellite appears to stay in one position in the sky?

A. HEO
B. Geostationary
C. Geomagnetic
D. LEO

A satellite in a ***geostationary*** orbit remains ***in one position*** above the Earth. No tracking necessary. You always know where it will be. **ANSWER B.**

Good To Know:

Geostationary or Geosynchronous?

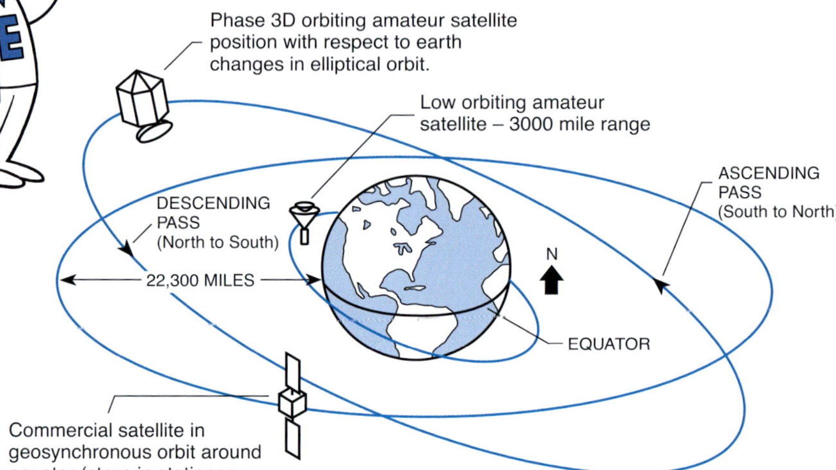

Orbiting Satellites

If you have worked any amateur satellites, you know that you have to track the satellite as it moves from horizon to horizon. The Keplerian Elements, or "Keps" are the positional data you need to "find" the satellite in its overhead pass. The ARRL and other amateur satellite entities faithfully broadcast the Keps or otherwise disseminate the data on a daily basis, which you can either insert into your automatic antenna steering program, or use as a rough guideline if you're using the "Armstrong Method" of manually tracking a bird.

Now, there is a bit of difference between geostationary orbit and geosynchronous orbit. A true geostationary orbit must be perfectly circular (or very nearly so) and directly above the equator. This is the orbit used for broadcast satellites, and is generally a very expensive proposition, initially, as the satellite needs to be launched in a very high orbit, around 26,199 miles. However, a true geostationary orbit allows you to dispense entirely with tracking, since the satellite is in a relative fixed position over the earth.

A geosynchronous orbit, on the other hand, can be highly elliptical. It can appear to "wobble" in its orbit around the earth as it moves closer or farther away from the equator. At least once a day, a geosynchronous orbit will appear over the same location. So, even with a fixed antenna, you will be able to regularly use the satellite.

The distinction between the two orbital types is actually not too important for amateurs in the long run, because geosynchronous orbits are almost always an intermediate stage on the way to geostationary orbit. Space agencies generally launch into a geosynchronous orbit first, and then begin the rather long process of "rounding out" to a true geostationary orbit.

E2A04 What is meant by the "mode" of an amateur radio satellite?

A. Whether the satellite is in a low earth or geostationary orbit
B. The satellite's uplink and downlink frequency bands
C. The satellite's orientation with respect to the Earth
D. Whether the satellite is in a polar or equatorial orbit

Working through amateur satellites is exciting, challenging, and fun! Transmit and receive frequencies for satellite operations are on separate bands. The satellite receives on one frequency band and transmits on another. Since two bands are involved, we have two ***mode designators*** to best describe the transmit and receive bands. Each band has its own designator. The Earth transmit band is called the "***uplink***," and the receive band is the "***downlink***." When we work a satellite, the term "mode" means the bands up and downlink, not the actual emission type. While emission types on HF, like SSB, CW, and DATA are also "modes", too, satellite MODE refers to uplink and downlink bands. And yes, satellites also have different emission types, too! **ANSWER B.**

Ready to hear a spacecraft data beacon? There are plenty on the air every day all over the world as they fly by.

E2A05 What do the letters in a satellite's mode designator specify?

A. Power limits for uplink and downlink transmissions
B. The location of the ground control station
C. The polarization of uplink and downlink signals
D. The uplink and downlink frequency ranges

The two ***mode designators indicate*** the transmit ***uplink frequency*** and the receive ***downlink frequency*** bands that the satellite uses. **ANSWER D.**

E2A09 What do the terms "L band" and "S band" specify?

A. The 23- and 13-centimeter bands
B. The 2-meter and 70-centimeter bands
C. FM and digital store-and-forward systems
D. Which sideband to use

You can work this out by recalling the MHz to meters and meters to MHz formula, but there are lots of conversions because we are working with centimeters and GHz. The L band at 1.26 GHz works out to be 23 centimeters. The S band, at 2.4 GHz, works out to be 13 centimeters. You can rule out Answer B as that would be VHF and UHF, with C and D having nothing to do with band specifics. ***23 centimeters is L band, and 13 centimeters is S band***. **ANSWER A.**

Frequency Bands	Frequency Range	Modes	Frequency Bands	Frequency Range	Modes
HF	21 - 30 MHz	Mode H	S band	2.4 - 2.45 GHz	Mode S
VHF	144 - 146 MHz	Mode V	C band	5.8 GHz	Mode C
UHF	435 - 438 MHz	Mode U	X band	10.4 GHz	Mode X
L band	1.26 - 1.27 GHz	Mode L	K band	24 GHz	Mode K

E2A07 Which of the following types of signals can be relayed through a linear transponder?

A. FM and CW
B. SSB and SSTV
C. PSK and packet
D. All these choices are correct

Thanks to AMSAT, many of our OSCAR satellites offer a "bent pipe" re-transmission of many transmission modes. If we uplink FM or CW, it comes back down as FM or CW, respectively, through the satellite's linear transponder. Same thing with single sideband, slow scan television, and the digital modes – ***all of these signals are relayed through*** the satellite's on-board ***linear transponder***. **ANSWER D.**

E2A08 Why should effective radiated power (ERP) be limited to a satellite that uses a linear transponder?

A. To prevent creating errors in the satellite telemetry
B. To avoid reducing the downlink power to all other users
C. To prevent the satellite from emitting out-of-band signals
D. To avoid interfering with terrestrial QSOs

Satellites are powered by on-board batteries charged by solar panels. On older satellites, ***any strong input signal is re-transmitted at the same strength. This quickly depletes satellite battery power***. Our modern constellation of low Earth orbit satellites re-transmits modest power signals at approximately the same output power to conserve battery life. An overly strong input signal can cause the satellite to reduce its output signal to protect the batteries from a high-power drain. Once the satellite cuts power back in response to a strong signal, all of its re- transmissions are done at a reduced power. Everyone suffers even if they are not using high power. So don't run any more power than it takes for you to hear your own signal coming back clearly. **ANSWER B.**

E2A02 Which of the following is characteristic of an inverted linear transponder?

A. Doppler shift is reduced because the uplink and downlink shifts are in opposite directions
B. Signal position in the band is reversed
C. Upper sideband on the uplink becomes lower sideband on the downlink, and vice versa
D. All these choices are correct

A linear transponder differs from a repeater in that there is a direct translation from one radio frequency to another without demodulating and re-modulating a signal. In a linear transponder, the output power is directly proportional to the input power. An INVERTING linear transponder (used on many satellites) is simply a frequency converter where the input and output frequencies are equally spaced on both sides of an incoming signal. This scheme cancels out any Doppler shift, but with the side effect that the sidebands are reversed on the ground station. ***All of these choices are correct***. **ANSWER D.**

E2A03 How is an upload signal processed by an inverting linear transponder?

A. The signal is detected and re-modulated on the reverse sideband
B. The signal is passed through a non-linear filter
C. The signal is reduced to I and Q components, and the Q component is filtered out
D. The signal is mixed with a local oscillator signal and the difference product is transmitted

An inverting linear transponder is incredibly simple – at least when compared to other "repeater" schemes. A mixer will always produce the sums and differences of two frequencies injected into it. One can use either the sum or the difference frequencies. ***Using the difference frequency inverts*** the sidebands of any incoming frequency. **ANSWER D.**

Good to Know:

Mixers and Frequency Inversion

You don't really need to know anything about the physical internal circuitry of a mixer to explain what it does to a radio signal. It is best to think of a mixer as a simple "black box" that performs a simple mathematical operation. The mixer takes any two input signals and produces two output signals, the sum of the two input frequencies, and the difference of the two frequencies. (A practical mixer can also produce other products, but the ones we're most interested in are the simple sums and differences).

If you put a fixed carrier frequency into one input of a mixer, and put an upper sideband signal which is above the fixed carrier frequency into the second input, the output will be an UPPER sideband signal ADDED to the fixed carrier frequency.

If the upper sideband signal is BELOW the fixed carrier frequency, the output signal will be a LOWER sideband signal.

Let's plug in some actual numbers to help clarify this. Let's take a properly aligned upper sideband SSB transmitter, operating on a carrier frequency of 14.2 MHz. We'll modulate the transmitter with two audio tones, one at 1000 Hz and one at 2000 Hz. The radiated signal from this transmitter will be two RF carriers, one at 14.201 MHz and another at 14.202 MHz.

Now, let us take the output from our transmitter and put it into one input of a mixer and inject a fixed carrier of 15 MHz into the other mixer input. What do we get out of the mixer? We actually get FOUR products out, two sum frequencies and two difference frequencies. Let's ignore the sum frequencies for now. What difference frequencies do we get?

15 MHz – 14.201Mhz =.799 MHz. 15 MHz – 14.202 MHz = .798 MHz

We see that our two input signals have swapped places, creating a lower sideband signal out of our original upper sideband signal. Now these frequencies aren't very realistic, but they do demonstrate the inversion concept.

The sum frequencies will be 29.201 MHz and 29.202 MHz, respectively and preserve their normal "order." In other words the SUM outputs will be an upper sideband signal.

E2A11 What type of antenna can be used to minimize the effects of spin modulation and Faraday rotation?

A. A linearly polarized antenna
B. A circularly polarized antenna
C. An isotropic antenna
D. A log-periodic dipole array

A circularly polarized antenna will give you the best results when working amateur satellites. Although you lose approximately 3 dB gain over a horizontal antenna, ***circular polarization*** minimizes fading. **ANSWER B.**

A small 3 element beam can easily hear a satellite signal, but continuously rotate the beam from vertical to horizontal to peak the signal. Good exercise!

Cross-polarized, dual-feed beams take the polarization challenge away for smooth copy.

E2A12 What is the purpose of digital store-and-forward functions on an amateur radio satellite?

A. To upload operational software for the transponder
B. To delay download of telemetry between satellites
C. To hold digital messages in the satellite for later download
D. To relay messages between satellites

Orbiting amateur radio satellites make great store-and-forward devices to ***hold messages that may be downloaded later*** by other stations. **ANSWER C.**

E1D07 Which group of HF amateur bands include allocations for space stations?

A. 40 meters, 20 meters, 15 meters, and 10 meters
B. 30 meters, 17 meters, and 10 meters
C. Only 10 meters
D. Satellite operation is permitted on all HF bands

The four most popular bands ***on High Frequency*** for space station operation just happen to be our most popular bands for voice and data, too. 40, 20, 15, and 10 meters is the correct answer. To avoid the crowd, most space station contacts are on most of the VHF, UHF, and SHF bands. But for HF space station allocations, it's 40, 20, 15, and 10 meters, if you like a crowd. **ANSWER A.**

E1D08 Which VHF amateur bands have frequencies authorized for space stations?

A. 6 meters and 2 meters
B. 6 meters, 2 meters, and 1.25 meters
C. 2 meters and 1.25 meters
D. 2 meters

On VHF (30-300 MHz), ***only the 2-meter band allows space station operation***. Space station operation is prohibited on 1.25 meters (the 222 MHz band) and 6 meters. Notice that the wrong answers include the other bands. [97.207] **ANSWER D.**

E1D09 Which UHF amateur bands have frequencies authorized for space stations?

A. 70 centimeters only
B. 70 centimeters and 13 centimeters
C. 70 centimeters and 33 centimeters
D. 33 centimeters and 13 centimeters

On UHF (300-3000 MHz), the 33 cm (900 MHz) band is NOT allowed for space station operations, but all other bands are allowed. This includes the ***70 cm (435 MHz) and 13 cm (2300 MHz) bands***. [97.207] **ANSWER B.**

E3A01 What is the approximate maximum separation measured along the surface of the Earth between two stations communicating by EME?

A. 2,000 miles, if the moon is at perigee
B. 2,000 miles, if the moon is at apogee
C. 5,000 miles, if the moon is at perigee
D. 12,000 miles, if the moon is "visible" by both stations

EME means Earth-Moon-Earth communications. It normally takes a minimum of a pair of long-boom Yagis, plus the maximum legal power output, to establish a Moon bounce QSO on CW, Digital JT8, or SSB phone. If you are working a station with a huge dish antenna, you might get by with a single long-boom Yagi and a good VHF or UHF kilowatt amplifier. Moon bounce is most common on the bottom edge of the 2-meter band. Communications over a great distance and even beyond the horizon are possible when ***both stations can see the Moon***. **ANSWER D.**

Good To Know:

Great news from AMSAT! They are working with the U.S. Air Force for a Geosynchronous digital satellite scheduled to lift off soon that may "park" at 74 degrees west over the Equator. As designed, this satellite will have a 12 foot wide "Field of View" seeing all of the U.S., Canada, Hawaii, and Mexico. In addition, it will also see a portion of lower Alaska, northern South America, the western edge of Europe and part of western Africa! The satellite will receive digital uplinks on 5.6 GHz, and downlink on 10 GHz. The on-board linear transponder may offer up to 100 channels of data capability! If this comes to pass, it will greatly simplify and expand ham satellite operation.

E3A02 What characterizes libration fading of an EME signal?

A. A slow change in the pitch of the CW signal
B. A fluttery, irregular fading
C. A gradual loss of signal as the sun rises
D. The returning echo is several hertz lower in frequency than the transmitted signal

Remember, EME means Earth-Moon-Earth and that the Moon is not a perfect reflector of radio signals. All those craters on the Moon will cause your ***return signal to sound fluttery, with rapid, irregular fading***. **ANSWER B.**

EME communications require specialized equipment and patience.

E3A03 When scheduling EME contacts, which of these conditions will generally result in the least path loss?

A. When the Moon is at perigee
B. When the Moon is full
C. When the Moon is at apogee
D. When the MUF is above 30 MHz

PERigee is the time when the Moon is closest to the Earth. This is the PERfect condition for EME to occur, and a PERfect day to schedule an EME contact. When the ***Moon is at perigee*** and just appearing above the horizon, your will get the ***best opportunity for Moon-bounce*** success. **ANSWER A.**

E2D06 Which of the following is a method for establishing EME contacts?

A. Time-synchronous transmissions alternating between stations
B. Storing and forwarding digital messages
C. Judging optimum transmission times by monitoring beacons reflected from the moon
D. High-speed CW identification to avoid fading

Although "in the wild" EME contacts have occurred, your chances of making an EME contact are much, much better if you have some means of scheduling ahead of time. The JT65 and QRA64 digital modes are ***scheduled modes*** especially suitable for EME contacts. **ANSWER A.**

E2D03 Which of the following digital modes is designed for EME communications?

A. MSK144 C. WSPR
B. PACTOR III D. Q65

Earth Moon Earth (EME or Moonbounce) communications have some stringent requirements, most significantly, very weak signals and ***libation fading***. Although many different digital modes have been used to achieve EME communications, Q65 is specifically designed to minimize the effects of libation fading. **ANSWER D.**

E2D05 What is the characteristic of the JT65 mode?

A. Uses only a 65 Hz bandwidth
B. Decodes signals with a very low signal-to-noise ratio
C. Symbol rate is 65 baud
D. Permits fast-scan TV transmissions over narrow bandwidth

Weak signal VHF/UHF and microwave operators know well the popular WSJT software that enables miraculous ***reception of digital signals well below a normal noise floor. JT65*** coding allows your radio to capture everything from digital conversations off the surface of the Moon! Just because you can't hear anything coming out of the speaker, many times your computer's sound card can deliver enough meaningful audio to the decoding programs to utterly amaze you! **ANSWER B.**

E2D09 What type of modulation is used by JT65?

A. Multi-tone AFSK C. RTTY
B. PSK D. QAM

Multi-tone modulation methods allow the use of a much narrower bandwidth (for a given data rate) than would ordinarily be possible with a single frequency shifted tone. JT65 uses ***multi-tone Audio Frequency Shift Keying (AFSK)***, which is converted to multiple radio frequency shift keying when applied to a single sideband transmitter. When this FSK signal is received by a single sideband receiver, it is converted back to AFSK. You can see all the different audio tones on a standard waterfall display. **ANSWER A.**

Good To Know:

Weak Signal Band Plan Recommendations

Over the past few years, tremendous advances have been made in weak-signal communications on the VHF and UHF bands using computer-generated data modes. These modes enable communications at or below the noise floor. Amateurs with modest stations are now making contacts over thousands of kilometers via EME and Meteor Scatter using these new modes. However, these very-weak data signals often cannot be detected by ear, which makes them incompatible with normal analog SSB or CW signals. There have been instances of interference caused by SSB operators who obviously can't hear the data signals. The recent EME Conference held in Germany recommended the creation of worldwide VHF and UHF sub-bands for JT-65, the weak signal data protocol. The suggested segments are 50.185 to 50.195 MHz on 6 meters; 144.115 to 144.135 MHz on 2 meters; 432.060 to 432.070 MHz on 70 centimeters; and 1296.060 to 1296.070 MHz.

E2E03 To what does the "4" in FT4 refer?

A. Multiple of 4 bits of user information
B. Four-tone continuous-phase frequency shift keying
C. Four transmit/receive cycles per minute
D. All these choices are correct

By far, the most popular HF digital mode is FT8, and running a close second is the somewhat faster FT4. Most of the "WSJT" modes use multi-tone baseband signals, each of which is frequency shifted with the data. FT-4 uses four different tones, hence the -4 suffix. Each of the four tones are in phase quadrature (90 degrees of phase shift between them) during their "resting" period, which is shifted in frequency slightly when digital data is present. **ANSWER B.**

Licensed local hams in Puerto Rico sometimes got to work the Arecibo radio telescope on ham bands, before its December 2020 collapse.

Good To Know:

WSJT-X Modes Explained

JT4, JT9, JT65, and QRA64 use nearly identical message structure and source encoding, through the efficient compression of standard messages used for short QSOs. These modes use timed 60-second T/R sequences synchronized with WWV.

- *FT4 is primarily used for VHF contests, although sometimes it is also used for HF DXing.*
- *JT65 and QRA64 were designed for moonbounce on VHF/UHF bands. JT65 has also proven popular and effective for worldwide QRP communications at HF and 6 meters.*
- *JT9 is optimized for the LF, MF, and HF bands. It is about 2 dB more sensitive than JT65 while using less than 10% of the bandwidth.*
- *FT8 is used for weak signal work on VHF/UHF using T/R cycles 15 seconds long. MSK144 is designed for Meteor Scatter on the VHF bands.*
- *WSJT-X Version 2.1 offers ten different protocols for those of you who have your computer tied in to your ham portable or base station.*

E2D04 What technology is used for real-time tracking of balloons carrying amateur radio transmitters?

A. FT8
B. Bandwidth compressed LORAN
C. APRS
D. PACTOR III

The Automatic Packet Reporting System (APRS) was developed in 1992 as an innovative way of tracking mobile amateur radio stations. It turns out to be a very practical way of tracking objects in the sky too! ***APRS*** has added a new dimension to ham radio science, allowing the ***precise tracking of high-altitude balloons*** carrying amateur radio and scientific equipment. **ANSWER C.**

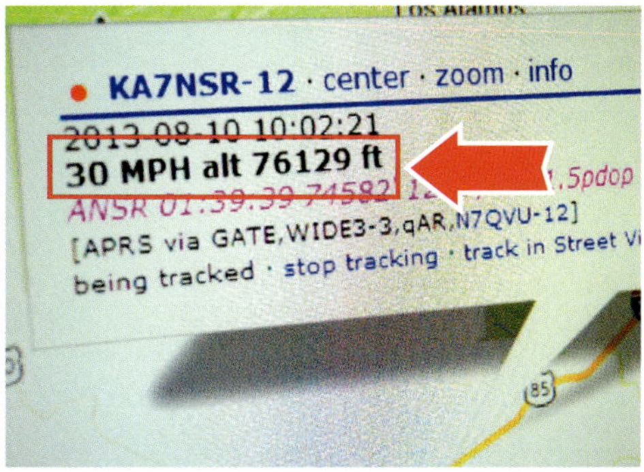

Look carefully at the balloon's altitude, 76,129 feet and drifting at 30 MPH, all reported via APRS!

E1D04 Which of the following is required in the identification transmissions from a balloon-borne telemetry station?

A. Call sign
B. The output power of the balloon transmitter
C. The station's six-character Maidenhead grid locator
D. All these choices are correct

Amateur Radio Station identification is almost always a very simple affair, and it's no different with an airborne balloon telemetry station. The only requirement is that you identify with your station's call sign, which may be transmitted either by voice or CW. Of course, you may add additional information to your ID, such as GPS coordinates, temperature, power output, and any other interesting information. But ***all that's required by FCC regulations is the call sign***. [97.119(a)] **ANSWER A.**

HERE ARE SOME WEB ADDRESSES OF USEFUL SITES RELATED TO OUTER SPACE COMMS

http://spaceweather.com/
http://prop.hfradio.org/
www.amsat.org/amsat-new/news/
http://hamwaves.com/propagation/
http://www.spacew.com/www/realtime.php
http://www.arrl.org/news/arrl-committee-seeks-microwave-band-plan-input
http://www.arrl.org/news/space-station-active-on-70-cm-packet
http://www.vhfdx.de/wsjt/
http://www.qsl.net/w8wn/hscw/hscwopen.html

Visuals & Video Modes

E2B04 How is color information sent in analog SSTV?

A. Color lines are sent sequentially
B. Color information is sent on a 2.8 kHz subcarrier
C. Color is sent in a color burst at the end of each line
D. Color is amplitude modulated on the frequency modulated intensity signal

Slow Scan Television (SSTV) is very different from broadcast television, so even if you're highly familiar with NTSC analog TV, it won't help you much with understanding amateur SSTV! Amateur SSTV is capable of transmitting high quality still images over a very narrow band of frequencies. A number of different SSTV protocols exist, but they all handle color information in essentially the same manner, ***sending one color at a time, one line at a time, sequentially***. **ANSWER A.**

E2B10 What aspect of an analog slow-scan television signal encodes the brightness of the picture?

A. Tone frequency
B. Tone amplitude frequency
C. Sync amplitude
D. Sync

Brightness of an SSTV picture depends on the ***tone frequencies*** received. The higher the frequency, the brighter the picture. **ANSWER A.**

While it takes about 30 to 45 seconds to download a photo, slow scan TV is popular on the HF bands, plus from the International Space Station on 2 meters!

E2B12 What signals SSTV receiving software to begin a new picture line?

A. Specific tone frequencies
B. Elapsed time
C. Specific tone amplitudes
D. A two-tone signal

In most SSTV receiving equipment, ***new picture lines are triggered when specific tone frequencies are received***. The software within the equipment knows what to listen for and exactly what to do when that tone is heard. **ANSWER A.**

E2B11 What is the function of the vertical interval signaling (VIS) code sent as part of an SSTV transmission?

A. To lock the color burst oscillator in color SSTV images
B. To identify the SSTV mode being used
C. To provide vertical synchronization
D. To identify the call sign of the station transmitting

There are over a dozen individual methods for sending a color slow scan television signal over HF and VHF airwaves. The most common are Martin and Scottie. These popular modes require the initial reception of ***the 300 ms VIS code, which is necessary for your computer to know which SSTV mode is being sent***. **ANSWER B.**

E2B09 What kind of receiver can be used to receive and decode SSTV using the Digital Radio Mondiale (DRM) protocol?

A. CDMA
B. AREDN
C. AM
D. SSB

The term "Digital Radio Mondiale" ***(DRM) is not a special kind of circuit*** with a big chip in the middle. Rather, DRM is a panel of commercial and non-commercial radio users who have developed the sole standard for sending broadcasts over worldwide low-frequency and high-frequency bands. SSTV is easy to hear at 14.230 MHz, and it's easy to capture a full-color, high resolution picture with just your HF transceiver and a suitably-equipped computer with a sound card running SSTV software. **ANSWER D.**

E1B02 Which of the following is an acceptable bandwidth for digital voice or slow-scan transmissions made on the HF amateur bands?

A. 3 kHz
B. 10 kHz
C. 15 kHz
D. 20 kHz

The restrictions for SSTV specify that the typical high-frequency, ***slow-scan signal*** can occupy no more room than a conventional single sideband voice signal, ***3 kHz or less***. [97.307(f)(2)] **ANSWER A.**

E2B08 What technique allows commercial analog TV receivers to be used for fast-scan TV operations on the 70-centimeter band?

A. Transmitting on channels shared with cable TV
B. Using converted satellite TV dishes
C. Transmitting on the abandoned TV channel 2
D. Using USB and demodulating the signal with a computer sound card

Unlike on HF where video is limited to narrowband SSTV methods, full broadcast-style TV is possible on amateur UHF frequencies. One of the last holdouts for NTSC TV (which has been replaced by digital TV in the commercial broadcast world) is fast scan amateur TV. It is a trivial matter to make NTSC equipment work on UHF amateur TV frequencies. ***Any commercial TV receiver capable of receiving Cable TV is capable of receiving amateur UHF transmissions***. Check out **www.atn-tv.com**. **ANSWER A.**

E2B01 In digital television, what does a coding rate of 3/4 mean?

A. 25% of the data sent is forward error correction data
B. Data compression reduces data rate by 3/4
C. 1/4 of the time interval is used as a guard interval
D. Three, four-bit words are used to transmit each pixel

Forward Error Correction (FEC) is a means of reducing the inevitable error that occurs when transmitting over a noisy signal channel. It involves sending part or all of a transmission more than once, so that a voting circuit on the receiving end can verify the data. Since this takes some time, it is best to not have to send EVERYTHING twice, but just enough information to perform some kind of check on the receiving end. A 3/4 coding rate means that three out of four bits of information are "useful" or "new" information, while the remainder, or 25%, is used for error correction. **ANSWER A.**

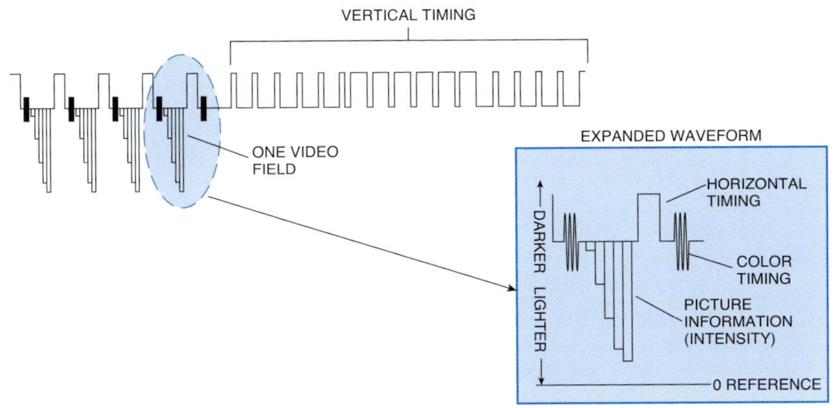

a. A TV Composite Video Signal

Source: *Using Video in Your Home,* G. McComb. © 1989, Master Publishing, Inc., Niles, IL.

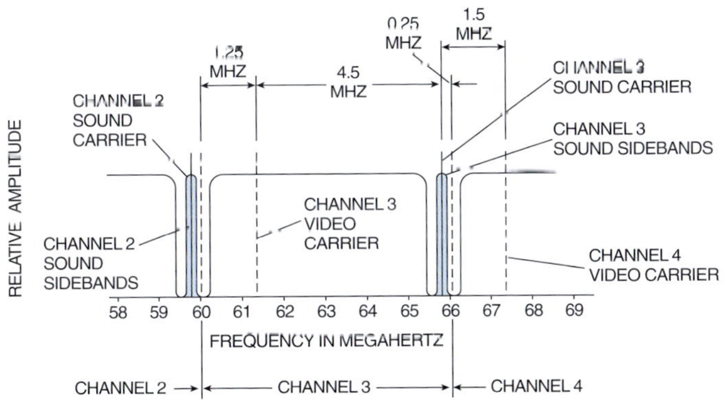

b. Bandwidth for a TV Channel

Source: *Antennas,* A. Evans, K. Britain. © 1988, Master Publishing, Inc., Niles, IL.

The video signal is composed of three major components: picture, timing, and color. Television sets "know" how to decode this stream of information into pictures that reproduce the changing scenes captured by the video camera.

E2B02 How many horizontal lines make up a fast-scan (NTSC) television frame?

A. 30
B. 60
C. 525
D. 1080

A total of ***525 lines from two fields make up a complete frame***. **ANSWER C.**

E2B03 How is an interlaced scanning pattern generated in a fast-scan (NTSC) television system?

A. By scanning two fields simultaneously
B. By scanning each field from bottom-to-top
C. By scanning lines from left-to-right in one field and right-to-left in the next
D. By scanning odd-numbered lines in one field and even-numbered lines in the next

Each field is 262.5 lines, and ***the frame is created by scanning even numbered lines in one field and odd numbered lines in the second field***. **ANSWER D.**

E2B07 Which types of modulation are used for amateur television DVB-T signals?

A. FM and FSK
B. QAM and QPSK
C. AM and OOK
D. All these choices are correct

Modern "fast scan" amateur radio television uses several different digital modulation schemes to achieve error correction and compression, the latter to decrease occupied bandwidth. ***DVB-T***, short for ***Digital Video Broadcasting – Terrestrial***, uses both Quadrature Amplitude Modulation (QAM) and Quadrature Phase Shift Keying (QPSK) to achieve these goals. This method was first used by commercial television broadcasters, and later adapted by radio amateurs. **ANSWER B.**

E2B06 What is vestigial sideband modulation?

A. Amplitude modulation in which one complete sideband and a portion of the other are transmitted
B. A type of modulation in which one sideband is inverted
C. Narrow-band FM modulation achieved by filtering one sideband from the audio before frequency modulating the carrier
D. Spread spectrum modulation achieved by applying FM modulation following single sideband amplitude modulation

Vestigial sideband modulation ***(VSB) is amplitude modulation of a carrier in which one complete sideband is transmitted*** (upper VSB is standard) ***and a vestigial sideband filter is used to attenuate the other sideband***. This helps reduce occupied bandwidth to 6 MHz and provides better spectrum efficiency than full double sideband amplitude modulation. **ANSWER A.**

Gordo is ready to operate video for his Wednesday evening Ham Nation pod-cast.

E2B05 Which of the following describes the use of vestigial sideband in analog fast-scan TV transmissions?

A. The vestigial sideband carries the audio information
B. The vestigial sideband contains chroma information
C. Vestigial sideband reduces the bandwidth while increasing the fidelity of low frequency video components
D. Vestigial sideband provides high frequency emphasis to sharpen the picture

Vestigial sideband (VSB) cuts off the lower color and sound sub-carriers at -3.58 and -4.50 MHz from the analog video carrier, respectively, as they are unnecessary. The frequencies below 1.25 MHz from the video carrier in a VSB transmitter will have very low energy radiated. This frees up room for more users of the band. Total VSB TV channel ***bandwidth*** is (14.75 ÷ 2) - 1.25 = 6 MHz. ***Simple analog video detectors are used*** in the TV receiver with most of the work being done by the VSB filter in the IF ahead of the detector. Even though commercial TV signals have switched from analog to digital, many ham fast-scan TV signals remain analog. New TVs still employ an analog tuner and vestigial sideband receiver with a simple video detector in addition to the digital system to allow compatibility with older VCRs and DVD players. **ANSWER C.**

Good To Know:

Amateur TV Goes Hi-Def!

In recent years, home television reception has gone from analog to digital, and the new high definition digital picture captured over the airwaves is phenomenal. Now, the Amateur Television Network is seeing ham stations around the country also making the switch to Hi-Def TV, although more gradually.

Most amateur systems use the mode DVB-T direct, transmitting through ATV Hi-Def repeaters and remote relay stations on UHF and microwave frequencies. Bandwidth and other parameters can be changed by a few key strokes on your computer. You can set up multiple channels with different bandwidths and parameters in the same exciter or receiver box.

For local hamfests, QAM-64 single carrier digital ATV (CATV) is a popular mode, as hams in their RVs can see the signal by hooking up their digital TV to an outside antenna and setting their TV source to the cable mode.

The Amateur Television Network publishes e-newsletters to keep both seasoned and ATV newcomers up to date on where to get the gear and how to go on the air with Hi-Def ATV. Learn about the new ATSC-3 format, which uses a modified form of DVB-T with extra layers that will work mobile ATV! New home TV sets are now being made with the ATSC-3 format inside. You can see the live ATV show and net every Tuesday evening on YouTube (https://www.youtube.com/amateurtelevision).

Get the full story on the switch to digital ATV at

→ **www.AmateurTelevisionNetwork.org**

→ **www.ATN-TV.com**

Digital Excitement with Computers & Radios

E2E01 Which of the following types of modulation is used for data emissions below 30 MHz?

A. DTMF tones modulating an FM signal
B. FSK
C. Pulse modulation
D. Spread spectrum

Frequency shift keying (FSK) is created by data pulses from your computer that shift the master oscillator back and forth to create two distinct tones in the other operator's receiver. Audio Frequency Shift Keying (AFSK) generates audio tones from your computer's sound card or multi-mode controller and is used for more complex digital signaling like PSK 31, MFSK, and CLOVER. These modes have more than two distinct frequency signals. Both ***Frequency Shift Keying (FSK)*** and Audio Frequency Shift Keying (AFSK) can be found on HF, ***below 30 MHz***. Listen to Gordo's audio CD course to hear the difference between these two modes. **ANSWER B.**

E2E11 What is the difference between direct FSK and audio FSK?

A. Direct FSK modulates to the transmitter VFO
B. Direct FSK occupies less bandwidth
C. Direct FSK can transmit higher baud rates
D. All these choices are correct

With ***direct frequency shift keying*** (FSK) the ***variable frequency oscillator*** in your transceiver ***changes the actual frequency*** in sync with the 1 and 0 data from your computer. With audio frequency shift keying (***AFSK***), your master oscillator stays on frequency and ***audio tones are generated*** by a sound card or modem to correspond precisely with the data being sent. **ANSWER A.**

E8D07 What is a common cause of overmodulation of AFSK signals?

A. Excessive numbers of retries
B. Excessive frequency deviation
C. Bit errors in the modem
D. Excessive transmit audio levels

It is important to know that many digital modes referred to as AFSK also include amplitude modulation components. If a mode is truly FSK, transmitter linearity is not an issue; you can run "plain vanilla" RTTY through a class C amplifier with no problem whatsoever (whether using a sound card mode, or direct FSK). However, if you are using a digital mode that has amplitude modulation as well, you MUST have a linear system from microphone to antenna. ***Overdriving the audio input*** of a receiving SSB transmitter is a sure way to create non-linearity; a signal that is difficult if not impossible to copy. **ANSWER D.**

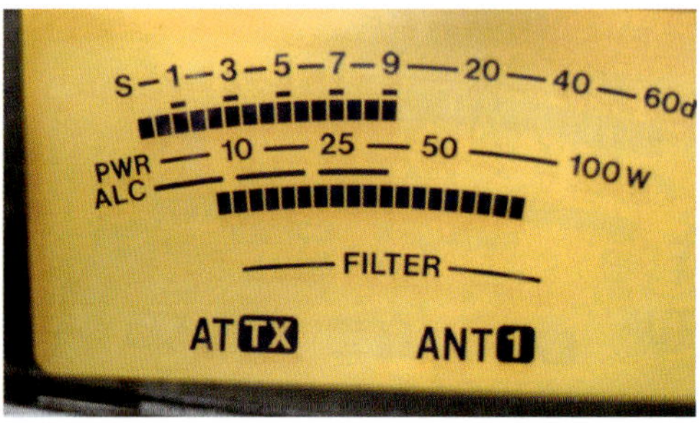

When you see an ALC reading that shoots to the right, turn down the digital drive level or your signal may be distorted.

E8D08 What parameter evaluates distortion of an AFSK signal caused by excessive input audio levels?

A. Signal-to-noise ratio
B. Baud error rate
C. Repeat Request Rate (RRR)
D. Intermodulation Distortion (IMD)

Intermodulation distortion is the primary reason why many digital modes do not decode properly. IMD doesn't always show itself on a waterfall display, although a signal with lots of IMD may appear wider than normal. Always be safe and run your audio well below the level that creates any ALC action on your transmitter. Intermodulation is not as much an issue when using sound card digital modes since they work with greatly reduced power. **ANSWER D.**

When operating digital, keep your ALC settings always in view on the meter.

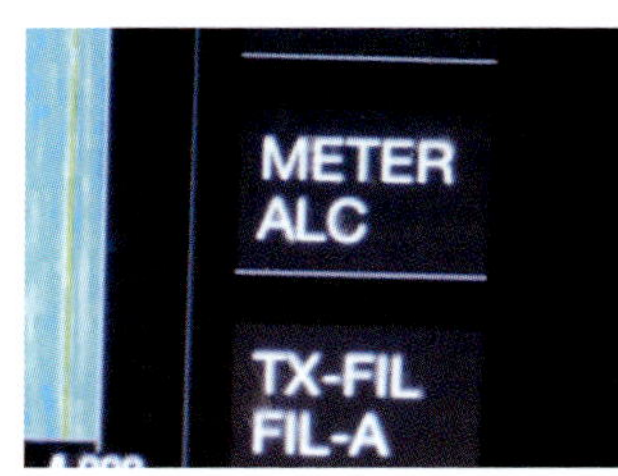

E8D09 What is considered an acceptable maximum IMD level for an idling PSK signal?

A. +5 dB
B. +10 dB
C. +15 dB
D. -30 dB

For IMD, ***the bigger the negative number the better. -30dB*** means that the sum of all intermodulation products is less than 1/1000th the power level of your desired signal. **ANSWER D.**

E8C03 Why should the phase of a PSK signal be changed at the zero crossing of the RF signal?

A. To minimize bandwidth
B. To simplify modulation
C. To improve carrier suppression
D. All these choices are correct

Zero crossing not only allows the ***least amount of spectrum*** to be occupied, but it also results in the least amount of confusion or "blurring" between symbols for a given rate. While commonly implemented in most digital modes, experimental modes such as coherent CW take advantage of this zero-crossing benefit. **ANSWER A.**

E2E07 How does Q65 differ from JT65?

A. Keyboard-to keyboard operation is supported
B. Quadrature modulation is used
C. Multiple receive cycles are averaged
D. All these choices are correct

Q65 is a recent addition to the WSJT-X suite of digital modes. It has superior weak signal performance over other modes, accomplished by averaging the data over several receive cycles. Averaging is a well-established means of handling rapidly changing signals such as those encountered with troposcatter or meteor scatter modes. **ANSWER C.**

E2E06 What is the length of an FT8 transmission cycle?

A. It varies with the amount data
B. 8 seconds
C. 15 seconds
D. 30 seconds

FT8 is a ***structured*** digital HF mode, which means both the data format and the timing are rigidly standardized. An FT8 transmission lasts 15 seconds, including the ***decode*** time at the receiving station. FT8 time slots are divided between even and odd multiples of 15 seconds, beginning at the top of the minute. Local computer time must be within about a second of absolute time in order for proper decoding to occur. **ANSWER C.**

E8C11 What is the relationship between symbol rate and baud?

A. They are the same
B. Baud is twice the symbol rate
C. Baud rate is half the symbol rate
D. The relationship depends on the specific code used

Symbol rate and baud rate are the same, and the best indication of the bandwidth necessary for transmission. The number of symbols per character may determine the ultimate "data rate" but not the bandwidth. **ANSWER A.**

E8C02 What is the definition of symbol rate in a digital transmission?

A. The number of control characters in a message packet
B. The maximum rate at which the forward error correction code can make corrections
C. The rate at which the waveform changes to convey information
D. The number of characters carried per second by the station-to-station link

A character can consist of several symbols. The simplest application of this is Morse code, where we have two basic symbols, a dot, and a dash, that are used in combinations to form characters. ***Symbol rate is the rate at which the waveform of a transmitted signal changes to convey information***. Depending on the complexity of the code, symbol rate and character rate can be two very different things! **ANSWER C.**

E2E13 Which of these digital modes has the highest data throughput under clear communication conditions?

A. MFSK16 C. FT8

B. 170 Hz shift, 45 baud RTTY D. PACTOR IV

The speed of data ***throughput*** depends on the quality of the transmission channel, generally decreasing in the presence of noise. Some modes are better under noisy conditions than others. PACTOR IV has a high data throughput under ideal, quiet, and non-fading conditions, but slows down considerably in the presence of noise or other channel disturbances. The reason we have so many different digital modes is that we have so many different radio conditions. It's great to have a choice! **ANSWER D.**

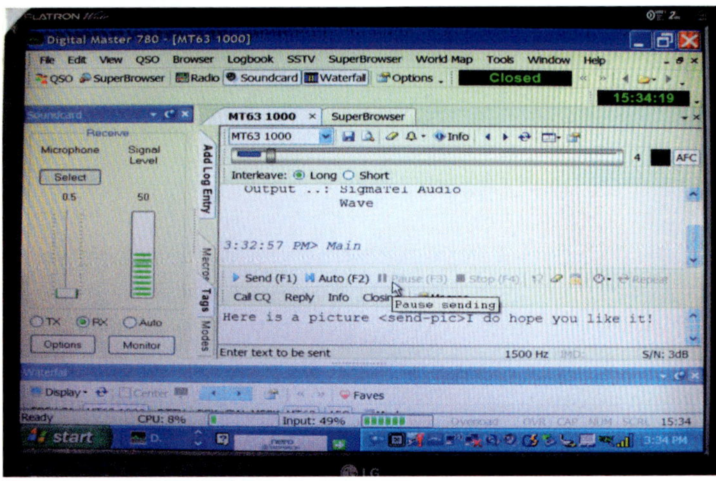

Going digital with your ham radio and computer will let you work stations you can't even hear above local noise!

E2E08 Which of the following HF digital modes can be used to transfer binary files?

A. PSK31 C. RTTY

B. PACTOR D. AMTOR

PACTOR transmissions sound like two birds chirping. Data is sent in blocks and the receiving station sends ACK for acknowledging everything sent or NAK indicating the request to re-send the block. ***PACTOR may be used to transfer binary files***, allowing you to download programs from a BBS and then work with the program on your computer. **ANSWER B.**

E8D10 What are some of the differences between the Baudot digital code and ASCII?

A. Baudot uses 4 data bits per character, ASCII uses 7 or 8; Baudot uses 1 character as a letters/figures shift code, ASCII has no letters/figures code

B. Baudot uses 5 data bits per character, ASCII uses 7 or 8; Baudot uses 2 characters as letters/figures shift codes, ASCII has no letters/figures shift code

C. Baudot uses 6 data bits per character, ASCII uses 7 or 8; Baudot has no letters/figures shift code, ASCII uses 2 letters/figures shift codes

D. Baudot uses 7 data bits per character, ASCII uses 8; Baudot has no letters/figures shift code, ASCII uses 2 letters/figures shift codes

---

Baudot is a 5-bit code; ASCII is a 7-bit code. In actuality, ASCII is an 8-bit code with only 7 bits uniformly defined and no shift code. There are ***2 shift codes in Baudot***: an up-shift and a down-shift. **ANSWER B.**

E2E02 Which of the following synchronizes WSJT-X digital mode transmit/receive timing?

A. Alignment of frequency shifts
B. Synchronization of computer clocks
C. Sync-field transmission
D. Sync-pulse timing

Most ***formatted*** digital amateur radio modes require fairly accurate synchronization between the transmitter and receiver. The time reference for the *WSJT-X* modes, such as WSPR, FT8, FT4, and the like is the computer clock on which the programs run. Network Time Protocol is the most convenient means of keeping your computer clock locked to an accurate, universal standard. An alternative means of synchronizing your computer clock is with the WWV broadcasts. **ANSWER B.**

E8C08 How does ARQ accomplish error correction?

A. Special binary codes provide automatic correction
B. Special polynomial codes provide automatic correction
C. If errors are detected, redundant data is substituted
D. If errors are detected, a retransmission is requested

ARQ (Automatic Repeat reQuest) is a digital protocol that acknowledges individual data blocks and ***requests a repeat of any missed block***. ARQ is part of the PACTOR and CLOVER digital signaling systems. **ANSWER D.**

E8C09 Which digital code allows only one bit to change between sequential code values?

A. Binary Coded Decimal Code
B. Extended Binary Coded Decimal Interchange Code
C. Extended ASCII
D. Gray code

In so-called coding theory, the ***Gray code*** has the shortest "distance" between any two symbols. This coding scheme reduces bandwidth and provides other benefits such as simpler error correction. **ANSWER D.**

E8B07 Orthogonal frequency division multiplexing (OFDM) is a technique used for which types of amateur communication?

A. Digital modes
B. Extremely low-power contacts
C. EME
D. OFDM signals are not allowed on amateur bands

OFDM uses the clever principle of frequency interleaving to allow ***much higher digital rates*** than "should" be possible on a given bandwidth channel. Weak signal VHF and UHF meteor and moonbounce digital exchanges use OFDM with great success. **ANSWER A.**

E8B08 What describes orthogonal frequency division multiplexing (OFDM)?

A. A frequency modulation technique that uses non-harmonically related frequencies
B. A bandwidth compression technique using Fourier transforms
C. A digital mode for narrow-band, slow-speed transmissions
D. A digital modulation technique using subcarriers at frequencies chosen to avoid intersymbol interference

OFDM requires precise timing and phasing of the component signals. By carefully choosing the spectrum generated by each of the modulated subcarriers, the ***signals can be interleaved without any mutual interference.*** **ANSWER D.**

E8C01 What is Quadrature Amplitude Modulation or QAM?

A. A technique for digital data compression used in digital television which removes redundancy in the data by comparing bit amplitudes
B. Transmission of data by modulating the amplitude of two carriers of the same frequency but 90 degrees out of phase
C. A method of performing single sideband modulation by shifting the phase of the carrier and modulation components of the signal
D. A technique for analog modulation of television video signals using phase modulation and compression

Two signals at 90-degree phase shift are known as Quadrature signals. Most modern receiving systems (and a good number of transmitting systems) use quadrature oscillators to modulate and demodulate RF signals. It can be shown mathematically that with just two quadrature oscillators you can synthesize ANY type of modulation: AM, FM, Phase Modulation, SSB, or any combination thereof. QAM is the amplitude modulation of two quadrature oscillators and is commonly used for both digital and analog modulation. **ANSWER B.**

E2E04 Which of the following is characteristic of the FST4 mode?

A. Four-tone Gaussian frequency shift keying
B. Variable transmit/receive periods
C. Seven different tone spacings
D. All these choices are correct

One of the latest modes in the WSJT-X suite of programs is FST4, which is particularly suitable for the weak signals on our two lowest frequency bands, 2200 meters and 630 meters. FST4 uses a number of “tricks” used for other weak signal digital modes, such as Four-tone Gaussian frequency shift keying, variable transmit/receive periods, and seven different tone spacings. **ANSWER D.**

E8D11 What is one advantage of using ASCII code for data communications?

A. It includes built-in error correction features
B. It contains fewer information bits per character than any other code
C. It is possible to transmit both uppercase and lower-case text
D. It uses one character as a shift code to send numeric and special characters

ASCII code allows upper case and lower-case text to be transmitted; Baudot does not. **ANSWER C.**

E8D06 What is the advantage of including parity bits in ASCII characters?

A. Faster transmission rate
B. Signal-to-noise ratio is improved
C. A larger character set is available
D. Some types of errors can be detected

The parity bit is used for error detection in a data transmission. A parity bit 1 or 0 might indicate the odd or even number of 1s or 0s in the data block. It is not foolproof – if a pair of 1s or a pair of 0s is missed, the error detection circuit might see no errors. Yet this simple error detection method is commonly used and works surprisingly well to confirm that the other station received the character intact or needs a repeat. **ANSWER D.**

E8C06 What is the bandwidth of an FT8 signal?

A. 10 Hz
B. 50 Hz
C. 600 Hz
D. 2.4 kHz

An FT8 ***baseband*** signal consists of 8 different audio tones, each separated by 6.25 Hz, giving a total bandwidth of 50 Hz. When the FT8 baseband is applied to a properly aligned single sideband transmitter, the radiated bandwidth remains the same, 50 Hz. (In practice, the transition between each of the 8 FT8 audio frequencies is quite gentle; if instantaneous frequency jumps were made between the tones the bandwidth would be infinite!) **ANSWER B.**

E8C07 What is the bandwidth of a 4,800-Hz frequency shift, 9,600-baud ASCII FM transmission?

A. 15.36 kHz
B. 9.6 kHz
C. 4.8 kHz
D. 5.76 kHz

Digital bandwidth is: BW = baud rate + (1.2 x frequency shift). In this problem, the baud rate is 9600 and th e frequency shift is 4800. ***Bandwidth = 9600 + (1.2 x 4800) = 15360***. The answer is 15,360 Hz, and when the decimal point is moved three places to the left, the bandwidth works out to be ***15.36 kHz***. **ANSWER A.**

E2E09 Which of the following HF digital modes uses variable-length character coding?

A. RTTY
B. PACTOR
C. MT63
D. PSK31

PSK31 stands for Phase Shift Keying with a bit rate slightly over 31. It is a little like Morse code, with the more common letters represented by a short count of bits: E (11), A (1011), T (101), or O (111), while less used characters such as Q and X, might contain up to 10 bits. Using PSK31, this ***varicode is transmitted*** with audio phase shifts which occupy a mere 31 Hz of bandwidth, leading to bandwidth efficiency. **ANSWER D.**

E8C10 How can data rate be increased without increasing bandwidth?

A. It is impossible
B. Increasing analog-to-digital conversion resolution
C. Using a more efficient digital code
D. Using forward error correction

The bandwidth of a digital signal of any kind is determined by the number of bits transmitted per second. However, not every "alphabet" has the same number of bits per character. For example, Baudot code only has 5 bits per character, while ASCII code has 8 bits per character. For the same information rate, (words per minute) Baudot code can consume considerably less bandwidth than ASCII. However, a number of ***advanced codes make very efficient use of information theory to increase the symbol to bandwidth ratio***. **ANSWER C.**

E8C04 What technique minimizes the bandwidth of a PSK31 signal?

A. Zero-sum character encoding
B. Reed-Solomon character encoding
C. Use of sinusoidal data pulses
D. Use of linear data pulses

Any digital mode, such as PSK31, that causes a sudden change of frequency, phase, or amplitude during data transmission will result in the generation of sidebands which increase the occupied bandwidth of the signal, sometimes dramatically. If the transitions between "mark" and "space" is made smoothly ***using sinusoidal data pulses,*** the ***bandwidth can be greatly reduced***. **ANSWER C.**

E2E10 Which of these digital modes has the narrowest bandwidth?

A. MFSK16 C. FT8
B. 170 Hz shift, 45-baud RTTY D. PACTOR IV

The bandwidth of a digital signal is determined both by the speed of the transmission and the particular coding. More complex codes require more bandwidth for a given ***baud rate***, the number of bits transmitted per second. Of the common text modes, FT8 has the narrowest occupied bandwidth, at the expense of somewhat slow information exchange. **ANSWER C.**

E2D07 What digital protocol is used by APRS?

A. PACTOR C. AX.25
B. QAM D. AMTOR

The digital packet protocol in use for ***APRS is AX.25***, the ham version of commercial packet networks operating X.25. The AX.25 protocol is recognized and authorized by the FCC with specific speed limitations depending on the frequency being used. **ANSWER C.**

E2D08 What type of packet frame is used to transmit APRS beacon data?

A. Acknowledgement C. Unnumbered Information
B. Burst D. Connect

The ***packet frames*** used to transmit APRS beacon data ***are called UIF – Unnumbered Information Frames***. Networking systems can allow an APRS position and short message to be transmitted and relayed throughout the country. Several manufacturers produce handheld and mobile radios with the APRS terminal node controller already built in. Handheld GPS equipment that outputs NMEA data is available for under $99. A mapping GPS can show ham call signs and locations. **ANSWER C.**

E2D11 How do APRS stations relay data?

A. By packet ACK/NAK relay
B. By C4FM repeaters
C. By DMR repeaters
D. By packet digipeaters

Since most VHF communications are primarily Line of Sight (LOS), it is usually necessary to have some sort of relay or repeater system to afford reliable, beyond-the-horizon communications. Packet Radio's success relies on a large number of packet ***digipeaters***, which relay packets on the same frequency on which they are received. This is in contrast with a standard FM repeater, which transmits on a different frequency than it receives. **ANSWER D.**

E2D10 What does the packet path WIDE3-1 designate?

A. Three stations are allowed on frequency, one transmitting at a time
B. Three subcarriers are permitted, subcarrier one is being used
C. Three digipeater hops are requested with one remaining
D. Three internet gateway stations may receive one transmission

Although most modern digital modes use the concept of ***packets*** to transmit information, there has been a resurgence of interest in "plain vanilla" packet radio. Radio packets can be traced along their routes to aid in troubleshooting. These packets include information in their ***overhead*** to aid in this tracing, which tells you how many hops, or "repeatings" the packet will take, as well as how many hops are remaining. The WIDE designation (for WIDE area network) shows the number of expected hops first, and how many are remaining until the final destination. The WIDE information can be retrieved at any point along a packet's path. **ANSWER C.**

E1F01 On what frequencies are spread spectrum transmissions permitted?

A. Only on amateur frequencies above 50 MHz
B. Only on amateur frequencies above 222 MHz
C. Only on amateur frequencies above 420 MHz
D. Only on amateur frequencies above 144 MHz

Spread spectrum is relatively new to ham radio operators and ***is confined to the 222 MHz band and above***. Spread spectrum is not allowed on 2 or 6 meters. And while spread spectrum is allowed on frequencies above 420 MHz, that answer leaves out the 220 MHz band. [97.305] **ANSWER B.**

E8D03 Which describes spread spectrum frequency hopping?

A. If interference is detected by the receiver, it will signal the transmitter to change frequencies
B. RF signals are clipped to generate a wide band of harmonics which provides redundancy to correct errors
C. A binary bit stream is used to shift the phase of an RF carrier very rapidly in a pseudorandom sequence
D. Rapidly varying the frequency of a transmitted signal according to a pseudorandom sequence

Frequency hopping depends on a particular signaling sequence, and the frequency of the RF carrier is changed so quickly that it is almost indistinguishable except to other spread spectrum receivers locked on to the same pseudorandom signaling sequence. **ANSWER D.**

E8D02 What spread spectrum communications technique uses a high-speed binary bit stream to shift the phase of an RF carrier?

A. Frequency hopping
B. Direct sequence
C. Binary phase-shift keying
D. Phase compandored spread spectrum

The term ***direct sequence*** is used to describe a system where a ***fast binary stream shifts*** the phase of an RF carrier. This is another spread spectrum communications technique. Using only a single carrier frequency, and then modulating the signal by shifting the phase in accordance with a high-speed binary bit stream, results in a signal whose pattern on a spectrum analyzer spreads out into a pattern of hills and valleys. This is called direct sequence spread spectrum. **ANSWER B.**

Spread spectrum phones are common!

E8D01 Why are received spread spectrum signals resistant to interference?

A. Signals not using the spread spectrum algorithm are suppressed in the receiver
B. The high power used by a spread spectrum transmitter keeps its signal from being easily overpowered
C. Built-in error correction codes minimize interference
D. If the receiver detects interference, it will signal the transmitter to change frequencies

Amateur operators transmitting and receiving spread spectrum signals are pioneering one more way to conserve our valuable frequency resources. Although spread spectrum may utilize an entire band, the hopping technique won't disturb other signals on the band, and all other ***signals not using the same spread spectrum algorithm are suppressed in the receiver***. So just when you thought there was no more room for any more signals on a particular band, spread spectrum gets through! **ANSWER A.**

E1F09 Which of the following cannot be transmitted over an amateur radio mesh network?

A. Third party traffic
B. Email
C. Messages encoded to obscure their meaning
D. All these choices are correct

Since the earliest days of Amateur Radio, there has been a strict prohibition against ***cyphers*** of any kind. A cypher is any means of obscuring or "secretizing" the meaning of any communication. This is in contrast with a ***code***, which is a perfectly legitimate means (such as Morse Code, or any number of other digital codes) of compressing information. A legitimate code has an established and ***known*** meaning. [FCC Part 97.113(a)(4)] **ANSWER C.**

E2C04 Which of the following frequencies can be used for amateur radio mesh networks?

A. HF frequencies where digital communications are permitted
B. Frequencies shared with various unlicensed wireless data services
C. Cable TV channels 41-43
D. The 60-meter band channel centered on 5373 kHz

A number of amateur UHF and microwave ***frequencies are shared with commercial broadband services, most of which are unlicensed***. Some hams experiment with modifying Wi-Fi routers to work on amateur radio frequencies, merging amateur radio seamlessly with the Internet. **ANSWER B.**

Take time to familiarize yourself with the Amateur Radio microwave band allocations listed below. All modes and licensees (except Novices) are authorized on the following bands [FCC Rules, Part 97.301(a)]:

2300-2310 MHz	*47.0-47.2 GHz*
2390-2450 MHz	*76.0-81.0 GHz**
3300-3450 MHz	*122.25 -123.00 GHz*
5650-5925 MHz	*134-141 GHz*
10.0-10.5 GHz	*241-250 GHz*
24.0-24.25 GHz	*All above 300 GHz*

**Amateur operation at 76-77 GHz has been suspended till the FCC can determine that interference will not be caused to vehicle radar systems.*

Commercial Wi-Fi (mesh) equipment is available in the shared 2390-2450 MHz band and less commonly in the 5650-5925 MHz band. Every ham should be aware that most of our microwave bands are unused and are wide open for experimentation and development!

E2C09 What type of equipment is commonly used to implement an amateur radio mesh network?

A. A 2-meter VHF transceiver with a 1,200-baud modom
B. A computer running EchoLink to provide interface from the radio to the internet
C. A wireless router running custom firmware
D. A 440 MHz transceiver with a 9,600-baud modem

The commercial frequencies in the 3.4 and 5 GHz bands actually overlap amateur radio frequencies in those bands. However, hams can use more power than the unlicensed wireless equipment, so you can do a few things the commercial folks can't when it comes to wireless networking. ***Off the shelf wireless routers*** can easily be configured to accommodate a wide area ***ham radio mesh network***. Recently, the FCC proposed restructuring the 3.5 GHz band, so hams may need to do some router frequency adjustments to those AREDN MESH systems currently on the air. **ANSWER C.**

E8C15 What technique do individual nodes use to form a mesh network?

A. Forward error correction and Viterbi codes
B. Acting as store-and-forward digipeaters
C. Discovery and link establishment protocols
D. Custom code plugs for the local trunking systems

Discovery and link establishment protocols are what allow "ad-hoc" mesh networks to assemble themselves. The concept is not entirely new; if you were involved in packet radio a few decades ago, there was a protocol called NetRom that did many of the functions of the more refined, modern, discovery and link establishment protocols. Fortunately, you don't need to look "under the hood" of these protocols to have a lot of fun with mesh networks...unless you really want to! **ANSWER C.**

E8C14 What type of addresses do nodes have in a mesh network?

A. Email
B. Trust server
C. Internet Protocol (IP)
D. Talk group

The TCP/IP internet protocol was so well conceived a few decades ago, that there have been few improvements on the basic format. Mesh networks, both amateur and commercial, take advantage of the tried and true TCP/IP to form ad-hoc "on-the-fly" networks almost instantly. In a mesh network, any node in the system can "talk" directly to any other node, unlike in a tree or star network, where packets often need to go through a central or semi-central node. Needless to say, there's some serious number-crunching that has to take place for this to all work. Fortunately, this is all invisible to the user! **ANSWER C.**

E2C12 What indicates the delay between a control operator action and the corresponding change in the transmitted signal?

A. Jitter
B. Hang time
C. Latency
D. Anti-VOX

Many processes beyond our control can slow down amateur radio communications. Although the RF path generally operates close to the speed of light, digital processing anywhere between "Point A and Point B" can slow down communications. Signal delay, or ***latency*** can occur in the actual computer used at the near and far ends of a communications link, but also at intermediate points such as digipeaters. All these delays add up to the overall latency of a communications link. **ANSWER C.**

E2E05 Which of these digital modes does not support keyboard-to-keyboard operation?

A. WSPR
B. RTTY
C. PSK31
D. MFSK16

You wouldn't think that a ham would talk and transmit to themselves, huh? With the weak signal propagation probe (WSPR), we do! Thanks to computers and software defined radios, we can actually see maps on our computer screens that our transmitted digital signal receives from locations all over the world, participating in this program! The question asks what is NOT a digital keyboard to keyboard ham-to-ham mode, and WSPR is an "echo" of your own signal all over the world. **ANSWER A.**

E8C12 What factors affect the bandwidth of a transmitted CW signal?

A. IF bandwidth and Q
B. Modulation index and output power
C. Keying speed and shape factor (rise and fall time)
D. All these choices are correct

While a continuous carrier RF wave has zero bandwidth, once you start or stop the wave, this no longer applies. During the rise and fall of a CW character, the bandwidth is no longer zero, but depends on the ***keying speed and the shape factor*** (rise and fall times) of the CW envelope. The "harder" the keying, the more bandwidth a CW signal will occupy. **ANSWER C.**

E8C05 What is the approximate bandwidth of a 13-WPM International Morse Code transmission?

A. 13 Hz
B. 26 Hz
C. 52 Hz
D. 104 Hz

To determine the bandwidth of a Morse code signal, use this formula:

BWcw = baud rate X wpm X fading factor

For CW, the baud rate is 0.8 and the fading factor is 5, therefore, ***multiply 0.8 times 13 wpm times 5. The answer works out to be 52 Hz***. **ANSWER C.**

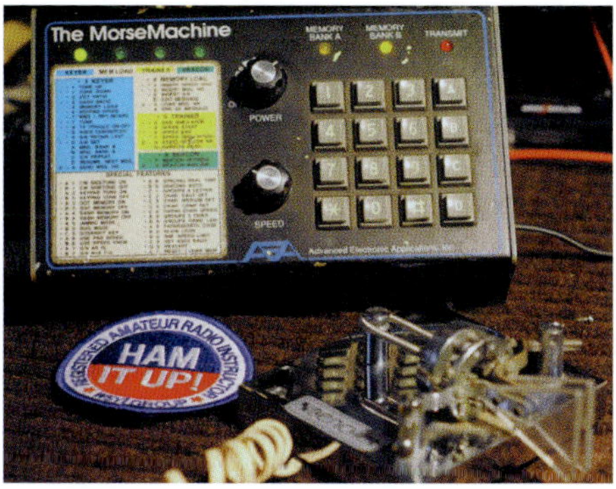

Morse code is more popular than ever these days on ham radio!

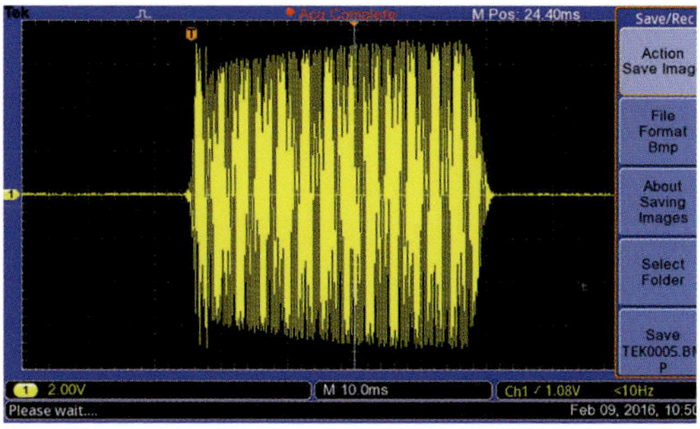

Waveform of a single CW dit.

E8D04 What is the primary effect of extremely short rise or fall time on a CW signal?

A. More difficult to copy
B. The generation of RF harmonics
C. The generation of key clicks
D. More difficult to tune

The process of keying a carrier on or off very quickly is identical to AM modulating it at a very high frequency; this generates sidebands resulting in ***key clicks***, which can be far removed from the carrier frequency. These sidebands can splatter across a large swath of an amateur band. Care should be applied to your keying wave shape for good CW operations. **ANSWER C.**

Morse code for low power (QRP) is just the ticket on a small, battery operated rig with built in paddles.

E8D05 What is the most common method of reducing key clicks?

A. Increase keying waveform rise and fall times
B. Insert low-pass filters at the transmitter output
C. Reduce keying waveform rise and fall times
D. Insert high-pass filters at the transmitter output

The CW envelope shape will greatly affect the emitted spectrum of a CW signal, as well as its sound. If the rise and fall times are too abrupt, it will emit key clicks, which are sidebands that can splatter across a large swath of an amateur band. If the keying is too soft, the tone can sound a bit mushy on the receiving end and high speed characters may overlap with succeeding characters. Linear amplifiers can alter the keying envelope too, so it's a great idea to be able to ***adjust both the rise and fall times of you CW "note."*** You might want to check out the December, 1984 QST article, "Try this Versatile CW Shaper" by Eric for ideas on how to do this on a variety of radios. **ANSWER A.**

Good To Know:

A New Mission for CW

For a few years now, knowing the International Radiotelegraph Code (usually synonymous with CW) has not been required for any class of amateur radio license. However, for the weak signal operator and the experimenter, CW still has a prominent role to play. Not only is a CW transmitter the simplest transmitter to build for any frequency from the lowest frequency medium wave bands clear up through our microwave bands, but CW has the best absolute weak signal performance of any mode. This includes exotic digital modes. Most of the activity on the experimental MF and lower frequency bands is by means of very slow CW, in ham vernacular QRSs... or QRSss...or QRSsss...depending on just how slow the CW is transmitted...sometimes as slow as a few characters per hour! By means of the lock-in amplifier, a special type of Direct Conversion receiver, signals as weak as 60 dB below the noise floor can be detected and decoded! It just takes a little time.

Amateur Radio still has exciting frontiers to explore, not only at the very highest frequencies of radio, but also at the very lowest frequencies of radio. We have a new amateur radio allocation at 2200 meters! One can rest assured that any successful communications achieved here will be by means of CW!

For the somewhat less-patient experimenter, high speed CW (QRR, QRRr, QRRrr, and QRRrrr) is still the preferred mode for fleeting phenomena like meteor burst, as well.

Until fairly recently, better-than-average CW proficiency was the hallmark of the Extra Class radio amateur. While this is no longer a compelling reason for most hams to become Extras, many many Extra Class radio amateurs still take great pride and joy in effortless CW operation...at any speed.

At one time or another, Eric has worked just about every mode of amateur radio communications there is, from CW, to SSB, to RTTY, to AMTOR, to PACKET, to SSTV, to exotic sound card digital modes that didn't even have a name at the time...not counting several military modes that you will never hear on the ham bands. But after 44 years on the air, CW remains his default mode.

SITES RELATED TO DIGITAL EXCITEMENT FOR YOUR COMPUTERS & RADIOS:

www.dstarinfo.com/
www.dstarinfo.com/repeater-lists.aspx
www.dstarusers.org/repeaters.php
www.cq-amateur-radio.com
www.dvdongle.com
www.rsgbiota.org/
www.dailydx.com
http://cluster.sdr-radio.com

Modulate Your Transmitters

E8B01 What is the modulation index of an FM signal?

A. The ratio of frequency deviation to modulating signal frequency
B. The ratio of modulating signal amplitude to frequency deviation
C. The modulating signal frequency divided by the bandwidth of the transmitted signal
D. The bandwidth of the transmitted signal divided by the modulating signal frequency

In an FM signal, the ***amount of deviation*** determines the "loudness" of the signal, while the ***modulating frequency*** determines the spacing between the sidebands. For any given modulating frequency, there is an optimum amount of deviation that can be accommodated by a given bandwidth. The relationship between deviation and modulating frequency is complex, described by ***Bessel Functions***. Different FM services are defined by specific modulation indices. **ANSWER A.**

E8B02 How does the modulation index of a phase-modulated emission vary with RF carrier frequency?

A. It increases as the RF carrier frequency increases
B. It decreases as the RF carrier frequency increases
C. It varies with the square root of the RF carrier frequency
D. It does not depend on the RF carrier frequency

This is one of those answers that has a "not" in it that makes it correct. When you calculate the ***modulation index***, you ***do NOT look at the actual RF carrier frequency*** as part of your calculations. **ANSWER D.**

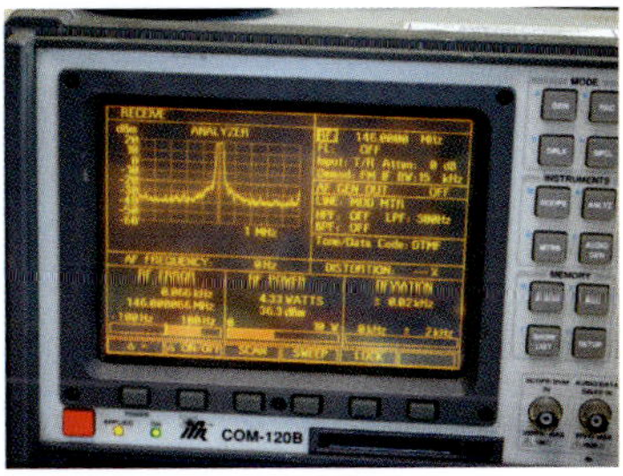

Close look at a clean FM signal on the scope.

FM versus AM

GREATER NUMBER OF SIDEBANDS

In FM modulation, one sine wave is used to modify another sine wave and multiple sums and differences result, depending on the deviation ratio. This is illustrated in Figure a. *Sidebands appear on either side of f_c at f_m, 2 f_m, 3 f_m, etc., depending upon the deviation ratio.* Figure b *shows that there are five significant sidebands for a deviation ratio of 3, and* Figure c *shows three significant sidebands for a deviation ratio of 1.67. The multiple sidebands are formed in both the upper and lower sideband positions, separated by the modulating frequency.*

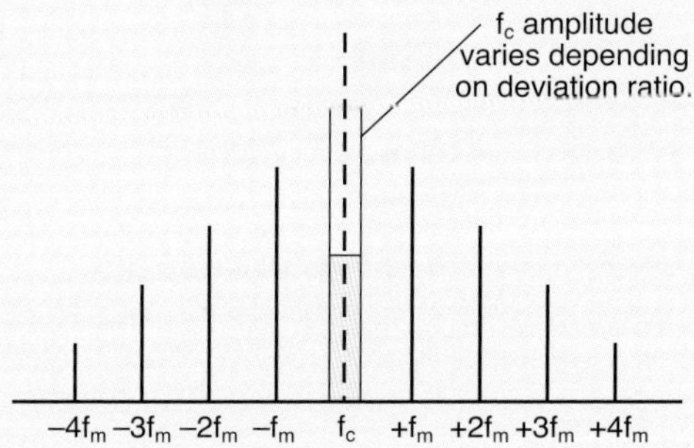

a. General Spectrum

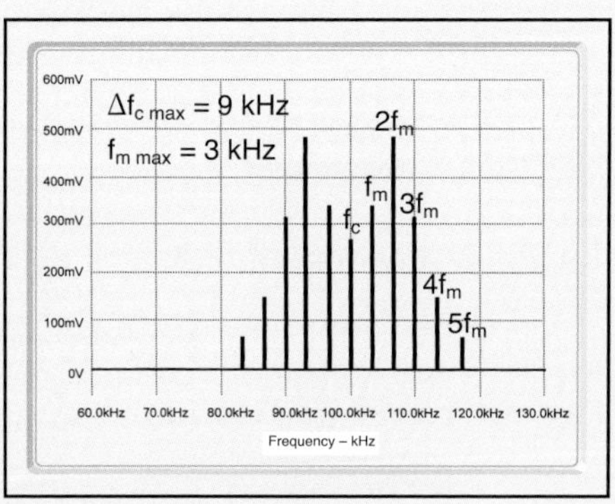

b. Deviation Ratio of 3

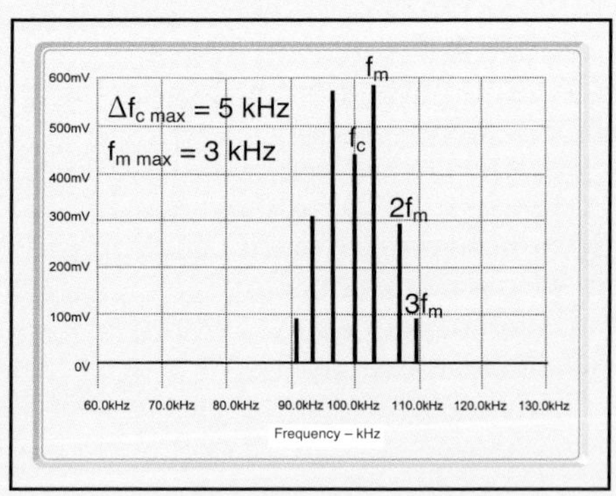

c. Deviation Ratio of 1.67

Frequency spectrum of FM signals.

Source: *Basic Communications Electronics*, Hudson & Luecke, © 1999 Master Publishing, Inc., Niles, IL

E8B03 What is the modulation index of an FM phone signal having a maximum frequency deviation of 3000 Hz either side of the carrier frequency if the highest modulating frequency is 1000 Hz?

A. 3 C. 6
B. 0.3 D. 0.6

This one is easy since both frequencies are given in Hz. Be careful of your units as you do this type of calculation. ***Simply divide 3000 by 1000 to get a modulation index of 3***. **ANSWER A.**

$$\text{Modulation Index } (\chi) = \frac{\text{Peak Deviation (D)}}{\text{Modulation Frequency } (m)} \text{ or } \chi = \frac{D}{m}$$

E8B04 What is the modulation index of an FM phone signal having a maximum carrier deviation of plus or minus 6 kHz if the highest modulating frequency is 2 kHz?

A. 0.3 C. 0.6
B. 3 D. 6

Again, easy because both frequencies are in kHz. Simply divide 6 by 2 to get a ***modulation index of 3***. **ANSWER B.**

E1C09 What is the highest modulation index permitted at the highest modulation frequency for angle modulation below 29.0 MHz?

A. 0.5 C. 2.0
B. 1.0 D. 3.0

With angle modulation, a sine wave carrier is varied by various angles to replicate the incoming modulation source. The ***highest modulation index permitted for angle modulation is exactly 1.0***. [97.307] **ANSWER B.**

E8B09 What is deviation ratio?

A. The ratio of the audio modulating frequency to the center carrier frequency
B. The ratio of the maximum carrier frequency deviation to the highest audio modulating frequency
C. The ratio of the carrier center frequency to the audio modulating frequency
D. The ratio of the highest audio modulating frequency to the average audio modulating frequency

It's important to set the deviation ratio properly on your FM transceiver to prevent splatter and a signal bandwidth that is too wide. ***Deviation ratio*** is the ratio of the ***maximum carrier swing to the highest audio modulating frequency*** when you speak into the microphone. **ANSWER B.**

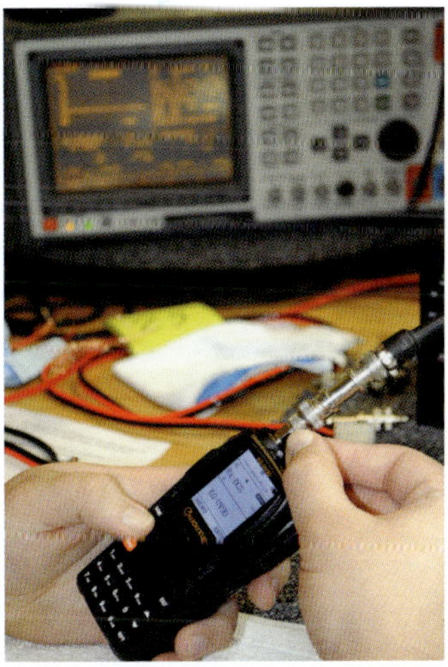

Check your deviation ratio by speaking loudly into the microphone while watching the scope.

E8B05 What is the deviation ratio of an FM phone signal having a maximum frequency swing of plus or minus 5 kHz if the highest modulation frequency is 3 kHz?

A. 6
B. 0.167
C. 0.6
D. 1.67

Deviation ratio is the ratio of the maximum carrier swing to the highest audio modulating frequency when you speak into the microphone. To calculate the deviation ratio, divide the maximum carrier frequency swing by the maximum modulation frequency. 5 kHz frequency swing divided by 3 kHz modulation frequency yields a ***deviation ratio of 1.67***. **ANSWER D.**

$$\text{Deviation Ratio} = \frac{\text{Maximum Carrier Frequency Deviation in kHz}}{\text{Maximum Modulation Frequency in kHz}}$$

Good To Know:

Modulation Index vs Deviation Ratio

Various FM services specify the Deviation Ratio of FM transmissions. The deviation ratio is a FIXED value of the maximum deviation allowed for a given modulating frequency, for a given service. There are various FCC assigned deviation ratios for, say FM broadcasting, or commercial VHF two-way radios.

Closely related to Deviation Ratio is the Modulation Index, which is the instantaneous value of FM deviation for a given applied modulating (audio) signal, and is continually changing. While Modulation Index is useful for analyzing FM signals with a spectrum analyzer or other instrument, when it comes to being legal, the Deviation Ratio is what you want to look at.

E8B06 What is the deviation ratio of an FM phone signal having a maximum frequency swing of plus or minus 7.5 kHz if the highest modulation frequency is 3.5 kHz?

A. 2.14
B. 0.214
C. 0.47
D. 47

To calculate deviation ratio of an FM signal, simply divide the maximum carrier frequency swing by the maximum modulation frequency. 7.5 kHz carrier frequency divided by 3.5 kHz modulation frequency yields a ***deviation ratio of 2.14***. Simple, huh? **ANSWER A.**

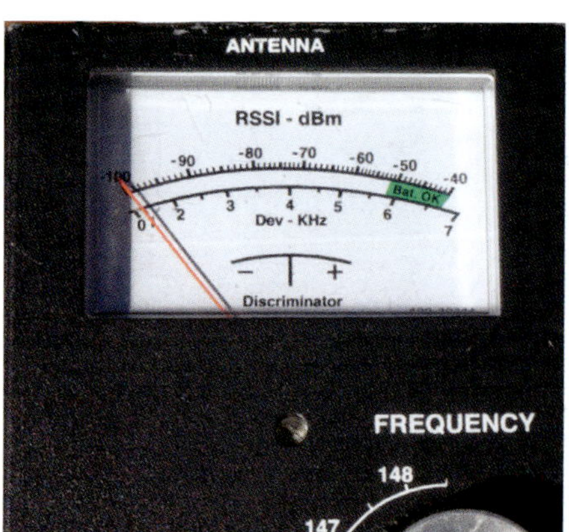

Check how many kHz of deviation you have on FM with this handy deviation meter.

E8B10 What is frequency division multiplexing (FDM)?

A. The transmitted signal jumps from band to band at a predetermined rate
B. Dividing the transmitted signal into separate frequency bands that each carry a different data stream
C. The transmitted signal is divided into packets of information
D. Two or more information streams are merged into a digital combiner, which then pulse position modulates the transmitter

With FDM, ***the transmit signal is divided into separate frequency streams***. Each information stream is modulated onto a sub-carrier (up-converting it), which is then used to modulate the carrier. In the receiver, the sub-carriers are removed by pass band filters, then each information stream is recovered by down-converting the sub-carrier to baseband. **ANSWER B.**

E8B11 What is digital time division multiplexing?

A. Two or more data streams are assigned to discrete sub-carriers on an FM transmitter
B. Two or more signals are arranged to share discrete time slots of a data transmission
C. Two or more data streams share the same channel by transmitting time of transmission as the sub-carrier
D. Two or more signals are quadrature modulated to increase bandwidth efficiency

Time division multiplexing (TDM) allows multiple users to share the same transmit and receive channel, divided into ***discreet time slots***, for sending data. This modern form of signaling is used most by commercial and telecommunications point-to- point broadcasting, but hams with the right equipment can take advantage of this frequency conserving technology. The two signals don't actually share discrete time slots – rather, they are separated by each using a different time slot. **ANSWER B.**

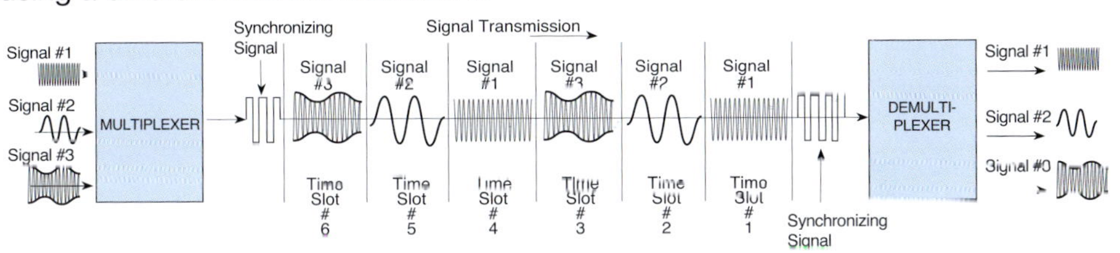

Time multiplexing and demultiplexing.

Source: , Hudson & Luecke, © 1999 Master Publishing, Inc., Niles, IL

E8A02 Which of the following is a type of analog-to-digital conversion?

A. Successive approximation
B. Harmonic regeneration
C. Level shifting
D. Phase reversal

Changing an analog signal into a digital signal takes place in the analog-to-digital converter through ***successive approximation*** of the waveform. Thousands of tiny audio slices are held in a sequential sample-and-hold circuit where they are converted into binary numbers. Some ADCs count with a staircase generator. Others convert analog voltage into digital values with multiple comparators, and others in DDS (Direct Digital Synthesis) use a phase accumulator to identify points on a specific waveform cycle. Successive approximation of the incoming waveform is the heart of converting analog signals to digital. **ANSWER A.**

E8A11 Which of the following is a measure of the quality of an analog-to-digital converter?

A. Total harmonic distortion
B. Peak envelope power
C. Reciprocal mixing
D. Power factor

The "front end" of an analog-to-digital converter is an analog amplifier (usually an op-amp), that is subject to all the requirements and characteristics of any amplifier. Any distortion or noise in the amplifier will be digitized as well as the desired analog signal. Digitizing an analog signal is not a "magic bullet" guaranteeing a high-quality communications circuit. A high-quality A/D converter must have an amplifier with very low ***total harmonic distortion*** and very low noise ahead of the digital sampling section. **ANSWER A.**

E8A01 What technique shows that a square wave is made up of a sine wave and its odd harmonics?

A. Fourier analysis
B. Vector analysis
C. Numerical analysis
D. Differential analysis

Fourier analysis was one of the greatest mathematical triumphs of the past millennium. It was so profound that it is the basis of all we do in digital signal processing. Joseph Fourier would be so proud. But alas, poor Joseph didn't have very good computers to work with in the 17th century, so he had to use a quill pen to do his calculations, which is why he had to settle for Really Slow Fourier Transforms (RSFT) instead of Fast Fourier Transforms (FFT). But somehow, he still got the job done. **ANSWER A.**

E8A03 Which of the following describes a signal in the time domain?

A. Power at intervals of phase
B. Amplitude at different times
C. Frequency at different times
D. Discrete impulses in time order

An oscilloscope is one of the most basic instruments in radio; it shows the amplitude (voltage) of a signal with respect to time. The ***domain*** of any mathematical function is generally shown on the horizontal axis of a graph; therefore, an oscilloscope shows amplitude in the ***time domain***. A spectrum analyzer, on the other hand, shows the amplitude of signals in the ***frequency domain***. **ANSWER B.**

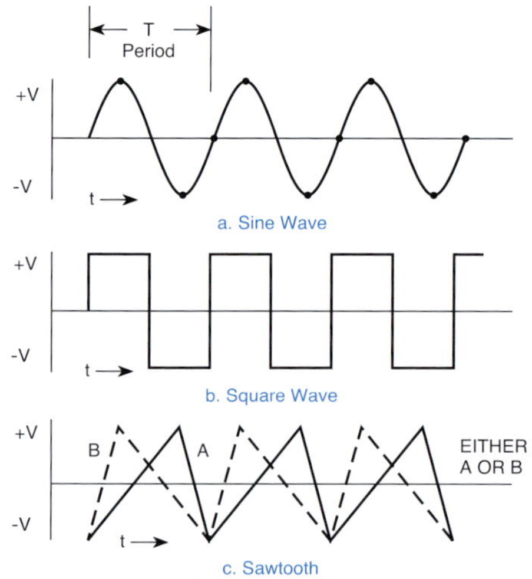

Sine, Square and Sawtooth Waveforms

E8A05 What is the benefit of making voltage measurements with a true-RMS calculating meter?

A. An inverse Fourier transform can be used
B. The signal's RMS noise factor is also calculate
C. The calculated RMS value can be converted directly into phasor form
D. RMS is measured for both sinusoidal and non-sinusoidal signals

The RMS voltage of an AC signal determines its "work producing" value; it is the equivalent heating value of a DC voltage of that same value. For a pure sine wave, the RMS value is always .707 times the peak voltage (or .3535 times the peak-to-peak value). However, if the AC wave is NOT a pure sine wave, all bets are off. ***A true RMS AC voltmeter will give you an accurate RMS voltage regardless of the waveshape***, a must when making accurate power measurements. **ANSWER D.**

E8A07 What determines the PEP-to-average power ratio of an unprocessed single-sideband phone signal?

A. The frequency of the modulating signal
B. Speech characteristics
C. The degree of carrier suppression
D. Amplifier gain

It's not all that easy to determine the ratio of peak envelope power to average power in your worldwide SSB set because your ***speech characteristics*** (the characteristics of the modulating signal) may vary from those of another operator when using the same equipment. Speech compression is specifically designed to reduce the peak- to-average power of an unprocessed SSB signal. **ANSWER B.**

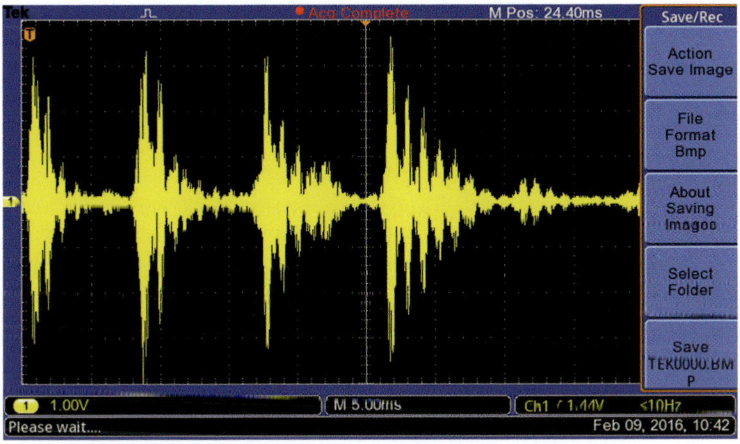

Average modulation can be increased by turning on speech processing, as seen on the right.

E8A06 What is the approximate ratio of PEP-to-average power in an unprocessed single-sideband phone signal?

A. 2.5 to 1
B. 25 to 1
C. 1 to 1
D. 13 to 1

When you talk about power output from your SSB worldwide set, we normally refer to peak envelope power. Depending on your modulation, ***the ratio of PEP to average power is about 2.5 to 1***, assuming no artificial speech compression is employed. You can remember 2.5 because that's very close to 2.5 kHz of bandwidth occupied by a properly modulated, unprocessed SSB signal. **ANSWER A.**

E7E01 Which of the following can be used to generate FM phone signals?

A. Balanced modulation of the audio amplifier
B. Reactance modulation of a local oscillator
C. Reactance modulation of the final amplifier
D. Balanced modulation of a local oscillator

The ***reactance modulator*** causes the ***oscillator*** to vary frequency in accordance with the modulation. It's the oscillator that is ***modulated***. **ANSWER B.**

E7E02 What is the function of a reactance modulator?

A. Produce PM or FM signals by varying resistance
B. Produce AM signals by varying an inductance
C. Produce AM signals by varying a resistance
D. Produce PM or FM signals by varying a capacitance

The ***reactance modulator*** is found in a frequency modulation transceiver, ***varying capacitance to produce*** the ***phase modulation (PM) or FM*** signal where the phase of the carrier current is varied with the frequency of the modulating voltage. **ANSWER D.**

E7E03 What is a frequency discriminator?

A. An FM generator circuit
B. A circuit for filtering closely adjacent signals
C. An automatic band-switching circuit
D. A circuit for detecting FM signals

Anytime you see the words ***frequency discriminator*** you know that you are dealing with ***signal detection or demodulation*** in an ***FM*** transceiver. **ANSWER D.**

E7E05 What is added to an FM speech channel to boost the higher audio frequencies?

A. A de-emphasis network
B. A harmonic enhancer
C. A heterodyne enhancer
D. A pre-emphasis network

Pre-emphasis networks in modern FM transmitters help improve the signal-to-noise ratio by proportionally attenuating the less-important lower audio frequencies for voice and ***boosting*** voice intelligence in the ***higher audio frequencies***. **ANSWER D.**

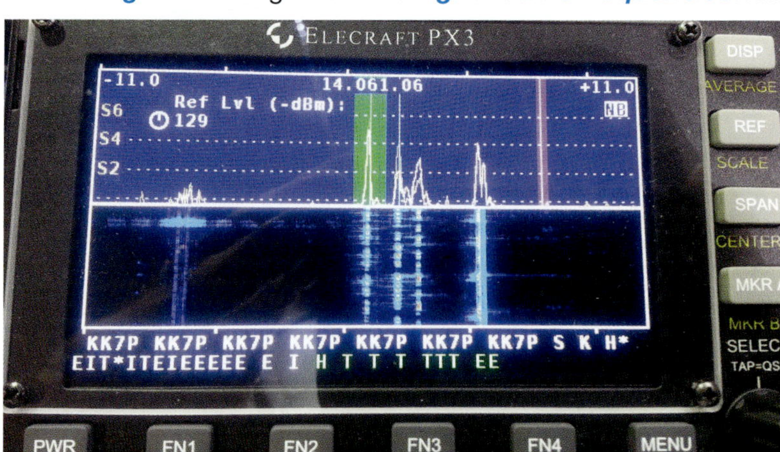

This modern spectrum scope lets you see the signal peaks in real time, plus the waterfall memory traces let you see signals that you might have missed. The waterfall memory won't let you miss any signal!

E7E04 What is one way to produce a single-sideband phone signal?

A. Use a balanced modulator followed by a filter
B. Use a reactance modulator followed by a mixer
C. Use a loop modulator followed by a mixer
D. Use a product detector with a DSB signal

In a ***single-sideband transceiver***, a ***lower or upper sideband filter*** removes the unwanted sideband. **ANSWER A.**

E7F04 Which method generates an SSB signal using digital signal processing?

A. Mixing products are converted to voltages and subtracted by adder circuits
B. A frequency synthesizer removes unwanted sidebands
C. Varying quartz crystal characteristics are emulated in digital form
D. Signals are combined in quadrature phase relationship

Digital signal processing has revolutionized ham radio transmitters and receivers and a host of after-market audio products by cleaning up signal reception. For SSB signal transmissions, ***DSP relies on phase-shifting*** the carrier and audio components by 90 degrees and sending this phased split to the SSB balanced modulator. The output is a DSP upper or lower sideband, with the carrier and unwanted sideband suppressed more than 40 dB. No heterodyning is required. All of this ***quadrature occurs in the DSP chip***. **ANSWER D.**

E7F03 What type of digital signal processing filter is used to generate an SSB signal?

A. An adaptive filter
B. A notch filter
C. A Hilbert-transform filter
D. An elliptical filter

A ***Hilbert transform filter*** is a ***DSP*** filter that performs an operation on an analytic signal, which is a signal containing both I and Q (In-phase and Quadrature) components. It is a "high tech" version of the old phasing method of ***single sideband generation***, popular before the crystal filter was readily available. All things old become new again in amateur radio! **ANSWER C.**

E7E07 What is meant by the term "baseband" in radio communications?

A. The lowest frequency band that the transmitter or receiver covers
B. The frequency range occupied by a message signal prior to modulation
C. The unmodulated bandwidth of the transmitted signal
D. The basic oscillator frequency in an FM transmitter that is multiplied to increase the deviation and carrier frequency

The term ***"baseband" refers to the frequency range occupied by a message signal prior to modulation***. For digital modes and CW, it could be as narrow as 30 Hz. For sideband voice and FM, baseband frequency response could range from 2.5 kHz to 25 kHz. **ANSWER B.**

E1B01 Which of the following constitutes a spurious emission?

A. An amateur station transmission made without the proper call sign identification
B. A signal transmitted to prevent its detection by any station other than the intended recipient
C. Any transmitted signal that unintentionally interferes with another licensed radio station and whose levels exceed 40 dB below the fundamental power level
D. An emission outside the signal's necessary bandwidth that can be reduced or eliminated without affecting the information transmitted

May your spurious emissions be minimum! The FCC says so. A ***spurious emission*** could include random splatter to adjacent frequencies, second harmonics, and any other microvolt or millivolt emission that serves no purpose ***outside of your signal's necessary bandwidth***. [97.3] **ANSWER D.**

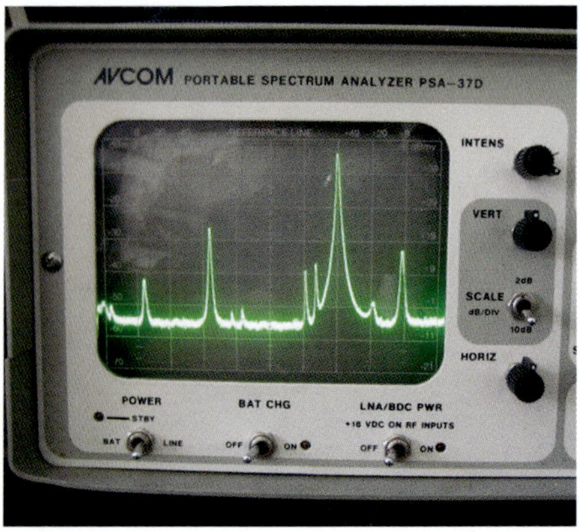

A close look at multiple spurious signals from a land mobile FM hand held not specifically tuned to the 2 meter ham band. All traces disappear when the hand held is un-keyed, proving these spurs are part of the main carrier.

E1B08 What limitations may the FCC place on an amateur station if its signal causes interference to domestic broadcast reception, assuming that the receivers involved are of good engineering design?

A. The amateur station must cease operation
B. The amateur station must cease operation on all frequencies below 30 MHz
C. The amateur station must cease operation on all frequencies above 30 MHz
D. The amateur station must avoid transmitting during certain hours on frequencies that cause the interference

If the FCC gets involved in an interference issue with your neighbor and it finds that your neighbor has a nice clean receiver of modern design, and your 1.5-kilowatt output is coming in over that receiver, ***the FCC could impose quiet hours on your transmissions on the specific band*** where you are causing the interference. Suggest you don't run the KW during ball games! [97.121] **ANSWER D.**

E7B12 What type of amplifier circuit is shown in Figure E7-1?

A. Common base
B. Common collector
C. Common emitter
D. Emitter follower

A common emitter amplifier is shown in Figure E7-1. It is the most common voltage amplifier circuit. While the emitter is not always connected directly to ground, it is also known as a grounded emitter amplifier. In a common emitter amplifier, the input signal is applied between the base and emitter while the output signal is taken between the collector and the emitter. The collector and the associated load resistor form a voltage divider, with the transistor acting as a variable resistor in series with the load resistance. At rest, the voltage at the collector of a properly functioning ***common emitter amplifier*** will be about half the supply voltage; this is a good quick test to perform when troubleshooting. **ANSWER C.**

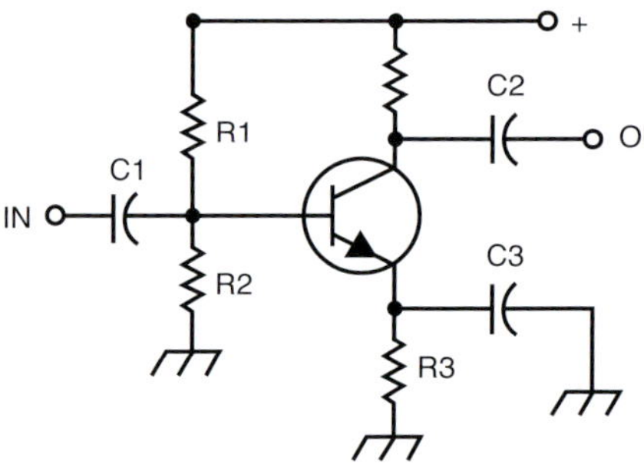

Figure E7-1

E7B10 In Figure E7-1, what is the purpose of R1 and R2?

A. Load resistors.
B. Voltage divider bias.
C. Self bias.
D. Feedback.

This reminds you a little bit of Thevenin's Theorem, doesn't it? ***R1 and R2 form a voltage divider*** to set the voltage and the current at the base of the transistor for voltage divider bias. **ANSWER B.**

E7B11 In Figure E7-1, what is the purpose of R3?

A. Fixed bias
B. Emitter bypass
C. Output load resistor
D. Self bias

Current through ***R3 produces*** a voltage drop that affects the base-emitter bias voltage. Since the transistor's collector and base current (the emitter current) determine the voltage at the emitter, it is called ***self-bias***. **ANSWER D.**

E7C01 How are the capacitors and inductors of a low-pass filter Pi-network arranged between the network's input and output?

A. Two inductors are in series between the input and output, and a capacitor is connected between the two inductors and ground
B. Two capacitors are in series between the input and output, and an inductor is connected between the two capacitors and ground
C. An inductor is connected between the input and ground, another inductor is connected between the output and ground, and a capacitor is connected between the input and output
D. A capacitor is connected between the input and ground, another capacitor is connected between the output and ground, and an inductor is connected between the input and output

The ***Pi-network features a capacitor in parallel with the input***, a second ***capacitor in parallel with the output***, and an ***inductor in series between the two***. It looks like the Greek letter π. **ANSWER D.**

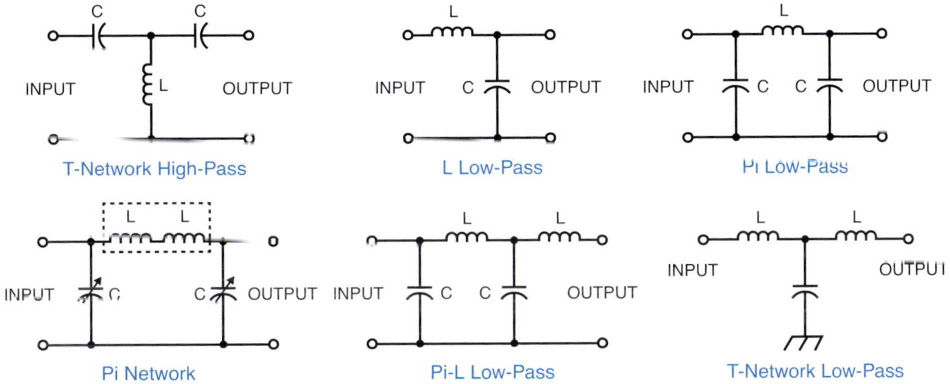

Matching Networks

E7C02 What is the frequency response of a T-network with series capacitors and a shunt inductor?

A. Low-pass
B. High-pass
C. Band-pass
D. Notch

While a ***T-network*** is commonly used as an antenna tuner or transmatch, it can inadvertently become a ***high-pass filter*** under certain conditions, which is something you generally want to avoid in this application. Remember that capacitors decrease reactance with increasing frequency; therefore, if these are in series with the signal path, they will pass higher frequencies with less attenuation. Likewise, an inductor has more reactance with increasing frequency, which means if they are in shunt parallel with the signal path, they will bypass less of the signal at higher frequencies. If you use a high-pass T-network as an antenna tuner, be sure your transmitter is clean, because the network will emphasize any harmonics you generate! **ANSWER B.**

E7C03 What is the purpose of adding an inductor to a Pi-network to create a Pi-L-network?

A. Greater harmonic suppression
B. Higher efficiency
C. To eliminate one capacitor
D. Greater transformation range

Worldwide ham sets transmitting into a multi-band antenna system can generate harmonics. Adding an inductor in series to a Pi-network will create a Pi-L-network, offering increased harmonic suppression for those of us still exercising our venerable tube rigs. **ANSWER A.**

E7C07 Which describes a Pi-L network?

A. A Phase Inverter Load network
B. A Pi-network with an additional output series inductor
C. A network with only three discrete parts
D. A matching network in which all components are isolated from ground

The Pi-L network is characterized by two series-adjustable inductors and two variable shunt-to-ground capacitors. You can spot the Pi-L network over a simple Pi network as the ***Pi-L network has an additional series inductor at the output***. **ANSWER B.**

E7B09 What is characteristic of an emitter follower (or common collector) amplifier?

A. Low input impedance and phase inversion from input to output
B. Differential inputs and single output
C. Acts as an OR circuit if one input is grounded
D. Input and output signals in-phase

A common collector amplifier, sometimes called an ***emitter follower***, has some unique and useful properties. As the name implies, the emitter follows the input voltage; it has the same phase and amplitude (nearly) as the input voltage. Since it has no ***voltage gain***, you might logically wonder what the purpose of the amplifier is. The emitter follower has very high input impedance and a low output impedance, which allows it to act as a buffer between two stages. It is often used in ***oscillator*** circuits to prevent "downstream" circuitry from ***loading*** the oscillator and pulling its frequency. It is also a main ingredient in ***active antennas***, to allow a very short antenna to have tremendous power gain. **ANSWER D.**

E7C04 How does an impedance-matching circuit transform a complex impedance to a resistive impedance?

A. It introduces negative resistance to cancel the resistive part of impedance
B. It introduces transconductance to cancel the reactive part of impedance
C. It cancels the reactive part of the impedance and changes the resistive part to the desired value
D. Reactive currents are dissipated in matched resistances

In older radios, you could vary the matching network to ***cancel the reactive (X) part of an impedance and change the value of the resistive part (R) of an impedance to a desired value***. Newer transistorized radios have a fixed output and there is no manual control of the fixed matching devices. However, manufacturers have come to the rescue in antenna matching by providing new radios with built-in automatic antenna tuners. **ANSWER C.**

E4D03 What creates intermodulation interference between two repeaters in close proximity?

A. The output signals cause feedback in the final amplifier of one or both transmitters
B. The output signals mix in the final amplifier of one or both transmitters
C. The input frequencies are harmonically related
D. The output frequencies are harmonically related

Intermod, in its true sense, ***occurs within the repeater's final amplifier***. When signals mix in one or both repeater amplifiers, sum and difference signals are heard on many different frequencies and as significant noise on the frequencies of the repeaters. **ANSWER B.**

E4D08 What causes intermodulation in an electronic circuit?

A. Negative feedback
B. Lack of neutralization
C. Nonlinear circuits or devices
D. Positive feedback

Intermodulation interference is minimized by a careful selection of linear circuits and components. If that newly designed transceiver for 2 meters or 440 MHz or that repaired repeater is experiencing high levels of intermodulation, the problem could be caused by ***non-linear circuits, devices, and components*** inside the receiver. **ANSWER C.**

E4D06 What is the term for the reduction in receiver sensitivity caused by a strong signal near the received frequency?

A. Reciprocal mixing
B. Quieting
C. Desensitization
D. Cross modulation interference

Two hams, close together working a public service event on the same band, like 2 meters, discovers when one transmits, the other ham says his received signal just vanished. This is ***desensitization***. Easy cure! Separate your stations at least 50 feet apart. **ANSWER C.**

E7C10 Which of the following filters is used in a 2-meter band repeater duplexer?

A. A crystal filter
B. A cavity filter
C. A DSP filter
D. An L-C filter

For a 2-meter repeater duplexer system, we need big filters. We call them ***cavity filters*** because they are hollowed out to form extremely narrow band tuned resonant circuits. **ANSWER B.**

E4D04 Which of the following is used to reduce or eliminate intermodulation interference in a repeater caused by a nearby transmitter?

A. A band-pass filter in the feed line between the transmitter and receiver
B. A properly terminated circulator at the output of the repeater's transmitter
C. Utilizing a Class C final amplifier
D. Utilizing a Class D final amplifier

Your ham friends who regularly trek to mountaintops to maintain ham radio VHF/UHF repeaters deserve your major support. Mountaintop repeaters are located in the worst RF environment possible – amid hundreds of nearby signals that may desensitize the receiver or create "garbage" on the input. Everything was fine when your team left the hill, but a new land-mobile radio system on the same tower is now causing ***intermodulation*** interference. One ***solution is to install a terminated circulator*** or ferrite isolator between your own transmitter and duplexer, which helps zero in on your own receive frequency and cancel out any other off-channel frequencies that seem to creep in only when your equipment is on transmit. **ANSWER B.**

The professional cavity filter system on a 2 meter repeater offers minimal attenuation to received signals in the pass band, and major attenuation to the repeater's transmitter frequency. It is a job to filter out your transmit signals only 600 kHz away! Make sure these filter cavities are in a dry, warm cabinet to prevent condensation on the gleaming copper insides!

E4E08 What current flows equally on all conductors of an unshielded multiconductor cable?

A. Differential-mode current
B. Common-mode current
C. Reactive current only
D. Magnetically-coupled current only

Common-mode currents are just what they sound like: currents that are equal in amplitude and phase on two or more conductors. The best way to reduce ***common-mode currents*** is to use a common-mode choke, such as passing the entire bundle through a single toroidal ferrite core or a string of ferrite beads. **ANSWER B.**

HERE ARE SOME WEB ADDRESSES OF USEFUL SITES RELATED TO MODULATE YOUR TRANSMITTERS:

www.dx-code.org/
www.qrz.com/
www.hamcall.net/call
www.dxzone.com/catalog/DX_Resources/Callsigns/

Amps & Power Supplies

E1F03 Under what circumstances may a dealer sell an external RF power amplifier capable of operation below 144 MHz if it has not been granted FCC certification?

A. Gain is less than 23 dB when driven by power of 10 watts or less
B. The equipment dealer assembled it from a kit
C. It was manufactured and certificated in a country which has a reciprocal certification agreement with the FCC
D. It was purchased in used condition from an amateur operator and is sold to another amateur operator for use at that operator's station

Amplifiers for frequencies below 144 MHz could be ***purchased in used condition*** by an equipment dealer and ***then sold by that dealer to another amateur operator*** for use at his or her station. This rule prevents the sale and use to hams of amplifiers intended for CB use that are constructed without the FCC's blessing. [97.315] **ANSWER D.**

E1F11 Which of the following best describes one of the standards that must be met by an external RF power amplifier if it is to qualify for a grant of FCC certification?

A. It must produce full legal output when driven by not more than 5 watts of mean RF input power
B. It must have received an Underwriters Laboratory certification for electrical safety as well as having met IEEE standard 14.101(B)
C. It must exhibit a gain of less than 23 dB when driven by 10 watts or less
D. It must satisfy the FCC's spurious emission standards when operated at the lesser of 1500 watts or its full output power

Recently, the FCC revised its policy on RF amplifiers. In order for an RF power amplifier to be granted FCC certification, ***the amp must meet tight spurious***

emission standards when it is operating at full power. Tube amplifiers with large band switching filter networks meet the specs more easily than smaller, solid-state amps. This may be why the solid-state amplifiers are still relatively expensive compared to tube amps. [97.317] **ANSWER D.**

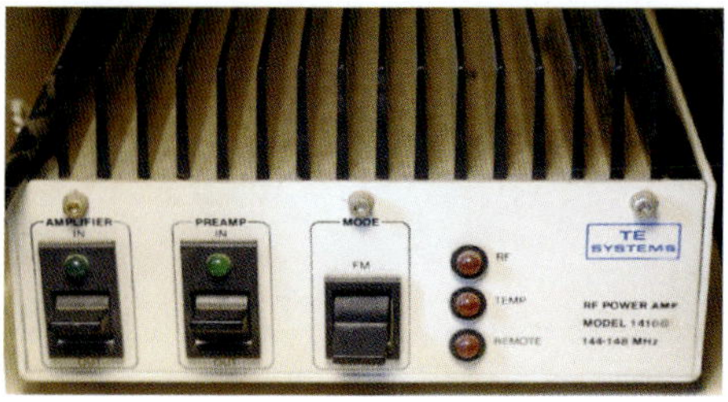

This little 2-meter linear power amplifier meets stringent FCC spurious emission standards.

E7B04 What is the operating point of a Class A common emitter amplifier?

A. Approximately halfway between saturation and cutoff
B. Approximately halfway between the emitter voltage and the base voltage
C. At a point where the bias resistor equals the load resistor
D. At a point where the load line intersects the zero bias current curve

The ***Class A amplifier*** offers the greatest linearity but provides only marginal efficiency, usually less than 25%. We normally find the Class A amplifier as a "small signal" amplifier. It is sometimes used in modulation circuits where perfect linearity is required. The input signal continuously varies the input current over the complete 360 degrees of the sine-wave signal, with ***bias normally set approximately half-way between saturation and cutoff***. **ANSWER A.**

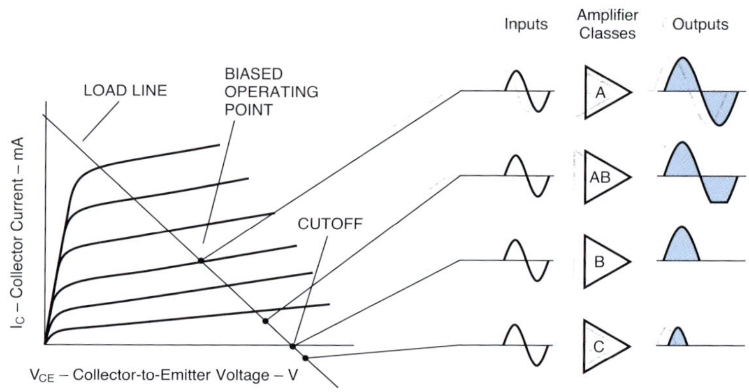

Various Classes of Transistorized Amplifiers

E7B01 For what portion of the signal cycle does each active element in a push-pull, Class AB amplifier conduct?

A. More than 180 degrees but less than 360 degrees
B. Exactly 180 degrees
C. The entire cycle
D. Less than 180 degrees

The ***Class AB amplifier*** has some of the properties of both the Class A and the Class B amplifier. It has output for ***more than 180 degrees*** of the cycle, but ***less than 360 degrees*** of the cycle. **ANSWER A.**

E7B06 What is a characteristic of a grounded-grid amplifier?

A. High power gain
B. Low input impedance
C. High electrostatic damage protection
D. Low bandwidth

Nowadays, most high-power vacuum tube linear amplifiers are of the grounded-grid topology. ***A grounded grid amplifier has a low input impedance,***

normally around 50 ohms, allowing it to be easily driven by a typical transceiver. Grounded grid amplifiers are inherently stable, and generally quite trouble free. Many common amplifier tubes, such as the 3-500 series, are specifically designed for grounded grid operation. **ANSWER B.**

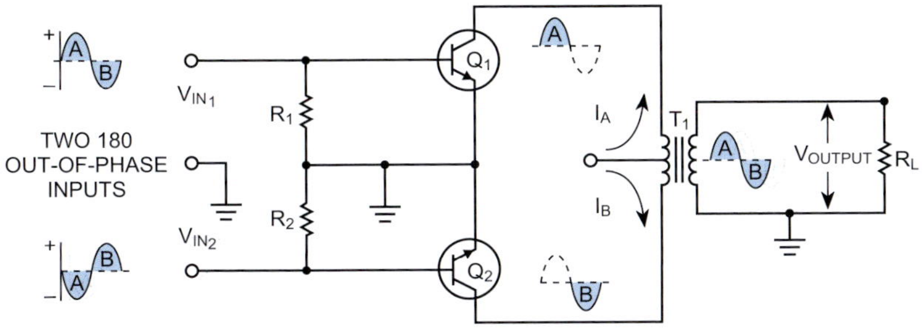

Transistorized Push-Pull Amplifier

E7B08 Why are switching amplifiers more efficient than linear amplifiers?

A. Switching amplifiers operate at higher voltages
B. The switching device is at saturation or cutoff most of the time
C. Linear amplifiers have high gain resulting in higher harmonic content
D. Switching amplifiers use push-pull circuits

In any linear amplifier the active device (a tube, a transistor, or an FET) functions as a voltage-controlled variable resistance and dissipates energy over the entire RF (or sometimes AUDIO) signal cycle. Theoretical efficiency for a resistive loaded Class A linear amplifier is only 25%! ***In a switching amplifier, the active device is either completely on or completely off.*** Power is only dissipated during the extremely short transition between the "fully on" and "fully off" state and can exceed 95% efficiency in very good designs. **ANSWER B.**

E7B05 What can be done to prevent unwanted oscillations in an RF power amplifier?

A. Tune the stage for minimum loading
B. Tune both the input and output for maximum power
C. Install parasitic suppressors and/or neutralize the stage
D. Use a phase inverter in the output filter

As a new Extra soon, you will likely be operating with a tower and beam tied into a power amplifier to boost your transceiver's 100-watt output signal. But cautiously work yourself into high power amplifier operation. Big amps may cause interference to your home electronics, and big amps need to be tuned precisely to minimize parasitic oscillations. When viewed on a spectrum analyzer, parasitic oscillations look like a forest of trees surrounding your transmitted signal. ***In tube amplifiers, RF chokes in the tube plate circuit and small-value bypass and suppressor capacitors in the grid circuit may help reduce unwanted oscillations***. With tube amplifiers, step-by-step neutralization procedures are detailed in the owner's manual. If the amplifier is new, you likely won't need to go through the neutralization process until you change out the final amplifier tube. **ANSWER C.**

E7B07 Which of the following is the likely result of using a Class C amplifier to amplify a single-sideband phone signal?

A. Reduced intermodulation products
B. Increased overall intelligibility
C. Reduced third-order intermodulation
D. Signal distortion and excessive bandwidth

The Class AB amplifier will take an input signal and cause a field effect transistor or bipolar transistor to operate near cutoff, but not into cutoff. This will lead to a nearly linear reproduction of the input signal and using Class AB is common with careful attention to setting the bias just above the point where the signal base current goes into cutoff at zero. Once the transistor is biased beyond cutoff, efficiency continues to climb but at the cost of distortion and excessive bandwidth. This is why satellite operators using SSB caution against the use of non-linear FM- only solid-state power amplifiers. The Class C amp is fine for data and FM, but for single sideband voice emissions, the ***Class C amplifier*** is operating outside of the linear portion of its operation. This ***causes the SSB voice signal to sound harsh, distorted, and occupying excessive bandwidth***. **ANSWER D.**

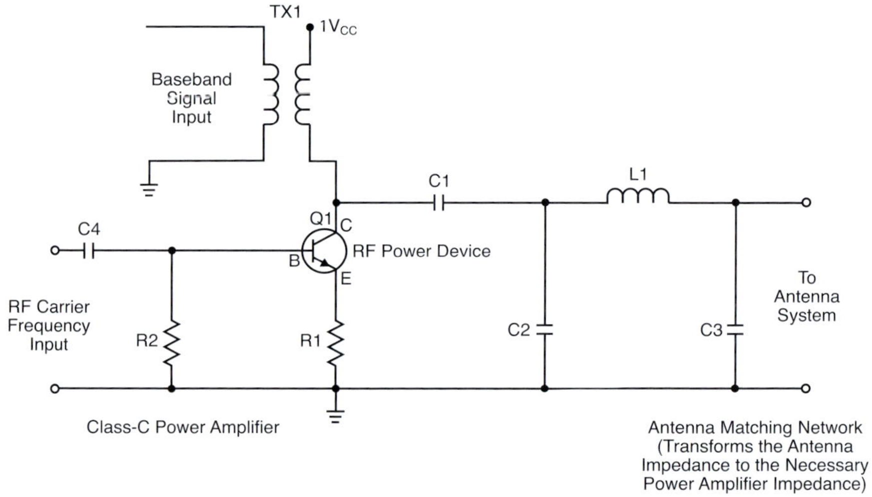

Class-C power amplifier-modulator being modulated with baseband signal.

Source: *Basic Communications Electronics*, Hudson & Luecke, © 1999 Master Publishing, Inc., Niles, IL

E7B02 What is a Class D amplifier?

A. An amplifier that uses switching technology to achieve high efficiency
B. A low power amplifier that uses a differential amplifier for improved linearity
C. An amplifier that uses drift-mode FETs for high efficiency
D. An amplifier biased to be relatively free from distortion

A less commonly known amplifier is called Class D. It is made up of two matched transistors driven with enough input power to produce a square wave output. The ***Class D amplifier employs switching technology to offer high efficiency*** from its balanced twin devices. **ANSWER A.**

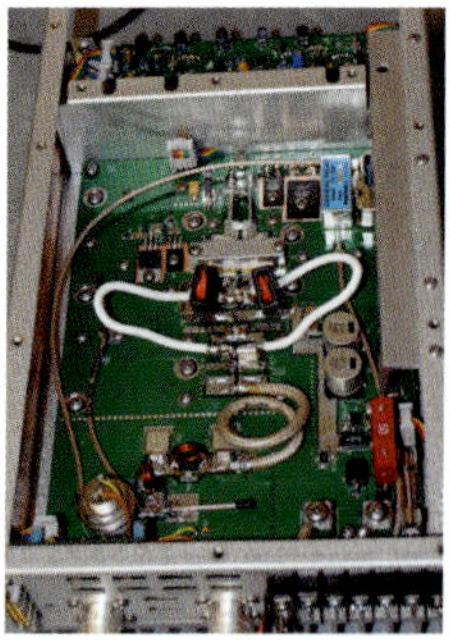

This small power amplifier from M2 Electronics puts out more than a kilowatt!

E7B03 What circuit is required at the output of an RF switching amplifier?

A. A filter to remove harmonic content

B. A high-pass filter to compensate for low gain at low frequencies

C. A matched load resistor to prevent damage by switching transients

D. A temperature compensating load resistor to improve linearity

Switching RF amplifiers are becoming especially popular on our lowest (2200 meter) amateur band where many audio amplifier circuits and methods are applicable. Many high-powered off-the-shelf audio amplifiers will work into the low radio frequency range, with excellent efficiency. Like any switching device, lots of harmonics are generated, which need to be eliminated before putting the amplifier on the air. A low pass filter, or very narrow (high-Q) bandpass filter, can accomplish this task nicely. **ANSWER A.**

E6D09 What devices are commonly used as VHF and UHF parasitic suppressors at the input and output terminals of a transistor HF amplifier?

A. Electrolytic capacitors

B. Butterworth filters

C. Ferrite beads

D. Steel-core toroids

If you look into the modern ***VHF/UHF*** single-band or dual-band ***amplifier***, you'll see many leads dressed with small ***ferrite beads*** to minimize parasitics coming down voltage or control lines. **ANSWER C.**

E7D12 What is the dropout voltage of a linear voltage regulator?

A. Minimum input voltage for rated power dissipation.

B. Maximum output voltage drop when the input voltage is varied over its specified range.

C. Minimum input-to-output voltage required to maintain regulation.

D. Maximum that the output voltage may decrease at rated load.

Unlike switching regulators, linear regulators can only reduce voltage from the unregulated value. This means that you need to have ***a bit more voltage going into the regulator than you want at the output***. Typically, you want the input voltage 30% to 50% greater than the output voltage for reliable regulation. **ANSWER C.**

E7D01 How does a linear electronic voltage regulator work?

A. It has a ramp voltage as its output

B. It eliminates the need for a pass transistor

C. The control element duty cycle is proportional to the line or load conditions

D. The conduction of a control element is varied to maintain a constant output voltage

A ***linear electronic voltage regulator varies the conduction of a circuit*** in direct proportion to variations in the line voltage to, or the load current from, the device. You will find a sophisticated voltage regulation circuit inside your base station power supply. **ANSWER D.**

A pair of three terminal regulators for a dual voltage power supply.

E7D03 What device is used as a stable voltage reference?

A. A Zener diode
B. A digital-to-analog converter
C. An SCR
D. An analog-to-digital converter

A ***Zener diode*** is primarily used as a ***voltage reference***. You can recognize the symbol of a Zener by the zig-zag bar used in its schematic symbol instead of the straight line designating the cathode of a standard diode. Remember: zig-zag = Zener! A Zener diode is normally operated *backwards*, in the normally non-conducting direction. At a very precise voltage, the Zener will conduct in the *reverse direction*, becoming a very low resistance. A Zener can serve as a voltage regulator by itself, or as a voltage reference for a more elaborate voltage regulator. Before Zener diodes were around, the voltage reference of choice was a gas filled tube. **ANSWER A.**

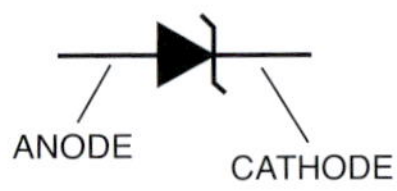

Here is the schematic symbol of a Zener diode. Since a diode only passes energy in one direction, look for that one-way arrow, plus a "Z" indicating it is a Zener diode. Doesn't that vertical line look like a tiny "Z".

Zener Diode

E7D02 How does a switchmode voltage regulator work?

A. By alternating the output between positive and negative voltages
B. By varying the duty cycle of pulses input to a filter
C. By varying the conductivity of a pass element
D. By switching between two Zener diode reference voltages

Switching regulators regulate the voltage by adjusting the duty cycle (on to off time ratio) of a square (or rectangular) wave. The output of the rectangular wave is filtered out or integrated to supply an output voltage equal to the AVERAGE voltage value of the rectangular wave string. The duty cycle controller is inside a closed loop, which samples the actual output voltage and adjusts the duty cycle appropriately. **ANSWER B.**

E7D04 Which of the following describes a three-terminal voltage regulator?

A. A series current source
B. A series regulator
C. A shunt regulator
D. A shunt current source

The very common ***series voltage regulator*** is used in many high-power ***linear power supplies***. With no load, the series regulator dissipates essentially zero power, while under full load it can dissipate up to 50% of the total power. By contrast, for lower power applications the *shunt regulator* is generally simpler. However, the shunt regulator dissipates the same amount of power regardless of the load, while the series regulator's power loss depends on the current draw. With no load, the shunt regulator is 0% efficient! **ANSWER B.**

E7D13 Which of the following calculates power dissipated by a series linear voltage regulator?

A. Input voltage multiplied by input current
B. Input voltage divided by output current
C. Voltage difference from input to output multiplied by output current
D. Output voltage multiplied by output current

Linear or ***analog*** voltage regulators are very clean, but they can be quite inefficient. A linear voltage regulator must always have a supply voltage significantly greater than the desired output voltage. In many high power regulated supplies, as much as half the supply voltage is dropped in the regulation circuitry, which typically includes large series pass transistors mounted on massive heat sinks. As in any ***series circuit the power dissipation of any component is equal to the voltage difference across that component times the output current***. **ANSWER C.**

E7D05 Which of the following types of linear voltage regulator operates by loading the unregulated voltage source?

A. A constant current source
B. A series regulator
C. A shunt current source
D. A shunt regulator

In circuits where the ***load on the unregulated input source must be kept constant***, we use the slightly less efficient ***shunt regulator***. These can sometimes get quite warm and are usually mounted to the chassis of the equipment, which acts as a heat sink. **ANSWER D.**

E7D06 What is the purpose of Q1 in the circuit shown in Figure E7-2?

A. It provides negative feedback to improve regulation
B. It provides a constant load for the voltage source
C. It controls the current to keep the output voltage constant
D. It provides regulation by switching or "chopping" the input DC voltage

Q1 is a transistor that can handle greater current than a Zener diode. They sometimes call this transistor "series-pass", and it ***is used to control the current supplied to the load***. Q1 will keep the output voltage constant. **ANSWER C.**

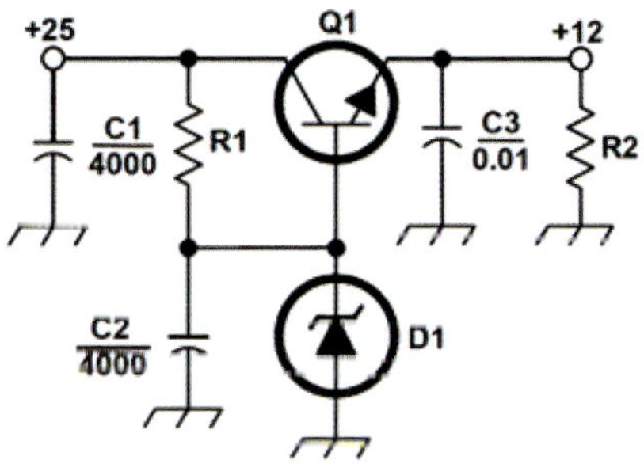

Figure E7-2

E7D07 What is the purpose of C2 in the circuit shown in Figure E7-2?

A. It bypasses rectifier output ripple around D1
B. It is a brute force filter for the output
C. To prevent self-oscillation
D. To provide fixed DC bias for Q1

C2 provides additional filtering for the ripple frequency. It ***bypasses voltage*** changes ***around D1***, the Zener diode, to keep the reference voltage stable. **ANSWER A.**

E7D08 What type of circuit is shown in Figure E7-2?

A. Switching voltage regulator
B. Common emitter amplifier
C. Linear voltage regulator
D. Common base amplifier

Components D1 and Q1 give away the answer – ***a linear voltage regulator***. This should give you a good, solid, regulated voltage output, even though the load current may be changing. **ANSWER C.**

E7D14 What is the purpose of connecting equal-value resistors across power supply filter capacitors connected in series?

A. Equalize the voltage across each capacitor
B. Discharge the capacitors when voltage is removed
C. Provide a minimum load on the supply
D. All these choices are correct

Those ***resistors*** in parallel across power supply electrolytic filter capacitors help ***equalize voltage drops*** across each of the capacitors, ***act as bleeder resistors*** when the power supply is turned off and ***add a constant load*** if there is minimal current required out of the power supply circuits. **ANSWER D.**

E7D15 What is the purpose of a step-start circuit in a high-voltage power supply?

A. To provide a dual-voltage output for reduced power applications
B. To compensate for variations of the incoming line voltage
C. To prevent arcing across the input power switch or relay contacts
D. To allow the filter capacitors to charge gradually

The modern, high-voltage power supply likely contains a series resistor and capacitor in parallel in the high-voltage plate circuit of the tube in order to buffer the initial in-rush of startup current. This ***allows the filter capacitors to gradually build their charges***, minimizing that dreaded arc sound when you turn on a high-voltage supply that has been turned off for several years! Some ham operators bring up high-voltage on older equipment with a variable AC transformer. **ANSWER D.**

E6D07 What is the current that flows in the primary winding of a transformer when there is no load on the secondary winding?

A. Stabilizing current
B. Direct current
C. Excitation current
D. Magnetizing current

With no load on the secondary of a transformer, the device works essentially as a pure inductor or choke. No power is dissipated except for the resistance of the primary winding and some core (or hysteresis) losses if a core is used. All the ***current is converted into a magnetic field***. Years ago, this is how I would bulk erase reel to reel audio tapes! **ANSWER D.**

E7D10 Why is a switching type power supply less expensive and lighter than an equivalent linear power supply?

A. The inverter design does not require an output filter circuit
B. The control circuitry uses less current, therefore smaller heat sinks are required
C. The high frequency inverter design uses much smaller transformers and filter components for an equivalent power output
D. It recovers power from the unused portion of the AC cycle, thus using fewer components

Conventional, heavy-transformer power supplies that convert home AC power down to 12 volts DC weigh in at about 2 lbs. for every 12-volt DC amp output. Gordo has plenty of 20-amp power supplies that weigh about 40 lbs. each. These power supplies are characterized by a huge step-down transformer, diodes, and big filter capacitors. The best efficiency you can hope for is about 50%. And ohhhh, do they get warm under load! The new power supply technology is called "switcher." These supplies are lightweight, small in size, give off little heat, and offer almost 90% efficiency. The output of a switching power supply (whether converting house power down to DC or switching house power up to various high- voltage levels) is clean and pure thanks to high-speed field-effect transistors taking the place of that big heavy transformer. New switching power supplies also have minimal EMI (RF interference), and some may even take any detectable emitted spur and let you adjust it out of your favorite listening spot. The ***switching power supply*** is about the same cost as the old, big-transformer supplies, and the ***very small transformer and filter components*** easily fit in a relatively small package, either inside or outside the radio equipment. **ANSWER C.**

High-frequency switching power supplies like this one weigh just a couple of pounds! Transformer type power supplies may weigh 10 times as much! But remember to keep inexpensive "no name" switchers away from coax lines and far away from your antenna system as these produce noise over HF frequencies. Modern name-brand switching power supplies are quiet on HF and VHF/UHF frequencies. The Jetstream power supply pictured here is nice and quiet on HF.

Receivers with Great Filters

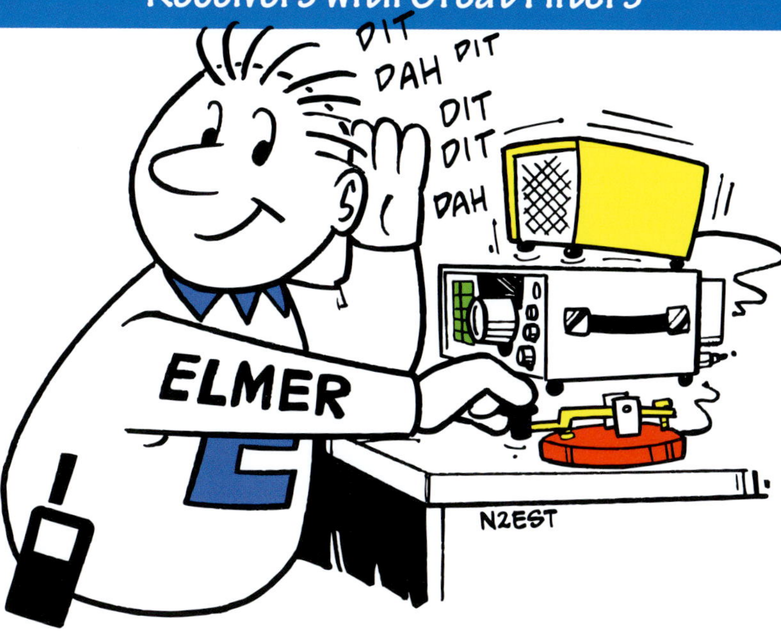

E4C03 What is the term for the suppression in an FM receiver of one signal by another stronger signal on the same frequency?

A. Desensitization
B. Cross-modulation interference
C. Capture effect
D. Frequency discrimination

A unique property of your FM receiver is the ability to hear only the strongest incoming FM signal. If a distant station is transmitting, and a strong local station comes on the air, you will only hear the local station. ***The stronger signal is "capturing" your receiver***. (The reason AM is still used on aircraft VHF radio is specifically to avoid the capture effect. Everybody needs to hear everybody else! This is the last major use of AM in the VHF world.) **ANSWER C.**

E4D12 What is the link margin in a system with a transmit power level of 10 W (+40 dBm), a system antenna gain of 10 dBi, a cable loss of 3 dB, a path loss of 136 dB, a receiver minimum discernable signal of -103 dBm, and a required signal-to-noise ratio of 6 dB?

A. -8dB
B. -14dB
C. +8dB
D. +14dB

In ANY complete communications link, the total path loss can be determined by simply adding all the losses and gains in dB. (We can usually assume in any REAL-WORLD system there will be a net path loss). In the case above, adding all the gains and losses we have +40 +10 -3 -136, which gives us a total ***path budget*** of -89dB, which is the signal level applied to the receiver's antenna terminals. If the minimum discernible signal is -103dB, our -89 dbm signal is 14 dB above that. If we subtract another 6dB to accommodate the required signal to noise ratio, we have 8dB left to work with. Since it is MORE than the minimum signal we need, we add a + sign, so it's +8dB. This means our signal can fade by 8dB and still be usable. (Some texts refer to link margin as ***fade margin***). **ANSWER C.**

E4D07 Which of the following reduces the likelihood of receiver desensitization?

A. Insert attenuation before the first RF stage
B. Raise the receiver's IF frequency
C. Increase the receiver's front-end gain
D. Switch from fast AGC to slow AGC

If you have a large HF/VHF/UHF base station transceiver, you'll notice that the insides are all enclosed in tiny metal cans. This shielding is necessary to keep the receiver from desensitizing on one band when you are transmitting on another band. Shielding is a good way to minimize internal desensitization. Desensitization occurs when nearby signals get into your receiver and cause the gain to be reduced. By ***decreasing the RF bandwidth of the receiver***, you reduce the likelihood of nearby signals affecting your receiver. Only signals that are within the passband of the RF are likely to have an effect. **ANSWER A.**

E7E06 Why is de-emphasis used in FM communications receivers?

A. For compatibility with transmitters using phase modulation
B. To reduce impulse noise reception
C. For higher efficiency
D. To remove third-order distortion products

De-emphasis in an FM receiver rolls off and attenuates the higher modulating frequency response. This compensates for the ***FM transmitter using phase modulation***, where the transmitter rolls off the lower frequencies. This provides ***compatibility*** in the FM communications systems. **ANSWER A.**

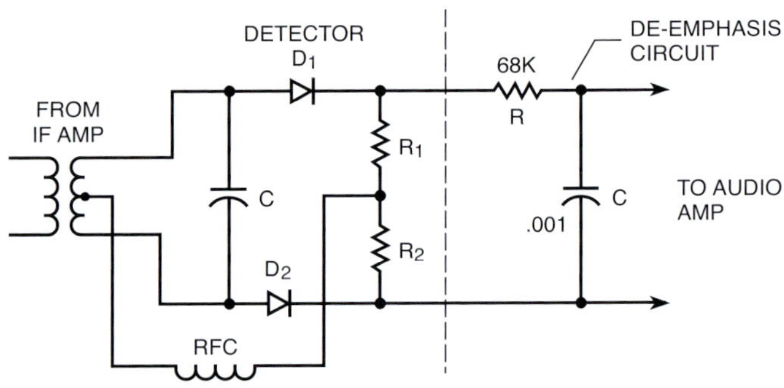

FM De-emphasis Circuit

E4D05 What transmitter frequencies would create an intermodulation-product signal in a receiver tuned to 146.70 MHz when a nearby station transmits on 146.52 MHz?

A. 146.34 MHz and 146.61 MHz
B. 146.88 MHz and 146.34 MHz
C. 146.10 MHz and 147.30 MHz
D. 146.30 MHz and 146.90 MHz

To calculate the possible sources of intermodulation interference, we check for the signals that lead to this type of problem that are common at base and repeater stations. There are four calculations to find the intermodulation frequencies that could be causing the problem – two that add frequencies and two that subtract the frequencies. The subtracted calculations are the only ones that produce frequencies close enough to cause significant interference. As the question states, the intermodulation product f_{IMD} is received on 146.70 MHz when a signal at 146.52 MHz is transmitted. The calculations when the frequencies are subtracted are:

$f_{IMD(2)} = 2f_1 - f_2$

Rearranging,

$f_2 = 2f_1 - f_{IMD(2)}$

$f_2 = 2(146.52) - 146.70$

$f_2 = 146.34$ MHz

$f_{IMD(4)} = 2f_2 - f_1$

Rearranging,

$2f_2 = f_{IMD(4)} + f_1$

$$f_2 = \frac{f_{IMD(4)} + f_1}{2}$$

$$f_2 = \frac{146.70 + 146.52}{2}$$

$f_2 = 146.61$ MHz

ANSWER A.

E4D02 Which of the following describes problems caused by poor dynamic range in a receiver?

A. Spurious signals caused by cross modulation and desensitization from strong adjacent signals
B. Oscillator instability requiring frequent retuning and loss of ability to recover the opposite sideband
C. Poor weak signal reception caused by insufficient local oscillator injection
D. Oscillator instability and severe audio distortion of all but the strongest received signals

In a receiver that does not have good dynamic range, ***cross-modulation*** of the desired signal can be encountered, and the receiver can be ***desensitized*** by strong adjacent frequency signals. **ANSWER A.**

E4C05 What does a receiver noise floor of -174 dBm represent?

A. The receiver noise is 6 dB above the theoretical minimum
B. The theoretical noise in a 1 Hz bandwidth at the input of a perfect receiver at room temperature
C. The noise figure of a 1 Hz bandwidth receiver
D. The receiver noise is 3 dB above theoretical minimum

Noise occurs any time objects are warmed above absolute zero. If you had a perfect receiver with no internal noise, it would still detect noise when connected to an antenna. ***At room temperature that noise level is about -174 dBm*** for every 1 Hz of bandwidth. In addition to the natural noise, man-made noise also occurs and is much more prevalent at some frequencies than others. On HF bands, man- made noise is so common that a receiver with a very good noise floor is not nearly as important as it would be for a receiver used at 10 GHz trying to detect an EME signal. **ANSWER B.**

E4D13 What is the received signal level with a transmit power of 10 W (+40 dBm), a transmit antenna gain of 6 dBi, a receive antenna gain of 3 dBi, and a path loss of 100 dB?

A. -51 dBm
B. -54 dBm
C. -57 dBm
D. -60 dBm

Here again, the dB is your friend; it makes path budget calculations a simple addition and subtraction problem. If we add all the pluses and minuses, we have +40 +6 +3 -100, which gives us -51 dBm at the receiver antenna terminals. What does the m mean? It means our dBs are referred to one milliwatt, a fairly common reference level in the communications business. Remember a dB is a power RATIO; without an actual power level to anchor it to, such as a milliwatt, it isn't too meaningful. **ANSWER A.**

E4D14 What power level does a receiver minimum discernible signal of -100 dBm represent?

A. 100 microwatts
B. 0.1 microwatt
C. 0.001 microwatts
D. 0.1 picowatts

In this problem, we work BACKWARDS from dB to power ratios. Remember the dB formula: dB=10 log(p1/p2). First, we divide -100 by 10 which gives us -10. Now we take the ***antilog*** of 10, which is 10 ^-10. But remember, this is a ratio. Our reference is 1 milliwatt, or 1/1000 watt. We have to move the decimal place 10 places to the left of .001 watts, which gives us 0.0000000000001 watts or .1 ***picowatts***. That's not a lotta watts! And yet this is a power level we routinely work with in ham radio communications. Are you impressed yet? You should be. **ANSWER D.**

E4C06 How much does increasing a receiver's bandwidth from 50 Hz to 1,000 Hz increase the receiver's noise floor?

A. 3 dB
B. 5 dB
C. 10 dB
D. 13 dB

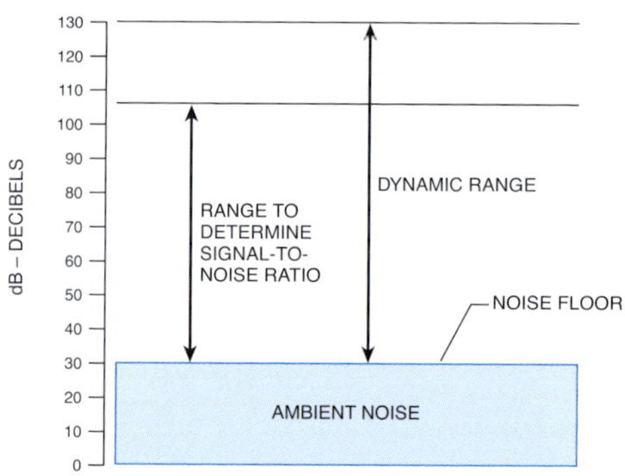

Noise Floor in a Receiver
Source: *Installing and Maintaining Sound Systems,* G. McComb, ©1996, Master Publishing, Inc.

Modern HF receivers are so sensitive that their internal noise floor can't be appreciated due to the many external noises picked up from your own home electronics.

In any electrical system the noise power is always proportional to the bandwidth. This is why signal-to-noise ratio increases as the IF bandwidth (or other selective stage bandwidth) becomes narrower. A bandwidth of 1,000 Hz is 20 times the bandwidth of 50Hz, and so there is 20 times the total power that can enter the system. A power increase of 20:1 is a 13 dB increase in power. Always remember and never forget the formula for dB: dB=10 log (P1/P2). **ANSWER D.**

E4C07 What does the MDS of a receiver represent?

A. The meter display sensitivity
B. The minimum discernible signal
C. The modulation distortion specification
D. The maximum detectable spectrum

The ***Minimum Discernible Signal*** is tested in the SSB or CW mode with internal RF gain full ON. Modern HF equipment has achieved extraordinarily low noise floors through product designs that minimize phase noise. **ANSWER B.**

E6E05 Which of the following noise figure values is typical of a low-noise UHF preamplifier?

A. 0.5 dB
B. -10 dB
C. 44 dBm
D. -20 dBm

At UHF frequencies and above, the sensitivity of a receiver is generally limited by the internal noise generated by the front-end amplifying device or devices. This is because any noise generated in the front end along with the desired signal is amplified by all the subsequent stages. The formula for Noise Figure is:

$$NF = 10\log_{10}(F) = 10\log_{10}\left(\frac{SNR_{in}}{SNR_{out}}\right) = SNR_{in\,(dB)} = SNR_{out\,(dB)}$$

where SNR is ***signal to noise*** ratio. When all the number crunching is done, we'll find that a pretty decent front end will have a noise figure of about ***0.5 dB***. **ANSWER A.**

E7C09 What is a crystal lattice filter?

A. A power supply filter made with interlaced quartz crystals
B. An audio filter made with four quartz crystals that resonate at 1 kHz intervals
C. A filter using lattice-shaped quartz crystals for high-Q performance
D. A filter for low-level signals made using quartz crystals

The new worldwide transceiver you purchase will probably have some good SSB and CW filters built in. The filters are made of ***quartz crystals designed for best response to low-level signals***. You can buy accessory filters for a tighter response, but many times these will make your received signals sound pinched. Unless you do a lot of CW work, stay with the filters that come with the set. While crystal ***lattice*** and ***ladder*** networks are similar in purpose, they have different topologies. Lattice networks use pairs of crystals where the series resonant mode of one crystal is matched to the parallel resonant mode of the second one, with the combined output forming the desired bandpass shape. **ANSWER D.**

E7C11 Which of the following measures a filter's ability to reject signals in adjacent channels?

A. Passband ripple
B. Phase response
C. Shape factor
D. Noise factor

In general, the quality of a filter is determined by how rapidly the response falls off at the cutoff frequency. A filter that very abruptly goes from "passing" to "stopping" a signal as the frequency changes is considered to have a good ***shape factor***. The response of the filter will have nearly vertical "walls" on each side of the passband. However, filter design is a balancing act of compromises; a very sharp filter may have some undesirable ***delay*** characteristics, which may or may not be a problem. In general, however, a filter used in, say, a single sideband filter should have as steep a shape factor as possible, usually defined as the ratio of the 60 dB attenuation to the 6 dB attenuation bandwidth. **ANSWER C.**

E7C08 Which of the following is most frequently used as a band-pass or notch filter in VHF and UHF transceivers?

A. A Sallen-Key filter
B. A helical filter
C. A swinging choke filter
D. A finite impulse response filter

At VHF and UHF frequencies, full size ***cavity resonators*** can be large and unwieldy. A good compromise between "lumped constants" and transmission line methods is used in the helical resonator, which is a cavity with inductive loading.

Good To Know:

Getting Your Filters In Shape

One of the most important parameters of a bandpass filter is the shape factor, sometimes called the skirt selectivity. An ideal bandpass filter response would look like a perfect rectangle if plotted on an amplitude versus frequency graph. Any signal within the two "walls" or "skirts" of the filter response would pass through with no attenuation, while any signal outside of the walls would be completely cut off. Although it is impossible to build a filter with a perfect shape factor (one having perfectly vertical walls), one can approach this response by using a filter with many poles (inductors or capacitors), or constructing a more modern digital filter.

The primary characteristic of a bandpass filter is defined as the bandwidth of the filter at the 3dB (half power) attenuation points. The shape factor is normally defined as the ratio of the bandwidth of the 60 dB attenuation points to the 3 dB attenuation points. (Sometimes different "endpoints" are defined such as 6dB and 40dB). In any case, a "perfect brick wall" filter would have a shape factor of 1:1, or just 1. In the simplest terms, the primary bandwidth tells you how far apart the walls are, while the shape factor tells you how vertical they are.

A simple single-tuned bandpass filter might have a very narrow primary bandwidth but a very poor shape factor, since single tuned circuits have a haystack shaped response curve. To achieve the more desirable rectangular shape factor, you might use a crystal lattice filter, or a multipole filter such as a mechanical filter to achieve a good shape factor [Figure 1]

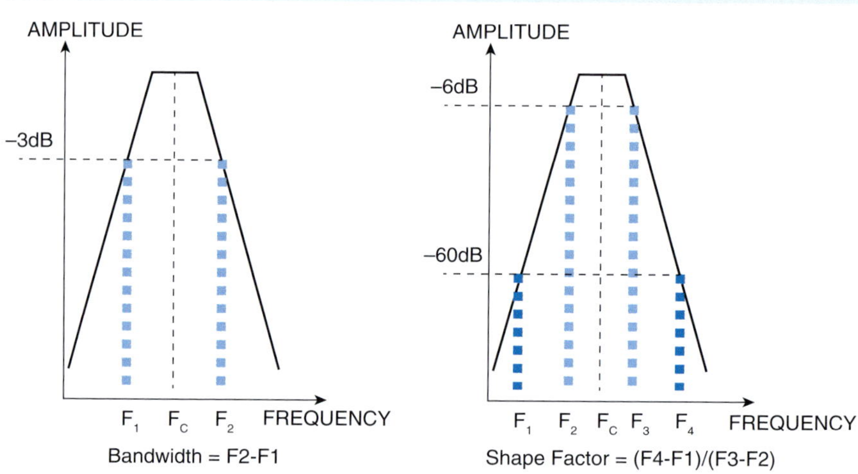

Helical resonators are used in the front end of most modern VHF receivers to achieve the necessary image rejection. They are also used in compact duplexers where full-sized cavities are impractical or impossible. **ANSWER B.**

E7C05 Which filter type has ripple in the passband and a sharp cutoff?

A. A Butterworth filter
B. An active LC filter
C. A passive op-amp filter
D. A Chebyshev filter

Look at the word ***Chebyshev***. It almost looks like it has a ***ripple***, doesn't it? **ANSWER D.**

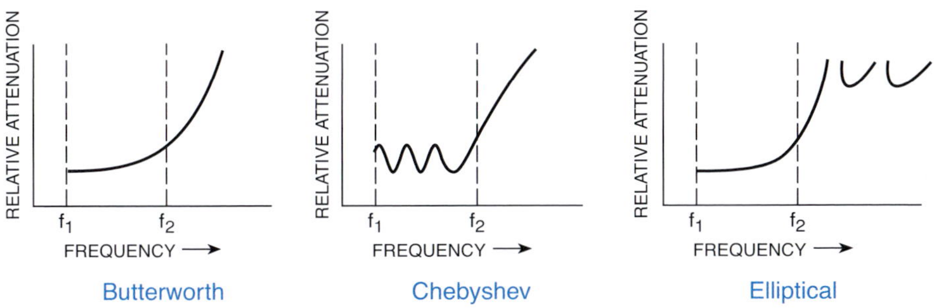

Comparison of Low-Pass Filter Circuits

E7C06 What are the characteristics of an elliptical filter?

A. Gradual passband rolloff with minimal stop-band ripple
B. Extremely flat response over its pass band with gradually rounded stop-band corners
C. Extremely sharp cutoff with one or more notches in the stop band
D. Gradual passband rolloff with extreme stop-band ripple

Different types of filters fulfill different priorities with regard to the ***cutoff*** or ***bandpass*** response. In audio applications, for instance, it's important to have very smooth transitions between bandpass and bandstop frequencies, with no abrupt phase changes. A Butterworth filter is commonly used for audio equalization. For radio frequency work, it's generally important to have abrupt cutoff response, consistent with not too much ripple within the desired bandpass. The ***elliptical filter*** has the best performance with regard to ***sharpness of cutoff*** but can have significant ripple in the bandpass. A less extreme response can be achieved with a Chebyshev type filter. Often, different types of filters are used in *cascade* to achieve the best overall response. **ANSWER C.**

E8A08 Why are direct or flash conversion analog-to-digital converters used for a software defined radio?

A. Very low power consumption decreases frequency drift
B. Immunity to out-of-sequence coding reduces spurious responses
C. Very high speed allows digitizing high frequencies
D. All these choices are correct

Besides being the simplest kind of A-to-D converter to understand, the flash A-to-D converter is fast, hence its name. In the ***flash A-to-D converter, every bit is digitized*** at the same time. **ANSWER C.**

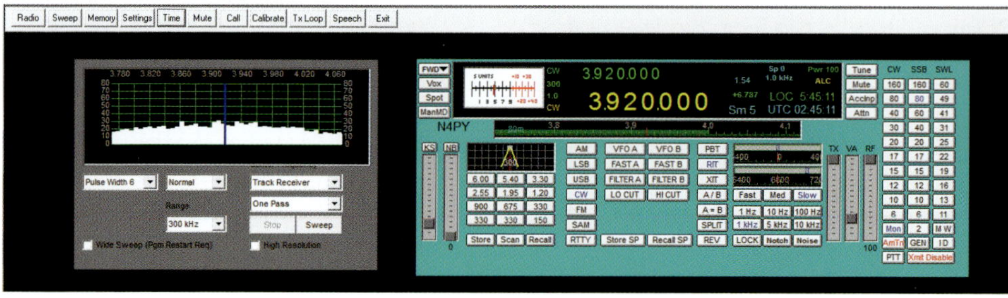

Software Defined Radio display.

E4C10 What is an advantage of having a variety of receiver bandwidths from which to select?

A. The noise figure of the RF amplifier can be adjusted to match the modulation type, thus increasing receiver sensitivity

B. Receiver power consumption can be reduced when wider bandwidth is not required

C. Receive bandwidth can be set to match the modulation bandwidth, maximizing signal-to-noise ratio and minimizing interference

D. Multiple frequencies can be received simultaneously if desired

The occupied bandwidth of any amateur radio signal depends on the mode of operation and the speed of data, among other factors. The best signal-to-noise ratio at the receiving end occurs when the ***receiver's bandwidth matches the bandwidth of the transmitted signal***. In most (but not all) modern receivers, it is the intermediate frequency (IF) stage that determines the receiver's bandwidth. Most high-quality communications receivers have a number of IF bandwidths available, or even a continuously variable IF bandwidth to allow the best signal-to-noise ratio for any given mode. **ANSWER C.**

E4C11 Why does input attenuation reduce receiver overload on the lower frequency HF bands with little or no impact on signal-to-noise ratio?

A. The attenuator has a low-pass filter to increase the strength of lower frequency signals

B. The attenuator has a noise filter to suppress interference

C. Signals are attenuated separately from the noise

D. Atmospheric noise is generally greater than internally generated noise even after attenuation

In many cases, it is advantageous to insert some RF attenuation in the "front end" of a receiver. Strong RF signals can overload a receiver, causing a number of undesirable effects, among them desensitizing the receiver to desired nearby weaker signals. Since most modern receivers have far more gain than is necessary, one can afford to sacrifice some gain to reduce the possibility of front-end overload. A good test for whether your receiver has adequate RF gain is whether you can hear ***atmospheric noise***. If you can, you have all the receiver gain you can use. Any gain beyond this point simply amplifies the atmospheric noise. **ANSWER D.**

E4C12 How does a narrow-band roofing filter affect receiver performance?

A. It improves sensitivity by reducing front-end noise

B. It improves intelligibility by using low Q circuitry to reduce ringing

C. It improves blocking dynamic range by attenuating strong signals near the receive frequency

D. All these choices are correct

Roofing filters in the IF minimize contesting receiver "swamping" when you have another ham running a kilowatt to chat with his buddy, well up the band. IF stage roofing filters, like automatically selected 15 kHz, 6 kHz, and 3 kHz filters, may feature 4-pole fundamental-mode monolithic crystal filter design to produce an ideal shape factor. The filters usually are positioned after the first mixer and significantly improve the 3rd-order intercept point performance for all stages that follow. Gordo appreciates the performance of roofing filters on 40 meters early in the morning when a nearby shortwave station numbs his receiver. ***Roofing filters improve the rig's blocking dynamic range***. **ANSWER C.**

E4C02 Which of the following receiver circuits can be effective in eliminating interference from strong out-of-band signals?

A. A front-end filter or preselector
B. A narrow IF filter
C. A notch filter
D. A properly adjusted product detector

An effective means of ***eliminating out-of-band image signal interference*** is with a ***front-end filter or pre-selector filter network***. These filters are much wider than typical IF filters and roofing filters. A higher IF also separates the image further from the incoming signal, thus easing the front-end filter requirements. The front-end filters are electrically close to the antenna input jack, as opposed to the IF and roofing filters, which are in the middle of the receiver section. The front-end filters must be band switched when there is a band change. **ANSWER A.**

E4C09 Which of the following choices is a good reason for selecting a high IF in a superheterodyne HF or VHF communications receiver?

A. Fewer components in the receiver
B. Reduced drift
C. Easier for front-end circuitry to eliminate image responses
D. Improved receiver noise figure

The intermediate frequency (IF) stage in a conventional receiver is where much of the amplification occurs. Selecting a high frequency for the IF allows for improved selectivity with minimum noise. This allows ***easier-to-build front end circuitry to eliminate image responses*** from nearby strong signals. **ANSWER C.**

E4C14 What is the purpose of the receiver IF Shift control?

A. To permit listening on a different frequency from the transmitting frequency
B. To change frequency rapidly
C. To reduce interference from stations transmitting on adjacent frequencies
D. To tune in stations slightly off frequency without changing the transmit frequency

One of the most useful developments in receiver design in the last century was the advent of IF Shift, also known as Passband Tuning, and a few other names by various manufacturers over the years. But the function is still the same. It moves the entire passband, regardless of the ***bandwidth*** chosen, up or down a few kilohertz. This allows you to reject adjacent frequency interference, especially when using single sideband. In years past, the same effect as passband tuning could be achieved by adjusting the ***BFO pitch***, and then readjusting the main VFO tuning. **ANSWER C.**

E4C13 What is reciprocal mixing?

A. Two out-of-band signals mixing to generate an in-band spurious signal
B. In-phase signals cancelling in a mixer resulting in loss of receiver sensitivity
C. Two digital signals combining from alternate time slots
D. Local oscillator phase noise mixing with adjacent strong signals to create interference to desired signals

The mixer in your new HF transceiver combines two input signals to produce an output signal that is equal to their sum or difference. Undesired ***phase noise at the local oscillator*** output may result in the potential masking of ultra-weak desired signals. Nearby strong signals will add to this problem, especially during HF contests when the bands are full. **ANSWER D.**

E7F02 What kind of digital signal processing audio filter is used to remove unwanted noise from a received SSB signal?

A. An adaptive filter
B. A crystal-lattice filter
C. A Hilbert-transform filter
D. A phase-inverting filter

The ***adaptive filter*** found in digital signal processing (DSP) circuits ***removes unwanted noise*** from incoming SSB signals. **ANSWER A.**

E4E01 What problem can occur when using an automatic notch filter (ANF) to remove interfering carriers while receiving CW signals?

A. Removal of the CW signal as well as the interfering carrier
B. Any nearby signal passing through the DSP system will overwhelm the desired signal
C. Excessive ringing
D. All these choices are correct

The automatic notch filter (ANF) on your new HF transceiver is designed to cancel an interfering steady carrier, as the "lid" across town tunes up without listening. Some transceivers even have more than one notch filter setting. On SSB 40 meters, 7 MHz foreign broadcast carriers can drive you nuts until you let the automatic notch filter find the steady tone and block it. Or perhaps you have local computer noise, with a steady tone right on your favorite HF frequency. With the ANF turned on, it will do the job. But if you are operating ***on CW or DATA***, it is best to turn ***the ANF*** circuit off, so it doesn't ***think a slow code message is just an almost steady tone to remove!*** The modern ANF filter will recognize CW that is faster than about 7 WPM but keep the ANF off during CW and DATA reception. **ANSWER A.**

E4C04 What is the noise figure of a receiver?

A. The ratio of atmospheric noise to phase noise
B. The ratio of the noise bandwidth in hertz to the theoretical bandwidth of a resistive network
C. The ratio in dB of the noise generated in the receiver to atmospheric noise
D. The ratio in dB of the noise generated by the receiver to the theoretical minimum noise

For VHF and UHF weak signal work on SSB and CW, you want to operate a receiver with an extremely low noise floor. An extremely low noise floor allows weak signals to be detected above the receiver's internal noise. A ***receiver with a good noise floor*** will yield nearly ***identical noise*** power ***when switched between a dummy load and an antenna***, measured 3 dB S (the desired signal power) + N/N, single tone, at 500 Hz and 2400 Hz. This works out to be a ratio, in dB, of the internal noise generated by the receiver itself compared to the theoretical minimum noise that the best receiver in the world would generate! Up on UHF, where we do a lot of weak signal operation like Moon bounce, this is an important consideration. But down on HF, atmospheric noise would drown out any slight differences in a receiver's own receiver-generated noise, so the noise floor is a less important ratio to consider in that new HF transceiver! **ANSWER D.**

E4D01 What is meant by the blocking dynamic range of a receiver?

A. The difference in dB between the noise floor and the level of an incoming signal that will cause 1 dB of gain compression
B. The minimum difference in dB between the levels of two FM signals that will cause one signal to block the other
C. The difference in dB between the noise floor and the third-order intercept point
D. The minimum difference in dB between two signals which produce third-order intermodulation products greater than the noise floor

When you are working a contest on HF and tune in to a rare station just above the noise floor, you may hear that station request responses to transmit 5 kHz "up." This is contest "split operation", and it leaves the station in the clear without people calling right on top of his ultra weak signal. Powerful stations just a few kilohertz away will sometimes cause the weak station's signal to disappear in the noise, because your receiver is saturated with powerful signals nearby on the band. A receiver with good dynamic range will minimize desensitization when nearby signals blast your front end circuitry. ***Blocking dynamic range*** is measured in decibels at your ***receiver noise floor***, with AGC turned off, and a ***nearby signal*** that leads to ***1 dB of gain compression in the receiver***, causing that weak signal to dip slightly into the noise. Blocking dynamic range tests are more fairly calculated at 100 kHz off from the desired signal. **ANSWER A.**

E4D09 What is the purpose of the preselector in a communications receiver?

A. To store frequencies that are often used
B. To provide broadband attenuation before the first RF stage to prevent Intermodulation
C. To increase the rejection of signals outside the band being received
D. To allow selection of the optimum RF amplifier device

You ask why those big contest HF transceivers are so large? On the inside, they may contain high-Q, motor-driven preselector coils that can automatically track, or may be manually adjusted, to your desired receive frequency. These preselector coils may achieve a Q of over 300 and provide steep passband filtering not available with smaller fixed components or older external bandpass boxes. The peak of these optional built-in ***preselector*** modules may even be adjusted away from your frequency to ***provide increased rejection of strong signals outside the band on which you are operating***. The Q is so high that each preselector on each band needs its own micro-motor drive assembly. But for big contesters, the big radio can really cut through the QRM. **ANSWER C.**

E4D10 What does a third-order intercept level of 40 dBm mean with respect to receiver performance?

A. Signals less than 40 dBm will not generate audible third-order intermodulation products
B. The receiver can tolerate signals up to 40 dB above the noise floor without producing third-order intermodulation products
C. A pair of 40 dBm input signals will theoretically generate a third-order intermodulation product that has the same output amplitude as either of the input signals
D. A pair of 1 mW input signals will produce a third-order intermodulation product that is 40 dB stronger than the input signal

If you input two tones that are f Hz apart to the receiver input, the mixer will output not only those two tones shifted down to the intermediate frequency but also third-order intermodulation tones at fHz above and below the original tones. At normal input signal levels, these third-order tones are much weaker than the original two tones. But as you increase the amplitude of the original two tones, the ***third-order tones*** increase in amplitude three times faster than the original tones. If you continue to increase the input tone signal levels, at some point the ***signal strength of the input tones and the third-order tones will be equal. This is called the third order intercept point***. A third-order intercept point of 40 dBm means that, if the input tones were 40 dBm, then the third order tones would also be 40 dBm. The 3rd order intercept point is not achievable in actual practice, as the receiver will saturate long before it reaches that point. **ANSWER C.**

E4D11 Why are odd-order intermodulation products, created within a receiver, of particular interest compared to other products?

A. Odd-order products of two signals in the band being received are also likely to be within the band
B. Odd-order products overload the IF filters
C. Odd-order products are an indication of poor image rejection
D. Odd-order intermodulation produces three products for every input signal within the band of interest

Your new high frequency transceiver boasts extraordinary receiver sensitivity, ultra-tight selectivity, and a 130 dBm odd order intercept point. Odd order intermodulation could occur in the receiver IF or RF stages when a pair of contest stations near you are hammering your "front end" just a few kHz away. Improved equipment designs minimize the ***odd order "phantom signals"*** that are non-desirable weak signal products from these other two stations, and are ***strong enough to interfere with the station you are trying to tune in***. Not only are you hammered by two strong signals of interference, but also artifacts of their clean transmitted signals. Physically larger inductors in the band pass filter stage may help, along with low-distortion diodes in the band pass frequency switching circuitry. Monolithic roofing filters, more expensive than overtone mode filters, will offer a better shape factor and are less susceptible to intermodulation distortion within your top-quality transceiver's RF and IF stages. If you are considering a contest-capable transceiver to feed into your 100-foot-long boom HF antenna, the multi-thousand-dollar high end HF transceivers offer the ultimate in the rejection of odd order phantom signals with a distortion-free, high-dynamic-range receiver system. **ANSWER A.**

E7E10 How does a diode envelope detector function?

A. By rectification and filtering of RF signals
B. By breakdown of the Zener voltage
C. By mixing signals with noise in the transition region of the diode
D. By sensing the change of reactance in the diode with respect to frequency

Since a ***diode*** only conducts for half of the AC signal, it ***may be used as a rectifier***. By ***filtering out the radio frequency energy*** after rectification, detection is accomplished – the modulated signal is all that remains. **ANSWER A.**

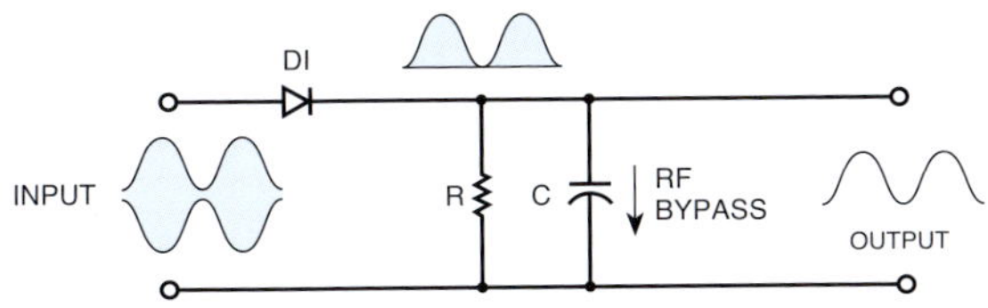

Diode Detector

E7E11 Which type of detector is used for demodulating SSB signals?

A. Discriminator
B. Phase detector
C. Product detector
D. Phase comparator

A ***product detector*** is found in ***SSB*** receivers. It mixes the incoming signal with a beat frequency oscillator signal that is a locally generated carrier, which is mixed with the incoming signal. **ANSWER C.**

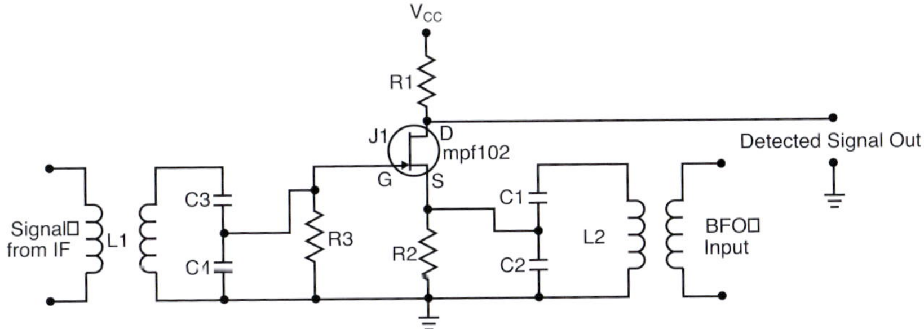

Product detector circuit used for SSB.

E7E08 What are the principal frequencies that appear at the output of a mixer?

A. Two and four times the input frequency
B. The square root of the product of input frequencies
C. The two input frequencies along with their sum and difference frequencies
D. 1.414 and 0.707 times the input frequency

The mixer produces your ***original two frequencies and the sum and difference frequencies***. The more elaborate the transceiver, the more mixing stages in the set. Mixing stages help filter out unwanted or phantom signals that could cause interference. **ANSWER C.**

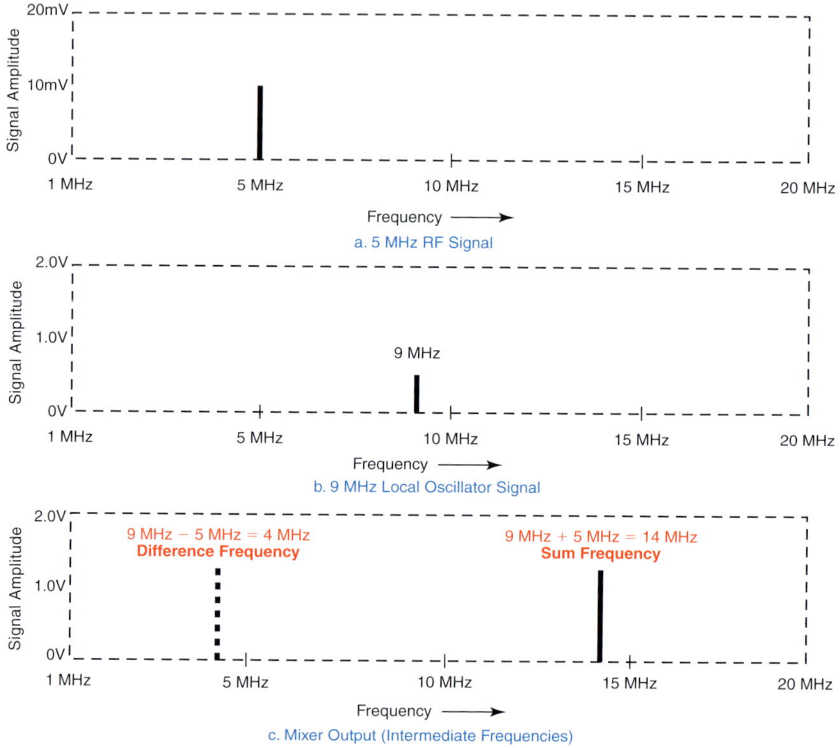

Frequency spectrum plot of mixer signals.

Source: *Basic Communications Electronics*, Hudson & Luecke, © 1999 Master Publishing, Inc., Niles, IL

E7E09 What occurs when the input signal levels to a mixer are too high?

A. Spurious mixer products are generated
B. Mixer blanking occurs
C. Automatic limiting occurs
D. Excessive AGC voltage levels are generated

Some VHF power amplifiers may incorporate a hefty pre-amplifier circuit for use on receiving that boosts incoming signal levels. When the pre-amp is turned on, it could generate ***spurious mixer products as a result of the high signal level***. The spurious products are not really signals that you are trying to receive. If you use a handheld with a power amp, turn the pre-amp off to avoid these unwanted signals. **ANSWER A.**

E4E03 Which of the following types of noise are removed by a noise blanker?

A. Broadband white noise
B. Impulse noise
C. Hum and buzz
D. All these choices are correct

Noise comes in all forms, and modern receivers can do a remarkable job of removing most forms of noise. Perhaps the oldest noise reduction system available to hams was the noise blanker, which is only effective for impulse noise, such as that from a vehicle ignition system. A ***noise blanker*** works by LIMITING or clipping the noise pulse that is above the average audio. Noise blankers always introduce a little distortion into the desired signal, but they are a lot better now in this regard than they used to be. Most modern rigs have adjustable threshold blankers which can be adjusted for best noise reduction consistent with minimum distortion. **ANSWER B.**

E4E02 Which of the following types of noise can often be reduced by a digital noise reduction?

A. Broadband white noise
B. Ignition noise
C. Power line noise
D. All these choices are correct

Most high frequency transceivers now include a digital signal processing filter. This filter will magically subtract broadband white noise, help minimize power line noise, may cancel out a steady frequency heterodyne signal, and help to reduce ignition noise. ***All of these may be addressed with a digital noise reduction filter***. But, you know, we always recommend solving the noise problem at the source, rather than trying to mask it, or subtract it, with DSP. **ANSWER D.**

Modern High Frequency rigs have builtin adaptive DSP filter options. If you have an old HF transceiver, you can add an external amplified DSP speaker and dial in your own level of noise elimination. Listen to this on Gordo's audio CD course.

E4E06 What type of electrical interference can be caused by computer network equipment?

A. A loud AC hum in the audio output of your station's receiver
B. A clicking noise at intervals of a few seconds
C. The appearance of unstable modulated or unmodulated signals at specific frequencies
D. A whining-type noise that continually pulses off and on

Personal computers and other digital equipment use square-wave signals called ***"clocks"*** that run at very high frequencies. These clocks can cause problems for radio receivers because they are composed of all the odd harmonics of the fundamental clock frequency. As you tune your receiver up and down the band, you can expect to find these harmonics. They may be ***unstable*** because clocks are turned on and off as different operations take place in the computer, and they may appear modulated when they are affected by other signals or operations of the computer. If you suspect interference from a digital device, find a frequency where the interference is happening and then simply turn off the suspected device. If the interference goes away, you've identified the culprit. To reduce the noise from such devices, try moving them farther away from the receiver, or try adding shielding around the device to reduce the level of interference. **ANSWER C.**

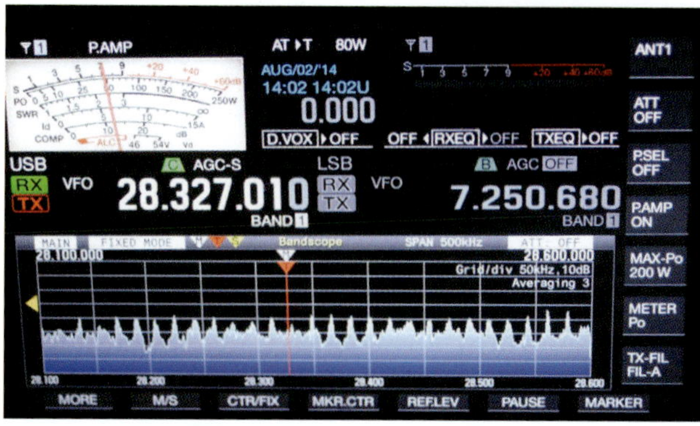

See the series of pulses on the spectrum scope, way up on 10 meters and 6 meters? You are seeing and hearing "birdies" generated by your home electronics.

E4E10 Which of the following can create intermittent loud roaring or buzzing AC line interference?

A. Arcing contacts in a thermostatically controlled device
B. A defective doorbell or doorbell transformer inside a nearby residence
C. A malfunctioning illuminated advertising display
D. All these choices are correct

With your new, big, directional HF antenna taking advantage of your new Extra Class privileges, you may pick up noise in one specific direction, likely when you're aimed at the local shopping center. ***All of these*** answers ***can generate a variety of interference*** that may come and go at intervals, just like that flashing neon sign down the street. There is no simple filter to cancel out many of these nuisance noise sources, other than aiming your beam to a point where the noise sources are minimized. **ANSWER D.**

E4E12 What causes interference received as a series of carriers at regular intervals across a wide frequency range?

A. Switch-mode power supplies
B. Radar transmitters
C. Wireless security camera transmitters
D. Electric fences

Despite the efficiency of a switch mode power supply, ***compared with an old-school linear power supply, one of these "gems" can be a cause of tremendous electrical interference.*** Most switching regulators have an oscillator somewhere in the 40 to 100 Khz frequency range, with many harmonics spread across the radio spectrum. If these units are not well shielded, they can wreak havoc with radio communications. Their "signature" is readily recognized on a spectrum or waterfall display, such as commonly used for digital HF communications, consisting of a large series of regularly spaced RF signals. **ANSWER A.**

E4E13 Where should a station AC surge protector be installed?

A. At the AC service panel
B. At an AC outlet
C. On the single point ground panel
D. On a ground rod outside the station

Although many "multi-outlet" power strips have surge protectors built in, these are not recommended for ham radio use. (Or any other use. A well-known ham and a good friend lost his entire ham shack and a good part of his house by a defective, inexpensive surge suppressing power strip.) The most effective and safest place to put a surge protector is right at the ***single point ground panel***. (You do have a single point ground panel, don't you?) The single point ground panel MAY be at the AC service panel, but since this is not always the case, the best answer is answer C. **ANSWER C.**

E4E09 What undesirable effect can occur when using a noise blanker?

A. Received audio in the speech range might have an echo effect
B. The audio frequency bandwidth of the received signal might be compressed
C. Strong signals may be distorted and appear to cause spurious emissions
D. FM signals can no longer be demodulated

For mobile applications, a high frequency transceiver will incorporate an IF noise blanker as well as digital noise reduction (DNR), which is part of the transceiver's digital signal processing. Some transceivers offer blanker on or blanker off with no adjustable amount of noise blanking. Others offer variable noise blanker settings. This is better. It is possible for the ***IF noise blanker to cause a very strong signal to sound distorted and may appear to cause spurious emissions*** when the noise blanker is set to maximum. It may also cause strong signals up to 30 kHz away to begin "chattering" your audio circuit on a frequency that is absolutely open. You may be tempted to tell the offending station up frequency that they are "splattering," but the actual problem is not their transmitter (well within specs), but the fact that you left your noise blanker turned on. **ANSWER C.**

E4E11 What could be the cause of local AM broadcast band signals combining to generate spurious signals on the MF or HF bands?

A. One or more of the broadcast stations is transmitting an over-modulated signal
B. Nearby corroded metal connections are mixing and re-radiating the broadcast signals
C. You are receiving skywave signals from a distant station
D. Your station receiver IF amplifier stage is overloaded

Your modern, high frequency transceiver employs filter networks, many specifically tuned to minimize powerful double sideband local AM broadcast signals coming in on your lower ham bands. But don't blame the AM broadcaster just yet – ***corroded connections*** on your tower or aluminum rain gutters might ***re-radiate the broadcast station signals***. This is why many hams with big towers use multiple insulators in their guy lines to minimize rectification of broadcast band signals due to corroded metal joints. **ANSWER B.**

E7F01 What is meant by "direct sampling" in software defined radios?

A. Software is converted from source code to object code during operation of the receiver
B. I and Q signals are generated by digital processing without the use of RF amplification
C. Incoming RF is digitized by an analog-to-digital converter without being mixed with a local oscillator signal
D. A switching mixer is used to generate I and Q signals directly from the RF input

Direct sampling in a software defined radio (SDR) is a method that has only been technologically feasible in very recent years, as ***the analog-to-digital sampling process is done right at the antenna input*** at radio frequencies. This method allows for extremely high performance and versatility but is subject to all the limitations of an A-to-D sampling scheme, such as aliasing. **ANSWER C.**

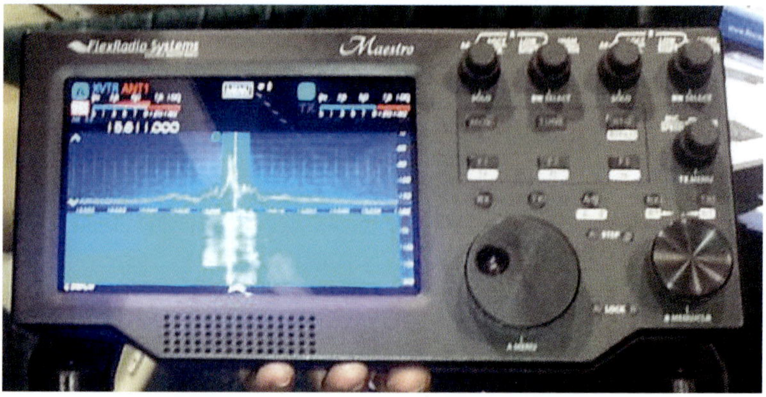

This HF radio head works Wi-Fi to the SDR black box in your shack. Great for lap top operation!

E8A04 What is "dither" with respect to analog-to-digital converters?

A. An abnormal condition where the converter cannot settle on a value to represent the signal
B. A small amount of noise added to the input signal to reduce quantization noise
C. An error caused by irregular quantization step size
D. A method of decimation by randomly skipping samples

In one of life's strange paradoxes, we can reduce noise in a digital sampler by introducing noise! In reality, this simply converts one type of noise into a much less obnoxious form. ***Adding a tiny bit of random noise may reduce quantization noise*** in any sampling system. **ANSWER B.**

E4C08 An SDR receiver is overloaded when input signals exceed what level?

A. One-half of the maximum sample rate
B. One-half of the maximum sampling buffer size
C. The maximum count value of the analog-to-digital converter
D. The reference voltage of the analog-to-digital converter

Unlike your conventional analog receiver, which tends to ***overload*** gradually in the presence of strong signals, your SDR receiver will hit a "brick wall" in the presence of overloading. The results of this are obvious and ugly. ***The A-D converter*** in your receiver's front end has nowhere to go when the "bit bucket" gets full, and ***simply ceases to function***. Happily, this state of being is only temporary, and as soon as you attenuate the signal to a proper level, your SDR receiver will be its old wonderful self again. **ANSWER D.**

Good To Know:

Improving your I/Q

If you've worked at all with Software Defined Radios (SDRs) or other forms of Digital Signal Processing (DSP), you have certainly heard the terms I and Q tossed around. What do these mean, and why are they so important? While I and Q modulation and demodulation are certainly integral to digital communications, I and Q methods have been around for ages in analog communications systems.

I and Q simply mean IN phase and QUADRATURE, which is a phase shift of 90 electrical degrees. Having two detectors in a receiver running in 90 degree quadrature allows you to unambiguously determine the phase angle of any radio signal, as opposed to a conventional detector that only gives you amplitude information. Why should you care?

An I/Q detector in a receiver allows you to demodulate ANY form of modulation: AM, FM, PM, SSB, and any combination of the above. In addition, an I/Q modulator in a transmitter will allow you to generate any of these signals, in any combination.

One of the very earliest methods for generating a single sideband signal was by means of the phasing method, which actually incorporated TWO I/Q generators, one for audio frequencies and one for radio frequencies. Early phasing transmitters were rather persnickety contraptions, but did work very well. But early on in the development of SSB, the phasing method was mostly replaced by the more common filter method.

However, with modern DSP methods, the phasing method of SSB generation has come back with a vengeance, because of its astonishing versatility in generating any imaginable kind of signal, analog or digital.

One color big screen does it all, from logging contacts, to checking propagation charts, checking for duplicate contacts, to working the radio's controls – all from one small SDR box hidden below the console.

Oscillate & Synthesize This!

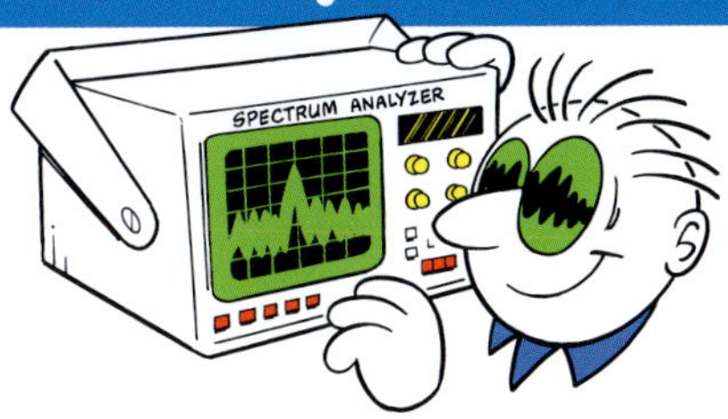

E7H01 What are three common oscillator circuits?

A. Taft, Pierce, and negative feedback
B. Pierce, Fenner, and Beane
C. Taft, Hartley, and Pierce
D. Colpitts, Hartley, and Pierce

Colpitts oscillators have a capacitor, just like C in the name. ***Hartley*** is tapped, and the ***Pierce*** oscillator uses a crystal. **ANSWER D.**

E7H03 What is a phase-locked loop?

A. An electronic servo loop consisting of a ratio detector, reactance modulator, and voltage-controlled oscillator
B. An electronic circuit also known as a monostable multivibrator
C. An electronic servo loop consisting of a phase detector, a low-pass filter, a voltage-controlled oscillator, and a stable reference oscillator
D. An electronic circuit consisting of a precision push-pull amplifier with a differential phase input

Few other circuits have changed the course of Amateur Radio equipment as much as the phase-locked loop (PLL) circuit. There is only one correct answer with the word "phase" in it, so ***look for the word phase when you read phase-locked loop*** in the answer choices. **ANSWER C.**

Phase-Locked Loop (PLL)

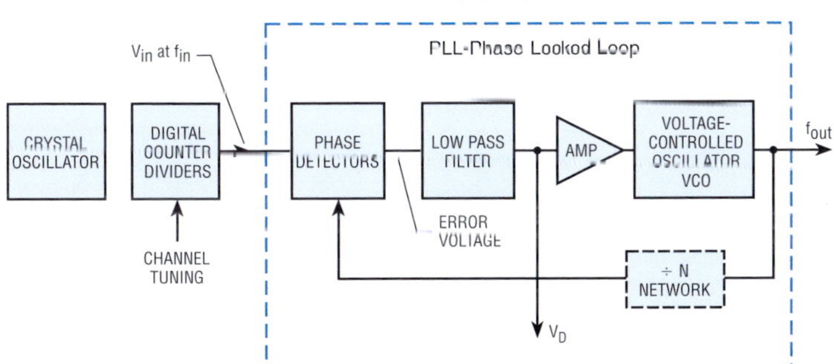

In a phase-locked loop the output frequency of the VCO is locked by phase and frequency to the frequency of the input signal V_{in}.

By using a very accurate crystal controlled source and dividing it down (or multiplying it up), the output frequency, f_{out}, of the PLL will be controlled very accurately and with rigid stability to the input frequency, f_{in}. The phase detector develops an error voltage that is determined by the frequencies and phase of the two inputs. The feedback system wants to reduce the error voltage to zero.

V_D is a voltage proportional to the frequency changes occuring in V_{in}; therefore, the PLL can be used as an FM demodulator.

With divide by N network, f_{out} will be f_{in} multiplied by N.

Channel tuning occurs in the Digital Counter Dividers.

E7H04 How is positive feedback supplied in a Colpitts oscillator?

A. Through a tapped coil
B. Through link coupling
C. Through a capacitive divider
D. Through a neutralizing capacitor

On the ***Colpitts*** oscillator, we use a ***capacitive divider*** to provide feedback. Look at the schematic on page 146. **ANSWER C.**

E7H05 How is positive feedback supplied in a Pierce oscillator?

A. Through a tapped coil
B. Through link coupling
C. Through a neutralizing capacitor
D. Through a quartz crystal

The ***Pierce*** oscillator uses a ***quartz crystal*** to obtain positive feedback. Look at the schematic on page 146. **ANSWER D.**

In a phase-locked loop the output frequency of the VCO is locked by phase and frequency to the frequency of the input signal V_{in}.

By using a very accurate crystal controlled source and dividing it down (or multiplying it up), the output frequency, f_{out}, of the PLL will be controlled very accurately and with rigid stability to the input frequency, f_{in}. The phase detector develops an error voltage that is determined by the frequencies and phase of the two inputs. The feedback system wants to reduce the error voltage to zero.

V_D is a voltage proportional to the frequency changes occurring in V_{in}; therefore, the PLL can be used as an FM demodulator.

With divide by N network, f_{out} will be f_{in} multiplied by N.

Channel tuning occurs in the Digital Counter D

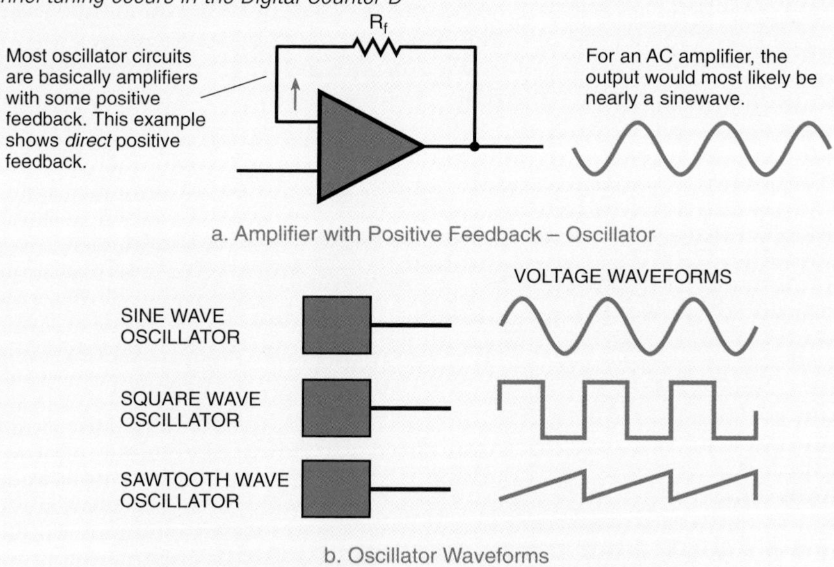

Oscillators – An oscillator is basically an amplifier with positive feedback from output to input.

Source: *Basic Electronics* © 1994, Master Publishing, Inc., Arlington, Texas

E7H12 Which of the following ensures that a crystal oscillator operates on the frequency specified by the crystal manufacturer?

A. Provide the crystal with a specified parallel inductance
B. Provide the crystal with a specified parallel capacitance
C. Bias the crystal at a specified voltage
D. Bias the crystal at a specified current

There is plenty to influence the ***precise frequency*** of a crystal in the oscillator circuit – crystal impedance, crystal aging, the method of holding the crystal in place inside the can, temperature and, finally, the ***crystal's natural parallel*** or ***holder capacitance*** (over which you have no control), usually in the order of 6 Pf. Your circuit's parallel capacitance, which you can control usually by means of a trimmer, is another influence to consider. **ANSWER B.**

Quartz crystals are still found in modern, synthesized ham radios. Some are enclosed in heated metal chambers to keep them temperature compensated for spot-on digital signal work. These two quartz crystals feed reference signals to various oscillator stages in the transceiver.

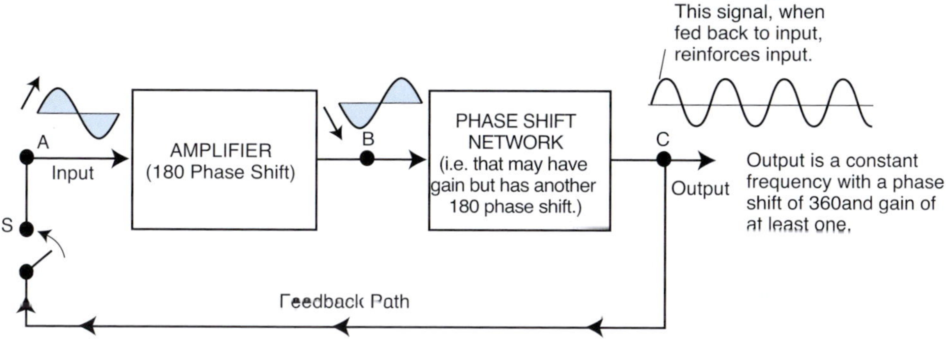

An oscillator outputs a signal of constant frequency.

Source: Basic *Communications Electronics*, Hudson & Luecke, © 1999 Master Publishing, Inc., Niles, IL

E7H08 Which of the following components can be used to reduce thermal drift in crystal oscillators?

A. NP0 capacitors
B. Toroidal inductors
C. Wirewound resistors
D. Non-inductive resistors

Well, this wobbles some old nostalgia glands. Lots of us homebrewers spent many hours attempting to stabilize oscillators by means of ***temperature compensating capacitors***. This is almost a lost art. An ***NP0 capacitor*** is one that has a *neutral* thermal coefficient; it drifts neither higher nor lower in capacitance as the temperature changes. NP0 actually stands for Negative/Positive 0. Other capacitors were designed with either a positive or negative temperature coefficient to compensate for other components with the opposite temperature coefficient. Are we having fun yet? **ANSWER A.**

E6D01 What is piezoelectricity?

A. The ability of materials to generate electromagnetic waves of a certain frequency when voltage is applied
B. A characteristic of materials that have an index of refraction which depends on the polarization of the electromagnetic wave passing through it
C. A characteristic of materials that generate a voltage when stressed and that flex when a voltage is applied
D. The ability of materials to generate voltage when an electromagnetic wave of a certain frequency is applied

Piezoelectricity is one of the greatest discoveries in all of electronics and is at the heart of all modern frequency and time standards, including the clock in your computer's microprocessor. The quartz crystal is the most commonly used piezoelectric material; if you stress the crystal, it generates an electrical voltage, and if you apply an electrical voltage, it deforms. It can be used as the heart of a radio frequency oscillator, or as part of an electrical filter of high quality. **ANSWER C.**

E6D02 What is the equivalent circuit of a quartz crystal?

A. Series RLC in parallel with a shunt C representing electrode and stray capacitance
B. Parallel RLC, where C is the parallel combination of resonance capacitance of the crystal and electrode and stray capacitance
C. Series RLC, where C is the parallel combination of resonance capacitance of the crystal and electrode and stray capacitance
D. Parallel RLC, where C is the series combination of resonance capacitance of the crystal and electrode and stray capacitance

Although the quartz crystal is actually an electromechanical device, its action can be described in purely electrical terms. It is primarily a series resonant RLC circuit, with a small amount of capacitance in parallel (sometimes called ***holder*** capacitance) which allows the device to resonate in two different modes, series or parallel. Either mode (and sometimes both!) can be used in crystal-based circuitry. **ANSWER A.**

E6D03 Which of the following is an aspect of the piezoelectric effect?

A. Mechanical deformation of material due to the application of a voltage
B. Mechanical deformation of material due to the application of a magnetic field
C. Generation of electrical energy in the presence of light
D. Increased conductivity in the presence of light

The piezoelectric effect is crucial to nearly all we do in radio. Every quartz crystal in your computer, cell-phone or ham rig relies on this property. In light of this, it's amazing that for a long time in early amateur radio we somehow managed to get by without even knowing about this phenomenon! When you apply a mechanical stress to a piezoelectric material, such as quartz, it generates a small DC voltage. And if you ***apply an electric field to the crystal, it mechanically deforms***. Quartz crystals are at the heart of nearly all precision oscillators and a great number of electrical filters. They can even be used as transducers in microphones or loudspeakers. **ANSWER A.**

E7H13 Which of the following is a technique for providing highly accurate and stable oscillators needed for microwave transmission and reception?

A. Use a GPS signal reference
B. Use a rubidium stabilized reference oscillator
C. Use a temperature-controlled high Q dielectric resonator
D. All these choices are correct

Up at 10 GHz (10,000 MHz) and 24 GHz (24,000 MHz) oscillator stability is of paramount importance. When Gordo is mountain topping with his 10 GHz setup, the sun warming his transmitter can throw him many kHz off frequency! Microwave operators who run beacons will many times use GPS satellite signals for a good stable reference, or may develop a Rubidium standard to keep the oscillator set on the nose. Also, a temperature-controlled high Q dielectric resonator can keep the beacon within a few cycles of being on frequency. ***All of these signal references are ways of providing good stability to your microwave system***. **ANSWER D.**

E7H02 What is a microphonic?

A. An IC used for amplifying microphone signals
B. Distortion caused by RF pickup on the microphone cable
C. Changes in oscillator frequency due to mechanical vibration
D. Excess loading of the microphone by an oscillator

Some of us old timers had a transmitter or two that you could modulate just by yelling at it… without a microphone! This was not a desirable condition, but it could be a lot of fun. Nevertheless, unless great care is taken, nearly any oscillator can become microphonic – this is where microscopic ***movements caused by vibrations*** of any frequency-determining components can change the value of that component and thus frequency modulate the thing. Oscillators aren't the only possibly microphonic component in a radio, but they're the most common. Well-designed oscillators are mechanically rigid and well isolated from their outside environment. **ANSWER C.**

E7H07 How can an oscillator's microphonic responses be reduced?

A. Use NP0 capacitors
B. Reduce noise on the oscillator's power supply
C. Increase the gain
D. Mechanically isolate the oscillator circuitry from its enclosure

Reducing microphonics requires removing any source of mechanical vibration by making everything more rigid, and physically ***isolating it***. **ANSWER D.**

E7H06 Which of these functions can be performed by a phase-locked loop?

A. Wide-band AF and RF power amplification
B. Frequency synthesis and FM demodulation
C. Photovoltaic conversion and optical coupling
D. Comparison of two digital input signals and digital pulse counting

When Gordo built his first external ***frequency synthesizer***, he was able to "synthesize" hundreds of channels with just a couple of crystals. How many of you remember being "rock bound" with an old rig? **ANSWER B.**

E7H09 What type of frequency synthesizer circuit uses a phase accumulator, lookup table, digital-to-analog converter, and a low-pass anti-alias filter?

A. A direct digital synthesizer
B. A hybrid synthesizer
C. A phase-locked loop synthesizer
D. A direct conversion synthesizer

Direct digital synthesis allows a single chip to handle multi-loop synthesizer circuits in your radio's local oscillator. Although the DDS does not output a pure sine wave, it offers fractions of a Hertz resolution with an extremely low noise floor. Within the chip is a ***phase accumulator, lookup table, digital-to-analog converter, and a low- pass anti-alias filter*** with a spectrally clean output. **ANSWER A.**

E7H10 What information is contained in the lookup table of a direct digital synthesizer (DDS)?

A. The phase relationship between a reference oscillator and the output waveform
B. Amplitude values that represent the desired waveform
C. The phase relationship between a voltage-controlled oscillator and the output waveform
D. Frequently used receiver and transmitter frequencies

Within the DDS, a lookup table will sample nearest available voltage levels in the digital-to-analog converter (DAC) ***that represent the desired waveform output***. **ANSWER B.**

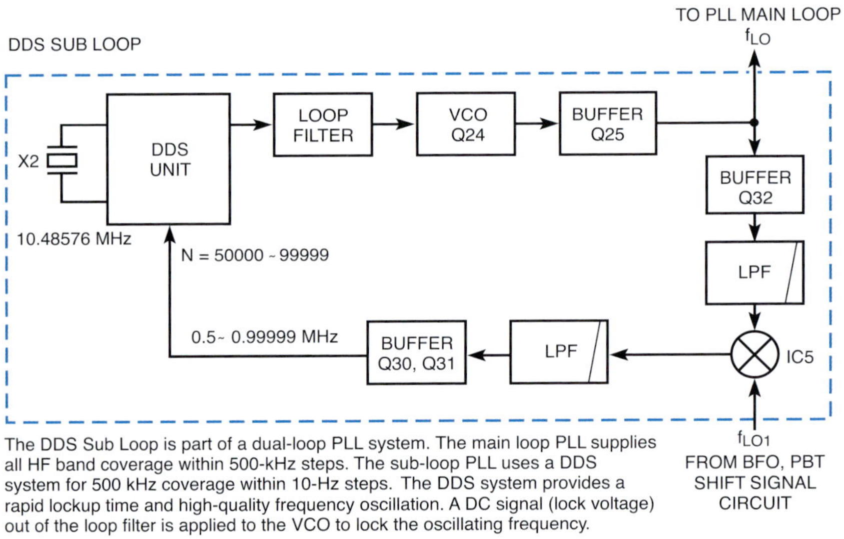

DDS Sub Loop of a Transmitter Dual-Loop PLL
Courtesy of ICOM America, Inc.

E7H11 What are the major spectral impurity components of direct digital synthesizers?

A. Broadband noise
B. Digital conversion noise
C. Spurious signals at discrete frequencies
D. Harmonics of the local oscillator

Unwanted components of a DDS output are spurs at discrete frequencies – unwanted emissions. These are easier to filter out than broadband noise. However, filtering these numerous spurious signals generated by the direct digital-frequency-synthesizer is tricky because they are spread over a wide frequency range. Some signals are called "alias signals," which are usually in multiples on each side of the desired frequency and may extend several MHz above and below the desired frequency. Many times a low pass filter may help mitigate the problem with some of these ***spurs at discreet frequencies***. **ANSWER C.**

E4C01 What is an effect of excessive phase noise in an SDR receiver's master clock oscillator?

A. It limits the receiver's ability to receive strong signals
B. It can affect the receiver's frequency calibration
C. It decreases the receiver's third-order intercept point
D. It can combine with strong signals on nearby frequencies to generate interference

Like every innovation in radio, each new concept can have undesirable side effects. The nearly universal adoption of Phase Locked Loop (PLL) technology greatly increased the versatility and stability of amateur radio receivers and transmitters, but created a few new problems in the process. All PLL circuits have a certain degree of ***phase noise***, which is a byproduct of the PLL's frequency correction process. If a PLL is used as a local oscillator in a receiver, any phase noise will also frequency modulate the incoming signals. This ***can interfere with reception of weak signals***, greatly reducing the usable sensitivity of the receiver. Phase noise is an important factor to consider when looking at the specifications of an amateur receiver or transceiver. Some transceivers are much better in this aspect than others. **ANSWER D.**

E7F05 How frequently must an analog signal be sampled to be accurately reproduced?

A. At least half the rate of the highest frequency component of the signal
B. At least twice the rate of the highest frequency component of the signal
C. At the same rate as the highest frequency component of the signal
D. At four times the rate of the highest frequency component of the signal

The Nyquist sampling theorem tells us that you need to ***sample the analog signal at twice the rate of the highest frequency signal***. Remember if you're sampling a *square wave,* you have odd harmonics well above the base frequency, and you need to include those too! **ANSWER B.**

Software Defined Radios provide rock-solid frequency stability with 1 Hz accuracy when operating in digital modes.

E7F06 What is the minimum number of bits required to sample a signal with a range of 1 volt at a resolution of 1 millivolt?

A. 4 bits
B. 6 bits
C. 8 bits
D. 10 bits

The dynamic range of a digital converter is equal to the power ratio between the maximum and minimum signals it can handle. Using the formula dB= 20 log (V_2/ V_1) where V_2 is the largest signal and V_1 is the smallest, we have a dynamic range of the "voltage dB" formula. This formula assumes the input and output impedances of the device are the same, which is a fair assumption in digital circuitry. Now, we still need to go one step further to answer the question. The formula we use to convert dynamic range to the number of bits is:

Dynamic Range (dB) = 20 × log (2n)

where: n = the number of bits in the ADC.

We know the dynamic range (60 dB) so we work this formula backwards.

$60/20 = \log(2^n)$

$3 = \log(2^n)$

$\text{Antilog}(3) = 1000 = 2^n$

$2^{10} = 1024$

So, if we plug in 10 for n, we know we can do the job with ***10 bits***, with just a tiny bit left over. **ANSWER D.**

E7F07 What function is performed by a Fast Fourier Transform?

A. Converting analog signals to digital form
B. Converting digital signals to analog form
C. Converting signals from the time domain to the frequency domain
D. Converting signals from the frequency domain to the time domain

The main "engine" of Digital Signal Processing is the Fast Fourier Transform which ***converts*** complicated ***frequency domain signals to*** simple ***time domains signals*** where the number crunching is much simpler. After the processing is done, the results are usually converted back to the frequency domain, using the inverse FFT function. **ANSWER C.**

E7F08 What is the function of decimation?

A. Converting data to binary-coded decimal form
B. Reducing the effective sample rate by removing samples
C. Attenuating the signal
D. Removing unnecessary significant digits

Digital sampling can be a labor-intensive process, at least for the computer doing the thinking. It's sometimes effective to ***remove redundant or unnecessary samples*** to speed up the process. **ANSWER B.**

E7F09 Why is an anti-aliasing filter required in a decimator?

A. It removes high-frequency signal components that would otherwise be reproduced as lower frequency components
B. It peaks the response of the decimator, improving bandwidth
C. It removes low-frequency signal components to eliminate the need for DC restoration
D. It notches out the sampling frequency to avoid sampling errors

The process of digitizing a signal is essentially identical to mixing in the analog universe and has the same byproduct, namely an ***image frequency***. In the A-to-D converter, the image ***is known as the alias***, but the physical process is identical. Images are usually not desirable and ***must be filtered out*** to avoid interfering with the "real" signal. **ANSWER A.**

E7F10 What aspect of receiver analog-to-digital conversion determines the maximum receive bandwidth of a direct-sampling software defined radio (SDR)?

A. Sample rate
B. Sample width in bits
C. Integral non-linearity
D. Differential non-linearity

In any digital processing scenario, whether in a digital oscilloscope or software defined radio, Nyquist sampling theorem reigns supreme. The ***bandwidth*** of the system ***is dictated by the sampling rate***, which must be twice the frequency of the highest frequency signal component. **ANSWER A.**

E7F11 What sets the minimum detectable signal level for a direct-sampling software defined receiver in the absence of atmospheric or thermal noise?

A. Sample clock phase noise
B. Reference voltage level and sample width in bits
C. Data storage transfer rate
D. Missing codes and jitter

As discussed earlier, the dynamic range is determined by the ***number of bits in the sample*** width, (more correctly termed *sample depth)*. The minimum detectable signal is the smallest "step" available in your Analog-to-Digital converter. **ANSWER B.**

E7F12 Which of the following is generally true of Finite Impulse Response (FIR) filters?

A. FIR filters can delay all frequency components of the signal by the same amount
B. FIR filters are easier to implement for a given set of passband rolloff requirements
C. FIR filters can respond faster to impulses
D. All these choices are correct

Digital Filters come in two general categories, the FIR (Finite Impulse Response), and IIR (Infinite Impulse Response). Unlike the more familiar analog filters that are generally defined by the number of poles, digital filters are characterized by the number of taps, which are logical interconnections. The IIR filter can re-use taps to obtain the effect of an equivalent analog filter with a great number of poles, but with a price: it can become unstable and difficult to design. The ***FIR filter,*** which does not re-use taps, ***has the same amount of delay for all frequencies of interest***, is much simpler to design, and more likely to be stable. **ANSWER A.**

E7F13 What is the function of taps in a digital signal processing filter?

A. To reduce excess signal pressure levels
B. Provide access for debugging software
C. Select the point at which baseband signals are generated
D. Provide incremental signal delays for filter algorithms

In a digital filter, ***taps*** are equivalent to poles in an analog filter and ***provide incremental signal delays*** for filter algorithms. You can synthesize the effect of a lot of physical components by using logical taps in a modern digital filter. **ANSWER D.**

E7F14 Which of the following would allow a digital signal processing filter to create a sharper filter response?

A. Higher data rate
B. More taps
C. Lower Q
D. Double-precision math routines

An analog filter with a lot of sections or poles can more closely approach the ideal frequency response than can a simpler filter. A digital filter can simulate a very complex (and expensive) analog filter through the use of ***more logical taps***, or iterations. **ANSWER B.**

Circuits for Resonance for All!

E5A01 What can cause the voltage across reactances in a series RLC circuit to be higher than the voltage applied to the entire circuit?

A. Resonance
B. Capacitance
C. Low quality factor (Q)
D. Resistance

Coils (having inductance) and capacitors (having capacitance) will be at resonance when the capacitive reactance equals the inductive reactance. ***At resonance, voltages higher than the applied voltage*** can be present across these components in the circuit. You can see this with a mobile whip antenna – if something were to touch the whip while you are transmitting, you might see quite a spark! **ANSWER A.**

Equation for Resonant Frequency

X_L = Inductive Reactance

$X_L = 2\pi fL$

X_C = Capacitive Reactance

$X_C = \frac{1}{2\pi fC}$

L = Inductance in henrys

C = Capacitance in farads

f = Frequency in hertz

At Resonance $X_L = X_C$

$$2\pi f_r L = \frac{1}{2\pi f_r C}$$

Solving for f_r

$$f_r^2 = \frac{1}{(2\pi)^2 LC}$$

The Resonant Frequency, f_r, is:

$$f_r = \frac{1}{2\pi\sqrt{LC}}$$

On-line calculator site:
www.calculatoredge.com/index.htm

E5A02 What is the resonant frequency of an RLC circuit if R is 22 ohms, L is 50 microhenries, and C is 40 picofarads?

A. 44.72 MHz
B. 22.36 MHz
C. 3.56 MHz
D. 1.78 MHz

The schematic shown in the diagram is a simple series resonant circuit. Don't panic. The following resonance formula will carry you through many questions on series and parallel circuits – one of which may be on your exam. First, you can throw away the resistance found in the parallel or series circuit. It is not needed for the resonance formula. Here is the resonance formula that you should memorize:

$$f_r = \frac{10^6}{2\pi\sqrt{LC}}$$

where: f_r is resonant frequency in **kHz**
$2\pi = 6.28$
L is inductance in **microhenries**
C is capacitance in **picofarads**

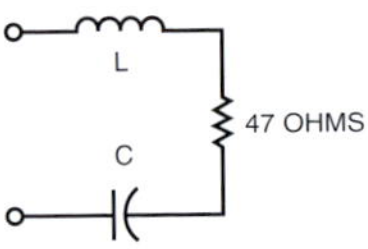

Series Resonant Circuit

First, multiply L ***X*** C. For this question, that is 50 ***X*** 40 = 2000. Now we need the square root of this, since in the resonance formula L and C are under the square root sign. With 2000 showing on your calculator, simply press the square root key. Presto, you have 44.72. Now multiply by 2π, which is 6.28 ***X*** 44.72 = 280.84. The final calculation for the formula is 1,000,000 ÷ 280.84. Press your clear button, and enter 1,000,000 ÷ 280.84 = 3560.7463 kHz. Your answer came out in kHz, but they are looking for MHz in the answer. No problem here – simply move the decimal point three places to the left, and you get answer C, 3.56 MHz. That's all there is to it! **ANSWER C.**

E5A10 What is the resonant frequency of an RLC circuit if R is 33 ohms, L is 50 microhenries, and C is 10 picofarads?

A. 7.12 MHz
B. 23.5 kHz
C. 7.12 kHz
D. 23.5 MHz

First multiply 50 microhenrys and 10 picofarads. Take the square root of the result. Then multiply by 6.28. Divide that answer into 1,000,000. Finally, change kHz to MHz. Easy! Here are the keystrokes: Clear, 50 × 10 = (square root symbol) 6.28 = 140.4. Remember 140.4. Clear, ***1,000,000 ÷ 140.4 = 7122 kHz which is 7.12 MHz***.**ANSWER A.**

E5A03 What is the magnitude of the impedance of a series RLC circuit at resonance?

A. High, compared to the circuit resistance
B. Approximately equal to capacitive reactance
C. Approximately equal to inductive reactance
D. Approximately equal to circuit resistance

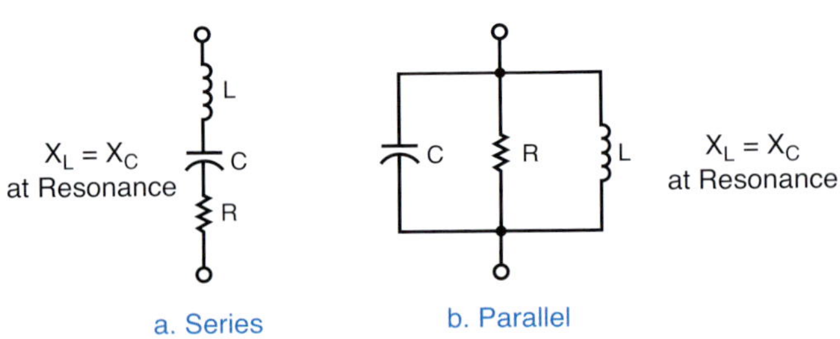

Series and Parallel Resonant Circuits

Did you ever wonder why the serious mobile ham always uses a gigantic loading coil for his series resonant antenna? The bigger the loading coil the less resistance. ***At resonance, the magnitude of the impedance is equal to the circuit resistance***. **ANSWER D.**

E5A04 What is the magnitude of the impedance of a parallel RLC circuit at resonance?

A. Approximately equal to circuit resistance
B. Approximately equal to inductive reactance
C. Low compared to the circuit resistance
D. High compared to the circuit resistance

Whether the system is parallel or series RLC resonant, the ***impedance will always be equal to the circuit resistance*** when the circuit is at resonance. **ANSWER A.**

E5A05 What is the result of increasing the Q of an impedance-matching circuit?

A. Matching bandwidth is decreased
B. Matching bandwidth is increased
C. Losses increase
D. Harmonics increase

One of the several definitions of Q is the "sharpness" of tuning. Since the purpose of a tuning circuit in the first place is to be *selective*, the better it is at being selective the higher the quality (Q) of the circuit. Quite logical, isn't it? Bandwidth is the reciprocal of selectivity, so ***the greater the Q, the lower the bandwidth***. **ANSWER A.**

E5A06 What is the magnitude of the circulating current within the components of a parallel LC circuit at resonance?

A. It is at a minimum
B. It is at a maximum
C. It equals 1 divided by the quantity 2 times pi, times the square root of (inductance L multiplied by capacitance C)
D. It equals 2 times pi, times the square root of (inductance L multiplied by capacitance C)

A ***parallel circuit at resonance*** is like a tuned trap in a multi-band trap antenna. At resonance, the trap keeps power from going any further. But be assured there is ***maximum circulating current*** within the parallel RLC circuit. Now don't get confused with this question – a parallel circuit has maximum current within it and minimum current going on to the next stage. That's why they call those parallel resonant circuits "traps." **ANSWER B.**

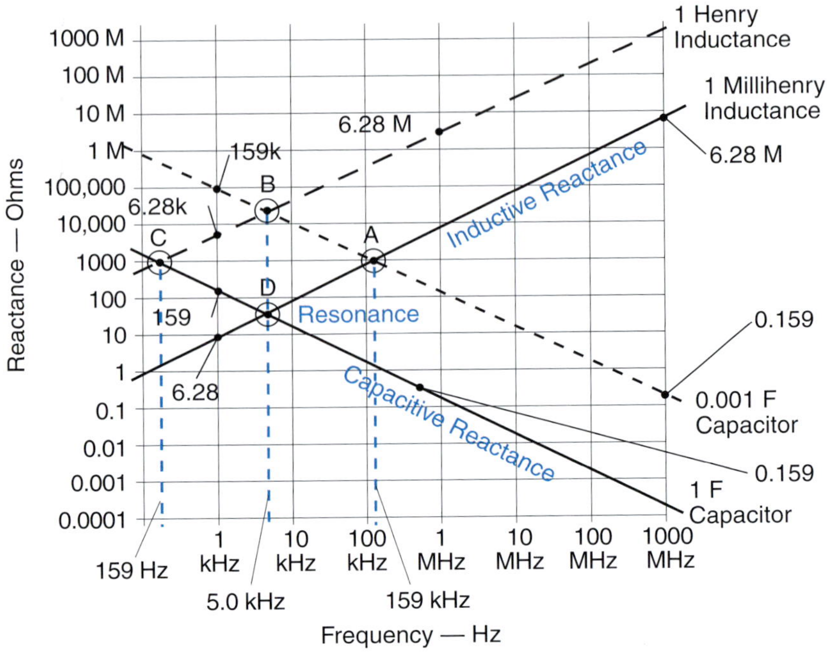

Variation of inductance and capacitive reactance with frequency (illustration not to exact log-log scale).

Source: Basic *Communications Electronics*, Hudson & Luecke, © 1999 Master Publishing, Inc., Niles, IL

E5A07 What is the magnitude of the current at the input of a parallel RLC circuit at resonance?

A. Minimum
B. Maximum
C. R/L
D. L/R

The total ***current into*** and out of a ***parallel RLC circuit at resonance is at a minimum***. On old radios and vacuum tube linear amplifiers, new or old, we would tune a plate control to see the current dip at resonance. **ANSWER A.**

E5A08 What is the phase relationship between the current through and the voltage across a series resonant circuit at resonance?

A. The voltage leads the current by 90 degrees
B. The current leads the voltage by 90 degrees
C. The voltage and current are in phase
D. The voltage and current are 180 degrees out of phase

At Resonance: Xl = Xc. When a circuit is ***at resonance, voltage and current are in phase***. **ANSWER C.**

Out of Resonance Examples

The two waveforms illustrate the relationship of voltage and current in an out-of-phase circuit.

ELI = voltage (E) leads current (I) in an inductive (L) circuit.
ICE = current (I) leads voltage (E) in a capacitive (C) circuit.

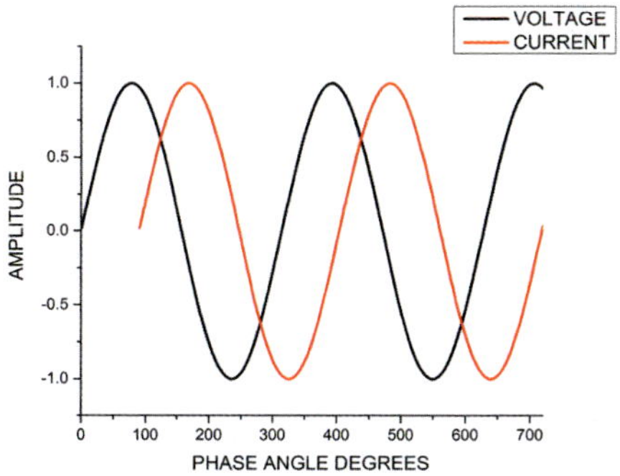

ELI voltage (E) leads current (I) (or current lags voltage)

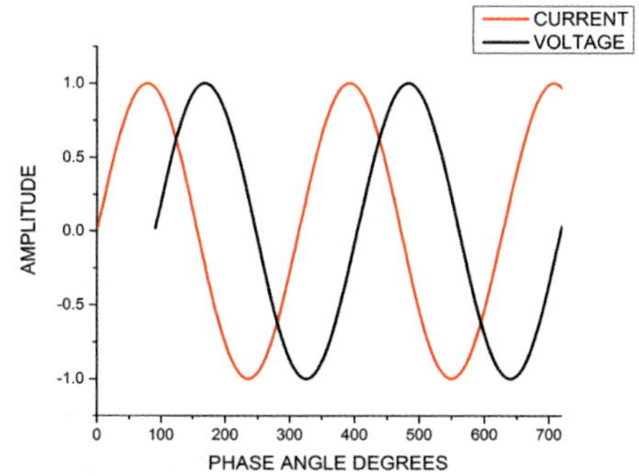

ICE current (I) leads voltage (E) (or voltage lags current)

E5B07 What is the phase angle between the voltage across and the current through a series RLC circuit if XC is 500 ohms, R is 1 kilohm, and XL is 250 ohms?

A. 68.2 degrees with the voltage leading the current
B. 14.0 degrees with the voltage leading the current
C. 14.0 degrees with the voltage lagging the current
D. 68.2 degrees with the voltage lagging the current

For any complex series circuit, the phase angle is the ***arctangent*** ($\tan^{-1}$) of the net reactance over the resistance. Also remember that in any circuit containing both inductive and capacitive reactance, you can always subtract the capacitive reactance from the inductive reactance. In this example, the NET reactance is 250-500 ohms, or -250 ohms. Dividing this by the resistance, we get -250/1000 or -.25. On MOST calculators, the key sequence is INV, TAN, (-.25), which gives ***-14 degrees***. The minus sign tells us the voltage lags the current (By ELI the ICE man).

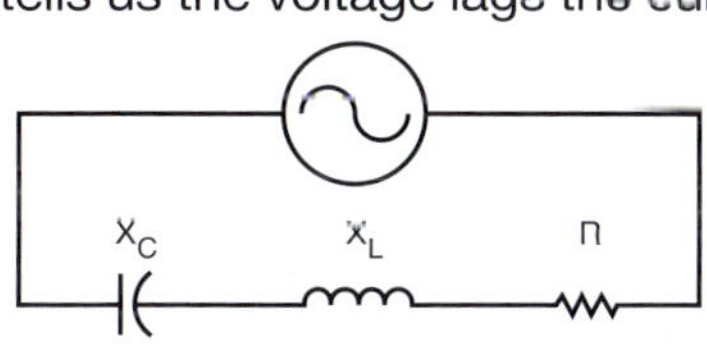

Series RLC Circuit Schematic for Phase Angle Calculations

Here is the formula, memorize it:

$$\text{Tan } \phi = \frac{X}{R}$$ where: ϕ is the phase angle in **degrees**
$X = (X_L - X_C)$ is total reactance in the circuit in ohms
R is circuit resistance in **ohms**

Since X_L and X_C are in opposition, we simply subtract one from the other. Now here's a fun way to remember the relationship of current and voltage in non-resonant circuits: ELI the ICE man! Say it out loud, "ELI the ICE man." In a more inductive (L) circuit, voltage (E) leads current (I) just like in the word "ELI" where E is before I. In a more capacitive (C) circuit, current (I) leads voltage (E), just like the word "ICE" where I comes before E. This is an ICE circuit, because X_C is 500 ohms and X_L is only 250 ohms. The total reactance (X) is 500 - 250 = 250. By dividing 1000 (the resistance) into 250, the tangent of ϕ comes out 0.25. A tangent 0.25 represents a phase angle of ***14 degrees***. This is an ICE circuit ***with voltage lagging current***. **ANSWER C.**

E5B08 What is the phase angle between the voltage across and the current through a series RLC circuit if XC is 300 ohms, R is 100 ohms, and XL is 100 ohms?

A. 63 degrees with the voltage lagging the current
B. 63 degrees with the voltage leading the current
C. 27 degrees with the voltage leading the current
D. 27 degrees with the voltage lagging the current

Always remember and never forget that ELI the ICE man is your friend! And, in an RLC circuit you can always subtract the smaller reactance from the larger reactance to obtain the NET reactance, in this case, X_C-X_L= 300-100 = 200 ohms CAPACITIVE. The phase angle of a complex series circuit is the ARC tangent (Tan^{-1}) of the reactance over the resistance. Keystrokes on MOST calculators will be ALT TAN (200/100) = ***63 degrees***. And we know that ***in a capacitive reactive circuit the voltage lags the current***. **ANSWER A.**

E5B11 What is the phase angle between the voltage across and the current through a series RLC circuit if XC is 25 ohms, R is 100 ohms, and XL is 75 ohms?

A. 27 degrees with the voltage lagging the current
B. 27 degrees with the voltage leading the current
C. 63 degrees with the voltage lagging the current
D. 63 degrees with the voltage leading the current

One of the things about complex impedances that makes things LESS complex is that you can always subtract an X_C from an X_L, to come out with the NET reactance. In this example, subtracting X_C from X_L gives us a net reactance of +50 ohms (the + indicating inductive reactance). Now here's another simplification as well as a sanity check. If the net reactance is greater than the net resistance, the phase angle is greater than 45 degrees. If the net reactance is LESS than the net resistance, the angle will be less than 45 degrees. Since the resistance of 100 ohms is greater than the reactance of +50 ohms, we can eliminate answers C and D right off the bat. Now, if we remember ELI the ICE man, we know that in an INDUCTIVE circuit, the voltage leads the current, which leaves Answer B as our only choice. **ANSWER B.**

A Neat Trick for Calculating Parallel Impedances

Graphical solutions to many electronics problems can often make things a lot simpler than working out complicated formulas, and they can also give you valuable insight into principles that you can't get just by "crunching numbers." Working out complex parallel impedances can be somewhat involved, and even seasoned electronics folks like us sometimes forget the formulas. By now, you should know how to work out series RLC networks using Cartesian coordinates. Working out parallel impedances is similar, but with an interesting twist. Let's figure out the total impedance (Z) and phase angle of a parallel 400 ohm resistor and 800 ohm inductor, using this clever graphical method. Take a look at the graph below.

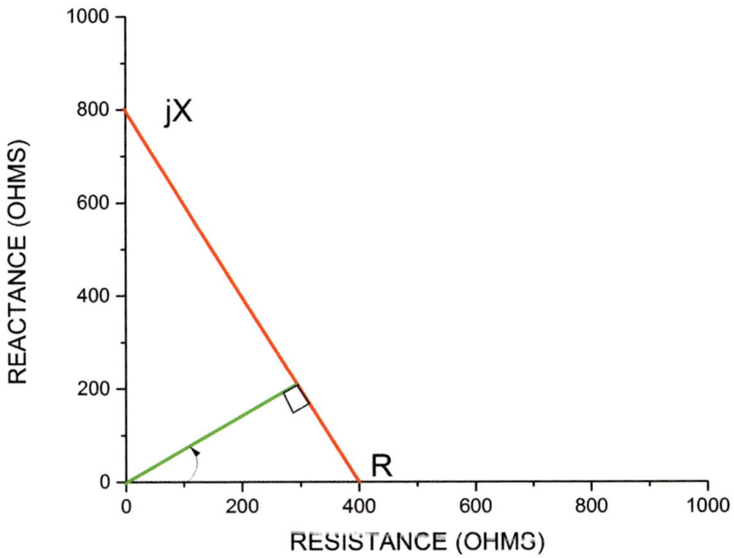

The first thing you'll need to do is plot the two component values on their representative axis. Make a small mark at 400 on the horizontal resistance axis, designated by R, and a small mark at 800 on the vertical reactance axis, designated jX. Remember the j prefix tells us the value is an imaginary number.

Now draw a straight line segment between R and jX, (shown in red). Next, draw a line segment from the origin (0,0) to the previous diagonal, at the point where the two segments are perpendicular, (shown in green). You'll need a square of some kind to do this...don't try to just eyeball it. The length of the green line segment is the impedance magnitude, and the angle the green segment makes with the horizontal axis is the phase angle. If you've drawn your graph carefully, the length of the green line should be 357. The phase angle should be 26.6 degrees. Nothing to it!

Now you can do the calculations the "hard" way and see if you get the same results. You should always know how to work a problem at least two different ways!

E5B09 What is the relationship between the AC current through a capacitor and the voltage across a capacitor?

A. Voltage and current are in phase
B. Voltage and current are 180 degrees out of phase
C. Voltage leads current by 90 degrees
D. Current leads voltage by 90 degrees

A ***capacitor*** is an "ICE" circuit where AC ***current leads voltage***. In the word "ICE" the letter "I" for current is leading the letter "E" for voltage, in a capacitive circuit. **ANSWER D.**

When re-capping an older tube transceiver, be sure and watch polarity on these types of capacitors. They make a one-time loud POOF BANG if you get them wrong!

E5B10 What is the relationship between the AC current through an inductor and the voltage across an inductor?

A. Voltage leads current by 90 degrees
B. Current leads voltage by 90 degrees
C. Voltage and current are 180 degrees out of phase
D. Voltage and current are in phase

An inductor is an "ELI" condition where voltage leads the current as we see in "ELI" where the letter "E" is ahead of the letter "I." ***Voltage leads current by 90 degrees***. **ANSWER A.**

Notice these coils are spaced apart, and the upper right ones oriented 90 degrees apart to prevent mutual coupling.

E5C04 What type of Y-axis scale is most often used for graphs of circuit frequency response?

A. Linear
B. Scatter
C. Random
D. Logarithmic

Frequency response of a circuit is displayed on a two-dimensional graph of amplitude vs frequency, with frequency on the X-axis and amplitude on the Y-axis. Since the amplitude values can vary over an extreme range, they are generally plotted on a ***logarithmic Y-axis***, or often in decibels. Sometimes the X-axis is also logarithmic if the frequency range is very large. **ANSWER D.**

E5C09 When using rectangular coordinates to graph the impedance of a circuit, what do the axes represent?

A. The X axis represents the resistive component, and the Y axis represents the reactive component
B. The X axis represents the reactive component, and the Y axis represents the resistive component
C. The X axis represents the phase angle, and the Y axis represents the magnitude
D. The X axis represents the magnitude, and the Y axis represents the phase

The horizontal ***(X) axis represents the resistive component*** of the impedance. The vertical ***(Y) axis represents the reactive component***. When the impedance is in a circuit and voltage is applied, there will be a current, and the voltage drop across the resistance is plotted on the horizontal axis. **ANSWER A.**

E5C01 Which of the following represents pure capacitive reactance of 100 ohms in rectangular notation?

A. 0 - j100
B. 0 + j100
C. 100 - j0
D. 100 + j0

When working with complex circuits (and by complex, we don't mean complicated, but simply circuits that contain resistance AND reactance), we plot the two components on two different axes. By convention, the resistance value, R, is plotted on the horizontal axis, and the reactance value, X, is plotted on the vertical axis, inductance values being UP and capacitive values being DOWN. The reactance value is prefixed with the letter "j" to indicate that this is the imaginary component of the impedance. The mathematicians amongst us use the letter "i" to indicate the ***imaginary component***, but we electronics folks need "i" for other things, so we use "j". **ANSWER A.**

Representing an AC Quantity

Rectangular Coordinates

A quantity can be represented by points that are measured on an X-Y plane with an X and Y axis perpendicular to each other. The point where the axes cross is the origin, the points on the X and Y axis are called coordinates, and the magnitude is represented by the length of a vector from the origin to the unique point defined by the *rectangular coordinates.*

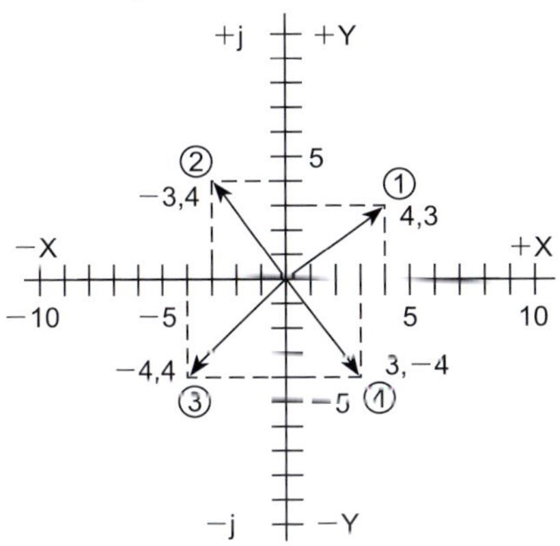

To be able to identify AC quantities and their phase angle, the X axis is called the real axis and the Y axis is called the imaginary axis. There are positive and negative real axis coordinates and positive and negative imaginary axis coordinates. The imaginary axis coordinates have a **j** operator to identify that they are imaginary coordinates.

In rectangular coordinates:

Vector 1 is 4 + j3
Vector 2 is −3 + j4
Vector 3 is −4 − j4
Vector 4 is 3 − j4

Polar Coordinates

A quantity can be represented by a vector starting at an origin, the length of which is the magnitude of the quantity, and rotated from a zero axis through 360°.

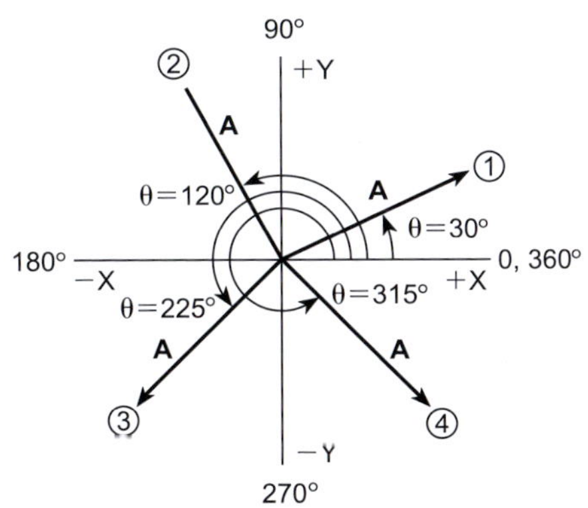

The position of the Vector A can be defined by the angle through which it is rotated. When we do so we are defining the AC quantity in *polar coordinates.*

In polar coordinates:

Vector 1 is A $\angle 30°$
Vector 2 is A $\angle 120°$
Vector 3 is A $\angle 225°$
Vector 4 is A $\angle 315°$

E5C05 What kind of diagram is used to show the phase relationship between impedances at a given frequency?

A. Venn diagram
B. Near field diagram
C. Phasor diagram
D. Far field diagram

A ***phasor is simply an arrow or vector that shows the magnitude and phase of an electrical current in a circuit***. The length of the phasor is the magnitude, and the direction of the arrow is the phase angle between voltage and current. **ANSWER C.**

E5C07 Where is the impedance of a pure resistance plotted on rectangular coordinates?

A. On the vertical axis
B. On a line through the origin, slanted at 45 degrees
C. On a horizontal line, offset vertically above the horizontal axis
D. On the horizontal axis

It's unfortunate that we use the term ***complex*** to describe values that are used all the time in electronics. Actually, they're quite simple once you've played around with them for a while. Impedance is always a complex value, because it contains both a resistance and a reactance (although any of these values can be zero. Always remember: zero is a number too!) Using Cartesian coordinates, the resistive values are plotted on the horizontal axis, while reactive components are plotted on the vertical axis. Any impedance that has positive resistance will show up on the right two quadrants of the graph; while any circuit with negative resistance (yes there is such a thing!) will show up somewhere on the left two quadrants. ***A pure resistance***, that is an impedance where the reactance is zero, ***will show up on the horizontal axis***. **ANSWER D.**

E5C06 What does the impedance 50 - j25 ohms represent?

A. 50 ohms resistance in series with 25 ohms inductive reactance
B. 50 ohms resistance in series with 25 ohms capacitive reactance
C. 25 ohms resistance in series with 50 ohms inductive reactance
D. 25 ohms resistance in series with 50 ohms capacitive reactance

In a complex impedance, the "j operator" always designates the reactive component. We know that + reactance is inductive and – reactance is capacitive. So, ***50 -j25 ohms is 50 ohms of resistance and 25 ohms of capacitive reactance***. **ANSWER B.**

E5C10 Which point on Figure E5-1 best represents the impedance of a series circuit consisting of a 400-ohm resistor and a 38-picofarad capacitor at 14 MHz?

A. Point 2
B. Point 4
C. Point 5
D. Point 6

Look at Figure E5-1 and notice it is a rectangular-coordinate system with axes at right angles to each other. The indicated points on the plane have X and Y coordinate values. In this question, the resistance of 400 ohms can be found along the +X axis; and since the circuit is capacitive, we go down to 300 on the -Y axis and project a line parallel to the X axis to intersect at ***Point 4*** a projection from 400 on the +X axis. You can work the problem all the way out, but in this question, you can visually spot it at a glance. **ANSWER B.**

To check yourself, solve the formula $X_C = 1/2\pi fC$ as follows:

$$X_c = \frac{1}{2\pi fC} = \frac{1}{6.28 \times 14 \times 10^6 \times 38 \times 10^{-12}}$$

$$= \frac{1}{3.341 \times 10^3 \times 10^{-6}} = 0.2993 \times 10^3 = 299.3 \text{ ohms or } 300 \text{ ohms}$$

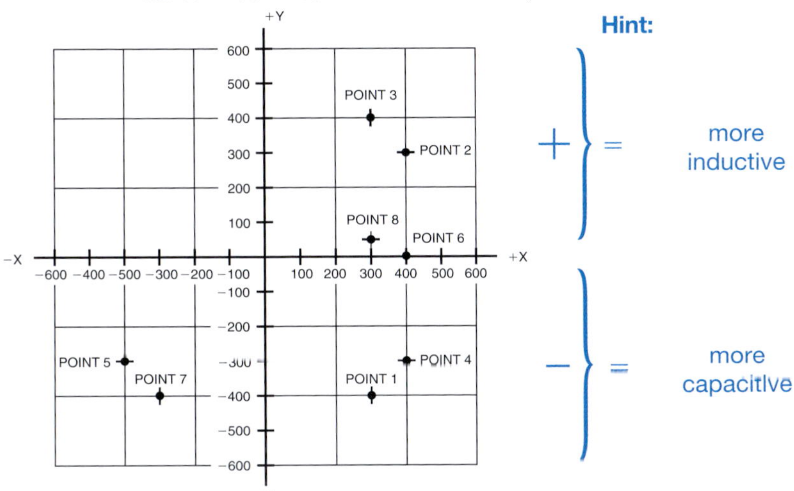

Figure E5-1

E5C11 Which point in Figure E5-1 best represents the impedance of a series circuit consisting of a 300-ohm resistor and an 18-microhenry inductor at 3.505 MHz?

A. Point 1
B. Point 3
C. Point 7
D. Point 8

The inductive reactance at 3.505 MHz is $X_L = 2\pi \times 3.505 \times 10^6 \times 18 \times 10^{-6} = 396.2$ or approximately 400 ohms of inductive reactance. ***Follow the +X axis out to 300*** for the resistance value. Move up on a line parallel to the ***+Y axis*** until you intersect the ***400 line***. You need to go in a positive Y direction because the circuit is inductive. ***Point 3*** is the answer. **ANSWER B.**

E5C12 Which point on Figure E5-1 best represents the impedance of a series circuit consisting of a 300-ohm resistor and a 19-picofarad capacitor at 21.200 MHz?

A. Point 1 C. Point 7
B. Point 3 D. Point 8

The capacitive reactance is $X_C = 1/2\pi fC = 1/6.28 \times 21.2 \times 10^6 \times 19 \times 10^{-12} = 1/2.53 \times 10^{-3} = 395$ ohms. We first move to the right along the ***+X axis*** to the value of the ***300-ohm resistor*** and move ***DOWN*** on a line perpendicular to the X axis (parallel to ***the -Y axis***) ***to Point 1*** at the intersection with the minus 400 line. 395 ohms is considered as 400 ohms. (If we went up, rather than down, we would be incorrect. Going up the +Y axis would mean the circuit is inductive. 19 picofarads makes the circuit capacitive, so we go down.) **ANSWER A.**

E5C08 What coordinate system is often used to display the phase angle of a circuit containing resistance, inductive, and/or capacitive reactance?

A. Maidenhead grid C. Elliptical coordinates
B. Faraday grid D. Polar coordinates

To display the impedance of a circuit with its phase angle, you use ***polar coordinates***. Polar coordinates provide a visual representation of the value of the impedance and its phase angle. **ANSWER D.**

E5C02 How are impedances described in polar coordinates?

A. By X and R values C. By magnitude and phase angle
B. By real and imaginary parts D. By Y and G values

Most electrical principles can be expressed in a number of ways. A complex impedance can be described either in rectangular coordinates or polar notation – just two ways of representing the same thing through different characteristics of it. A rectangular coordinate system uses resistance and total reactance to represent inductance. ***In polar notation, impedance is expressed by magnitude and phase angle***. **ANSWER C.**

E5C03 Which of the following represents a pure inductive reactance in polar coordinates?

A. A positive 45 degree phase angle
B. A negative 45 degree phase angle
C. A positive 90 degree phase angle
D. A negative 90 degree phase angle

Electrical values containing both resistance and reactance are best displayed on a two-dimensional graph, with the horizontal axis being resistance, and the vertical axis being reactance. By convention, inductive reactance is UP, and capacitive reactance is DOWN. The phase angle is measured counterclockwise, beginning from the positive resistive axis. This means ***a pure inductive reactance is straight up, a POSITIVE 90 degrees from a pure resistance***. **ANSWER C.**

E5B12 What is admittance?

A. The inverse of impedance
B. The term for the gain of a field effect transistor
C. The inverse of reactance
D. The term for the on-impedance of a field effect transistor

In circuits containing many parallel components, it can be much easier to do calculations based on the ***reciprocal values of*** resistance, reactance, or ***impedance***, which are conductance, susceptance, and ***admittance,*** respectively. The unit of conductance, susceptance, or admittance is the siemens, designated by the upside-down OMEGA symbol. Before 1971, the MHO (ohm spelled backwards) was the official unit of conductance, susceptance, and admittance, and many of us old timers still believe mho is mo' better. It's easy to remember that with mo' MHO, you have mo' current flow! But siemens (spelled with a small s) is now the official term. **ANSWER A.**

E5B03 How is impedance in polar form converted to an equivalent admittance?

A. Take the reciprocal of the angle and change the sign of the magnitude
B. Take the reciprocal of the magnitude and change the sign of the angle
C. Take the square root of the magnitude and add 180 degrees to the angle
D. Square the magnitude and subtract 90 degrees from the angle

Admittance is the reciprocal of impedance, and its unit is the mho (ohm spelled backwards), or, more recently, the ***siemens***. Admittance is more convenient to use than impedance in complex PARALLEL circuits. But you should learn to be comfortable with both ways of expressing the problem. It's a snap to convert impedance to admittance by simply ***taking the reciprocal of the magnitude and flipping the sign of the phase angle***. Or, if you're using a Smith Chart, just flip the chart upside down! **ANSWER B.**

E5B06 What is susceptance?

A. The magnetic impedance of a circuit
B. The ratio of magnetic field to electric field
C. The imaginary part of admittance
D. A measure of the efficiency of a transformer

A circuit or component has susceptance because it is susceptible to current flow. ***Susceptance*** is measured in siemens (symbol "B"). It is the ***imaginary reactance component of admittance***. **ANSWER C.**

E5B02 What letter is commonly used to represent susceptance?

A. G
B. X
C. Y
D. B

It's easy to confuse units with symbols. The ***symbol for susceptance is B***, while the unit of susceptance is the mho (or siemens). **ANSWER D.**

E5B05 What is the effect on the magnitude of pure reactance when it is converted to a susceptance?

A. It is unchanged
B. The sign is reversed
C. It is shifted by 90 degrees
D. It is replaced by its reciprocal

In electronics, we have several reciprocal functions (1/original value), which are particularly useful when working with parallel circuits. The three most common are ***conductance***, the reciprocal of resistance, ***susceptance***, the reciprocal of reactance, and ***admittance***, the reciprocal of impedance. All three reciprocal values are in units of ***mhos***, or more recently, ***siemens***. Most of us old timers prefer the venerable term MHO, because it's easy to remember that with mo' mho, you have mo' current flow! **ANSWER D.**

E5A09 How is the Q of an RLC parallel resonant circuit calculated?

A. Reactance of either the inductance or capacitance divided by the resistance

B. Reactance of either the inductance or capacitance multiplied by the resistance

C. Resistance divided by the reactance of either the inductance or capacitance

D. Reactance of the inductance multiplied by the reactance of the capacitance

By definition, in a resonant circuit, inductive reactance and capacitive reactance are equal, so either can be used for the calculation of Q. In a ***parallel circuit, Q = R/X*** where R is resistance and X is inductance. **ANSWER C.**

E5A13 What is an effect of increasing Q in a series resonant circuit?

A. Fewer components are needed for the same performance

B. Parasitic effects are minimized

C. Internal voltages increase

D. Phase shift can become uncontrolled

The Q of a resonant circuit is defined as the ratio of reactance to resistance (X/R). A resonant circuit with no internal resistance can have infinite circulating current, as well as infinite voltage across the inductor or capacitor. In the real world, you will never have this situation, but you can still have some ***very high voltage increases***, especially in Hi Q mobile antenna series circuits, where high voltage sparks may sometimes emit from the whip tip! Never park where your HF antenna whip tip could be touching a tree! **ANSWER C.**

Half-Power Bandwidth

An amplifier's voltage gain will vary with frequency as shown. At the cutoff frequencies, the voltage gain has dropped to 0.707 of what it is in the mid-band. These frequencies, f_1 and f_2, are sometimes called the half-power frequencies.

If the output voltage is 10 volts across a 100-ohm load when the gain is A at the mid-band, then the power output, P_O, at mid-band is:

$$P_O = \frac{E^2}{R} = \frac{10^2}{100} = \frac{100}{100} = 1 \text{ Watt}$$

At the cutoff frequency, the output voltage will be 0.707 of what it is at the mid-band; therefore, 7.07 volts. The power output is:

$$P_{0.707} = \frac{(7.07)^2}{100} = \frac{50}{100} = 0.5 \text{ Watt}$$

The power output at the cutoff frequency points is one-half the mid-band power. The half-power bandwidth is between the frequencies f_1 and f_2.

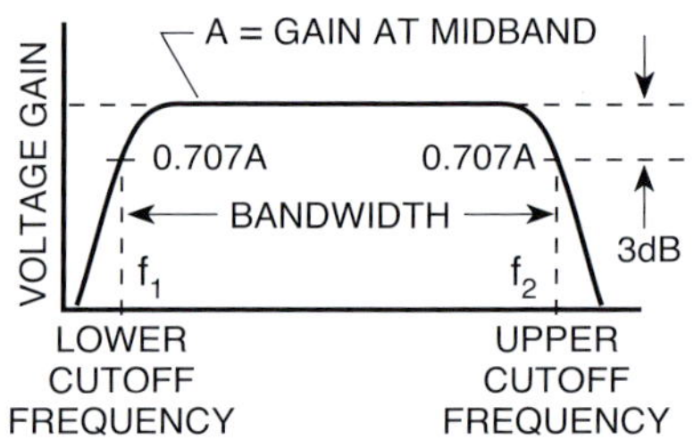

Power Ratio in dB

$$P_{dB} = 10 \log_{10}\left(\frac{P_O}{P_{0.707}}\right)$$
$$= 10 \log_{10}\left(\frac{1}{0.5}\right)$$
$$= 10 \log_{10} 2$$
$$= 10 \text{ X } 0.301$$
$$= 3 \text{ db}$$

The power output at the 0.707 frequencies is 3 db down from the mid-band power.

E5A11 What is the half-power bandwidth of a resonant circuit that has a resonant frequency of 7.1 MHz and a Q of 150?

A. 157.8 Hz
B. 315.6 Hz
C. 47.3 kHz
D. 23.67 kHz

To find the half-power bandwidth, divide the resonant frequency (in kHz) by Q. 7100 kHz divided by 150 equals 47.33. Remember, to convert MHz to kHz, move the decimal three places to the right. On your calculator, press clear, ***7100 ÷ 150 = 47.33***. **ANSWER C.**

Here is the formula:

$$\text{Half-power bandwidth} = \frac{f_r}{Q}$$

where: f_r is the resonant frequency in **kHz**
Q is a **quality factor** of the circuit

E5A12 What is the half-power bandwidth of a resonant circuit that has a resonant frequency of 3.7 MHz and a Q of 118?

A. 436.6 kHz
B. 218.3 kHz
C. 31.4 kHz
D. 15.7 kHz

With this question, go from MHz to kHz, and then divide it by the Q. Clear, enter ***3700 ÷ 118 = 31.36***. Round the answer to 31.4. **ANSWER C.**

E5B01 What is the term for the time required for the capacitor in an RC circuit to be charged to 63.2% of the applied voltage or to discharge to 36.8% of its initial voltage?

A. An exponential rate of one.
B. One time constant.
C. One exponential period.
D. A time factor of one.

In an RC circuit, assuming there is no initial charge on the capacitor, it takes ***one time constant to charge a capacitor to 63.2 percent*** of its final supply voltage value. **ANSWER B.**

E5B04 What is the time constant of a circuit having two 220-microfarad capacitors and two 1-megohm resistors, all in parallel?

A. 55 seconds
B. 110 seconds
C. 440 seconds
D. 220 seconds

Here is an RC circuit with all components in parallel. It will be easy if you remember that two equal capacitors in parallel have a total capacitance of twice the value of one of the capacitors, and that two equal value resistors in parallel have a total resistance equal to one-half the value of one of the resistors. More good news – "megs" (10^6) are being multiplied by "micros" (10^{-6}) so the powers of ten are cancelled in the calculation. Two 220 microfarad capacitors in parallel equal a larger 440 microfarad capacitor. Two 1-megohm resistors in parallel equal a resistance of 0.5 megohms. ***Solve for the time constant using the simple formula T = R times C***. Multiply ***0.5 × 440 and you end up with 220 seconds*** as the correct answer. The calculator keystrokes are: Clear, 0.5 × 440 = 220. **ANSWER D.**

E5D01 What is the result of conductor skin effect?

A. Resistance increases as frequency increases because RF current flows closer to the surface
B. Resistance decreases as frequency increases because electron mobility increases
C. Resistance increases as temperature increases because of the change in thermal coefficient
D. Resistance decreases as temperature increases because of the change in thermal coefficient

If you ever had a chance to inspect an old wireless station, you would have seen that the antenna "plumbing" leading out of the transmitter was usually constructed of hollow copper tubing. ***Radio frequency current always travels along the thinner outside layer of a conductor***, and as frequency increases, there is almost no current in the center of the conductor. The higher the frequency the greater the ***skin effect***. **ANSWER A.**

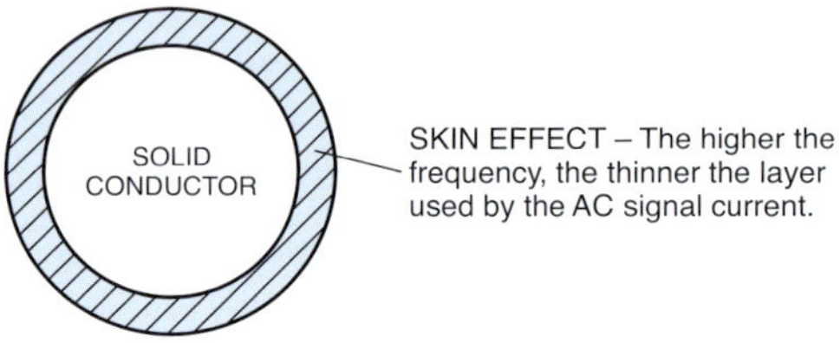

Cross-Section of Wire Conductor Illustrating Skin Effect

E5D02 Why is it important to keep lead lengths short for components used in circuits for VHF and above?

A. To increase the thermal time constant
B. To minimize inductive reactance
C. To maintain component lifetime
D. All these choices are correct

VHF circuitry can sometimes look pretty ugly. This is because a straight line between two points isn't always the prettiest path. But for VHF circuitry, it is the path that works the best. Any time a piece of wire is a significant fraction of a wavelength long, it can become an antenna. Or, short of that, it can act as an unwelcome inductor. So, ***to minimize unwanted inductance, keep all your leads in a VHF circuit as short as possible***. If you can do this and make it look pretty, you're doing very well! **ANSWER B.**

Notice no long lead lengths on this VHF amplifier wire-wound toroidal inductor.

E5D04 Why are short connections used at microwave frequencies?

A. To increase neutralizing resistance
B. To reduce phase shift along the connection
C. To increase compensating capacitance
D. To reduce noise figure

A transmission line, whether in the form of a run of coaxial cable or a copper trace on a circuit board, will introduce some degree of phase shift since it has a finite velocity factor. Whether the phase shift causes a problem or not depends on the application. For a given length of transmission line, the ***phase shift*** will be greater as the frequency increases. It's usually safest to keep transmission lines as short as possible unless there is a compelling reason to do otherwise. **ANSWER B.**

E5D06 What parasitic characteristic creates an inductor's self-resonance?

A. Skin effect
B. Dielectric loss
C. Coupling
D. Inter-turn capacitance

The inter-turn capacitance of an inductor can be significant, depending on the construction. Close wound inductors can have substantial inter-turn capacitance, which can greatly alter their self-resonant frequency, or even worse, create multiple self-resonant frequencies. One solution to this is the UNIVERSAL WINDING form, where different turns only cross each other at right angles to minimize the inter-turn capacitance. There was a time, long ago, when just about every ham had a universal coil winding machine.... sort of a modified fishing reel arrangement. (I wish I still had mine.) Here is a wonderful video of a beautiful home-brew coil winder: **https://www.youtube.com/watch?v=bnGfSruSddk** **ANSWER D.**

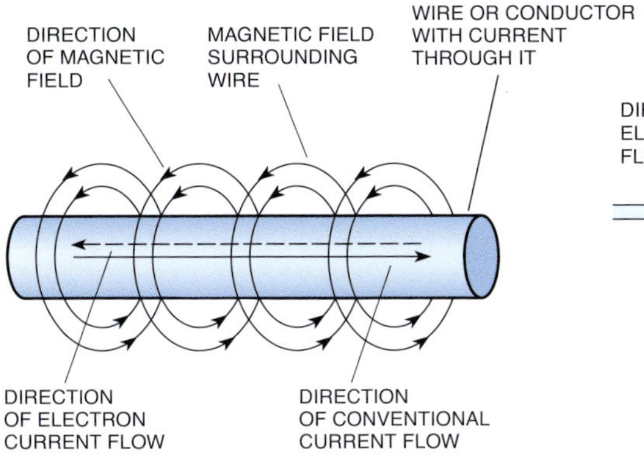

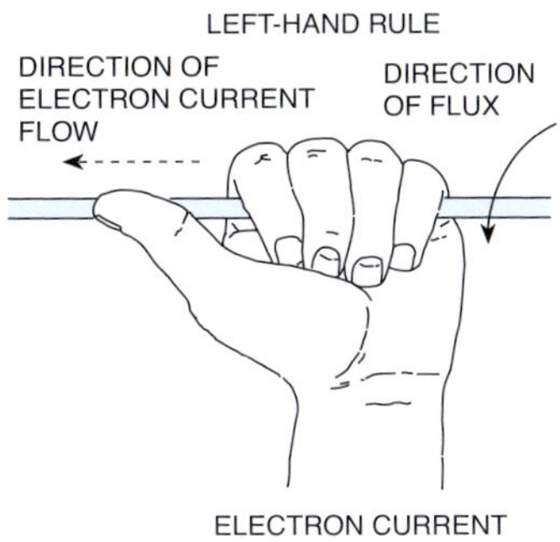

Left-Hand Rule

E6D04 Why are cores of inductors and transformers sometimes constructed of thin layers?

A. To simplify assembly during manufacturing
B. To reduce power loss from eddy currents in the core
C. To increase the cutoff frequency by reducing capacitance
D. To save cost by reducing the amount of magnetic material

An iron core transformer is ***laminated*** in order to break the magnetic path of eddy currents, which are currents perpendicular to the desired magnetic currents, and do little except waste energy. Powdered iron or ferrite cores expand the lamination idea to the extreme; the core consists of billions of magnetic domains completely insulated from one another. **ANSWER B.**

E6D08 Which of the following materials has the highest temperature stability of its magnetic characteristics?

A. Brass
B. Powdered iron
C. Ferrite
D. Aluminum

Ferrite materials have the highest permeability (mu) of any commonly used cores, affording high values of inductance for a small number of turns. However, ferrites are highly temperature sensitive, and can saturate easily. Powdered iron, while having only about 1/100th the ***permeability*** of the highest mu ferrites, are much more temperature stable. Powdered iron is preferable for tuned circuits that need to maintain high frequency stability and high-power applications where saturation could cause signal distortion. You'll need more windings for a powdered iron core than a ferrite core, but it's usually worth the extra effort (and copper!) for the effort. **ANSWER B.**

E6D11 Which type of core material decreases inductance when inserted into a coil?

A. Ceramic
B. Brass
C. Ferrite
D. Aluminum

Brass has a negative permeability and ***can be used to reduce the inductance*** of an inductor. While you are not too likely to see brass slugs in amateur radio gear, high powered broadcast transmitters often used moveable brass slugs in the tuning networks. **ANSWER B.**

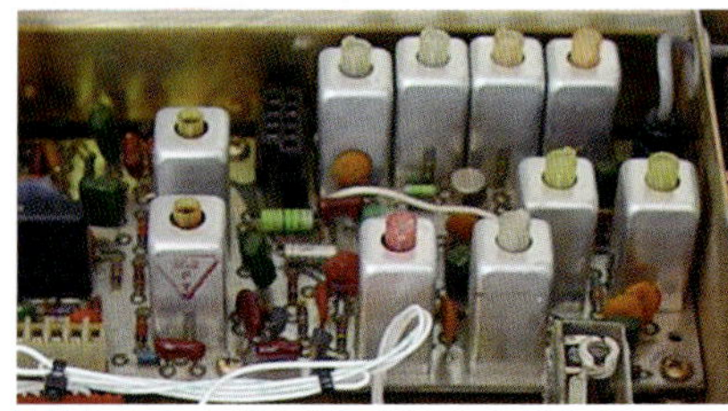

Veteran land mobile radio guys know to carefully tune these inductors, as the core "slugs" are fragile! Only the proper plastic tuning tool will do!

E6D12 What causes inductor saturation?

A. Operation at too high a frequency
B. Selecting a core with low permeability
C. Operation at excessive magnetic flux
D. Selecting a core with excessive permittivity

While ferrous materials of any kind will increase the permeability of an inductor, they are all subject to ***saturation***, where an increase in coil current results in no further increase in the magnetic field. When this happens, the permeability (and inductance) can drop precipitously, and depending on the application can create distortion, overheating, or frequency shift and drift. Most of the time saturation is not a good thing, but there are some circuits that employ ***saturable reactors***. Before the Varactor diode was invented, saturable reactors were often used to create an electrically tuned oscillator or frequency modulator. **ANSWER C.**

E5D05 What parasitic characteristic causes electrolytic capacitors to be unstable for use at RF?

A. Skin effect
B. Shunt capacitance
C. Inductance
D. Dielectric leakage

Any real-world component has some parasitic characteristics. Every resistor has some capacitance and inductance, every inductor has some capacitance and resistance, and every capacitor has some resistance and inductance. Electrolytic capacitors in particular have a great deal of parasitic inductance, which makes their series impedance much higher than the capacitive reactance alone might suggest. RF bypass capacitors, especially, need to have very low impedance to ground at the operating frequency. Parasitic inductance can make the capacitors look more like an inductor than a capacitor, entirely defeating their purpose. Power bus RF bypassing often uses a combination of electrolytic and low inductance capacitors in parallel. **ANSWER C.**

E5D07 What combines to create the self-resonance of a component?

A. The component's resistance and reactance
B. The component's nominal and parasitic reactance
C. The component's inductance and capacitance
D. The component's electrical length and impedance

It would be nice if every component were absolutely pure. Unfortunately, all electrical components have some parasitic components. For instance, an inductor always has some parasitic resistance, because of the DC resistance of the wire (or other conductor) it's made from. Also, every capacitor has some parasitic inductance, and every inductor has some parasitic capacitance, which means every inductor or capacitor has a self-resonant frequency (or sometimes multiple frequencies). Normally, resistance has no effect on the resonant frequency of a component, which is determined only by its capacitance and inductance. **ANSWER B.**

E5D09 What happens to reactive power in ideal inductors and capacitors?

A. It is dissipated as heat in the circuit
B. Energy is stored in magnetic or electric fields, but power is not dissipated
C. It is canceled by Coulomb forces in the capacitor and inductor
D. It is dissipated in the formation of inductive and capacitive fields

In a circuit that has both capacitors and inductors, the ***reactive power*** oscillates back and forth between the magnetic and electric fields and ***is not dissipated***. **ANSWER B.**

E5D03 What is the phase relationship between current and voltage for reactive power?

A. They are out of phase
B. They are in phase
C. They are 90 degrees out of phase
D. They are 45 degrees out of phase

It's never a bad idea to review some of the basics like ELI the ICE Man, which tells us the relationship between voltage and current in an inductor or a capacitor. In any complex circuit, reactive or "wattless" power has a ***phase difference of 90 degrees (leading OR lagging) between voltage and current***. Reactive power doesn't do anyone much good, but it is necessary to know in order to determine the voltage or current ratings of the various components in a circuit. **ANSWER C.**

E5D11 How much real power is consumed in a circuit consisting of a 100-ohm resistor in series with a 100-ohm inductive reactance drawing 1 ampere?

A. 70.7 watts
B. 100 watts
C. 141.4 watts
D. 200 watts

The 100 ohms inductive reactance in the coil will not be a factor in this problem, as a pure inductance will not consume power in this circuit. Ohm's Law, $P = I^2 \times R$, will result in 1 × 1 = 1, times R of 100 ohms resistance, equal to ***100 watts*** of power consumed. **ANSWER B.**

$P = I^2 \times R$	Where: P = Power
$P = (1 \times 1) \times 100$	I = Amps
$P = 1 \times 100$	R = Resistance in ohms
$P = 100$ watts	

E5D08 What is the primary cause of loss in film capacitors at RF?

A. Inductance
B. Dielectric loss
C. Self-discharge
D. Skin effect

Different types of capacitors have different loss processes, depending on both the construction and frequency of operation. Film capacitors, while having little dielectric loss or parasitic inductance, have a frequency limit imposed by skin effect, where the currents only travel on the outer surface of a conductor. Skin effect makes the conductors look a lot "skinnier" than they do at DC and incur greater I^2R loss. **ANSWER D.**

E5D12 What is reactive power?

A. Power consumed in circuit Q
B. Power consumed by an inductor's wire resistance
C. The power consumed in inductors and capacitors
D. Wattless, nonproductive power

The letters AC are found in the word ***reactive*** (meaning out-of-phase). It indicates ***non-productive power*** produced in circuits containing inductors and capacitors. Since out-of-phase power in inductors will tend to cancel out-of-phase power in capacitors, reactive power is ***wattless***. It is not converted to heat and dissipated. It is energy being stored temporarily in a field and then returned to the circuit. It circulates back and forth in a coil's magnetic field and/or a capacitor's electrostatic field. **ANSWER D.**

E4E04 How can conducted noise from an automobile battery charging system be suppressed?

A. By installing filter capacitors in series with the alternator leads
B. By installing a noise suppression resistor and a blocking capacitor at the battery
C. By installing a high-pass filter in series with the radio's power lead and a low-pass filter in parallel with the antenna feed line
D. By installing ferrite chokes on the charging system leads

Connect your radio's red and black leads ***directly to your battery's*** positive and negative terminals using the shortest possible run. Be sure you put a fuse right next to the battery's positive pick-off point for safety. Also, it is recommended that you ***install ferrite chokes on the charging system leads***. **ANSWER D.**

E4E05 What is used to suppress radio frequency interference from a line-driven AC motor?

A. A high-pass filter in series with the motor's power leads
B. A brute-force AC-line filter in series with the motor's power leads
C. A bypass capacitor in series with the motor's field winding
D. A bypass choke in parallel with the motor's field winding

One way of reducing radio frequency interference (RFI) coming up the line from an electric motor is to ***install an AC-line filter in series with the motor leads***. The incorrect answers have the right components, but the wrong usage. RF beads are used to keep RF off of specific circuits, and bypass capacitors in parallel are used to bypass noise. As for the ground-fault interrupter, this won't help with motor noise. Go with the ***brute- force AC-line filter in series***. **ANSWER B.**

E6D06 What core material property determines the inductance of an inductor?

A. Permittivity
B. Resistance
C. Reactivity
D. Permeability

The key words in the question are ***core material property***. The material that determines the ***inductance of a toroid is the permeability*** of the iron donut. **ANSWER D.**

E6D05 How do ferrite and powdered iron compare for use in an inductor core?

A. Ferrite cores generally have lower initial permeability
B. Ferrite cores generally have better temperature stability
C. Ferrite cores generally require fewer turns to produce a given inductance value
D. Ferrite cores are easier to use with surface mount technology

A sound reason to use ferrite cores rather than powdered-iron cores is the fact that ***ferrite cores require fewer turns of wire to produce a given inductance*** value. **ANSWER C.**

Don't be tempted to adjust any coil core unless you have the shop manual for the radio on hand. Technical shop manuals are sold separately from your radio. Unless you have all the scopes and test equipment on hand, stay away from doing any coil turning!

E6D10 What is a primary advantage of using a toroidal core instead of a solenoidal core in an inductor?

A. Toroidal cores confine most of the magnetic field within the core material
B. Toroidal cores make it easier to couple the magnetic energy into other components
C. Toroidal cores exhibit greater hysteresis
D. Toroidal cores have lower Q characteristics.

The ***toroidal core inductor concentrates the magnetic field within the core material***, protecting other nearby components from stray magnetic fields. **ANSWER A.**

GOOD TO KNOW:

To calculate the percentage of the supply voltage after two time constants in an RC circuit, write down the percent after a single time constant as 63.2 percent. The remaining percent of charge that the capacitor must charge is 36.8 percent. It is found by deducting 63.2 percent from 100 percent. Since 36.8 percent is the final value to which the capacitor will charge after it has charged to 63.2 percent, in the next time constant the capacitor will charge to 63.2 percent of the 36.8 percent. In other words, another percent charge is added to the capacitor in the second time constant equal to:

36.8% x 63.2% = 23.26% (rounded to 23.3%)

This really needs to be calculated in decimal format as:

0.368 x 0.632 = 0.2326

Percent values are converted to decimals by moving the decimal point two places to the left. Decimal values are converted to percent values by multiplying by 100 (moving decimal point to right two places.)

Since the final percent of charge in two time constants is desired, it is only necessary to add the two percent charges together:

	Percent	Decimal
Charge in first time constant =	63.2%	0.632
Charge in second time constant =	23.3%	0.233
Total charge after two time constants	86.5%	0.865

The process can be continued for the third time constant. Since the capacitor has charged to 86.5 percent in two time constants, if you start from that point the value of the third charge would be another 13.5 percent. Therefore, in the next time constant, the third, the capacitor would charge another:

13.5% x 63.2% = 8.53% (rounded to 8.5%)

After the third time constant the total charge is:

86.5% + 8.5% = 95%

If the process is continued through five time constants, the capacitor charge is 99.3 percent. Therefore, in electronic calculations, the capacitor is considered to be fully charged (or discharged) after five time constants

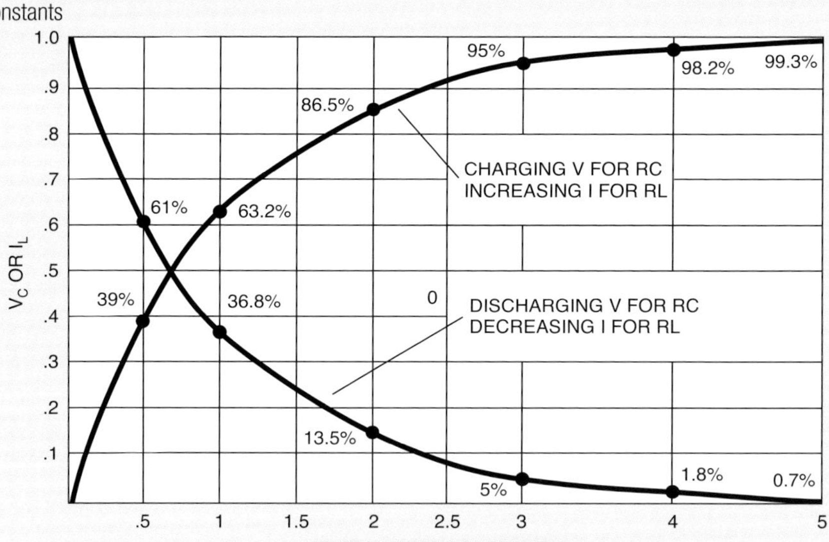

Time Constants Showing V and I Percentages

Components in Your New Rig

E6A01 In what application is gallium arsenide used as a semiconductor material?

A. In high-current rectifier circuits
B. In high-power audio circuits
C. In microwave circuits
D. In very low-frequency RF circuits

Here's a new word for your amateur radio vocabulary: GaAsFET. The GaAs stands for ***gallium arsenide***, and FET stands for field effect transistor. We use the ***GaAsFET*** transistor in VHF, UHF, and ***microwave receivers***. This transistor offers excellent signal gain with an extremely low noise figure. **ANSWER C.**

E6A02 Which of the following semiconductor materials contains excess free electrons?

A. N-type
B. P-type
C. Bipolar
D. Insulated gate

There are more ***free electrons*** orbiting the nucleus in an ***N-type*** material. **ANSWER A.**

You will have one question from this group about semiconductor materials. Remember this phrase, "ANFREE" for an N-type material where antimony atoms or arsenic atoms have been added, and there are more free electrons in an N-type semiconductor. "PIUMH" (pronounced like sneezing) is a P-type material produced by adding either galIIUM or indIUM, producing more "Potholes" than free electrons. Also, anything that PROVIDES is a donor, and an Acceptor impurity will Add holes. Got it? Good reading about semiconductors is found in the latest edition of the American Radio Relay League Handbook for Radio Amateurs.

E6A04 What is the name given to an impurity atom that adds holes to a semiconductor crystal structure?

A. Insulator impurity
B. N-type impurity
C. Acceptor impurity
D. Donor impurity

The ***impurity atom that adds holes*** to a semiconductor crystal structure is called an ***acceptor impurity***. It's an acceptable impurity. **ANSWER C.**

E6A05 How does DC input impedance at the gate of a field-effect transistor (FET) compare with that of a bipolar transistor?

A. They are both low impedance
B. An FET has lower input impedance
C. An FET has higher input impedance
D. They are both high impedance

A field effect transistor, ***FET***, is much easier to work with in circuits than the simple bipolar transistor. The FET's ***higher input impedance*** is less likely to load down a circuit to which it is attached, and the FET can handle big swings in signal levels. **ANSWER C.**

E6A06 What is the beta of a bipolar junction transistor?

A. The frequency at which the current gain is reduced to 0.707
B. The change in collector current with respect to base current
C. The breakdown voltage of the base-to-collector junction
D. The switching speed

Another similar figure is the ***beta*** of the transistor, which ***is the ratio of collector current change to the change in base current***, in the common emitter configuration. While the common emitter or ***grounded emitter*** configuration is more familiar to most hams, in the early days of transistors the common base configuration was more common, so the Alpha of a transistor was deemed more "basic." However, for most voltage amplifiers the common emitter is more prevalent, so the beta figure is more useful. In either case, alpha or beta, it is a measurement of the effective current gain of the transistor, sometimes loosely defined as "amplification factor." **ANSWER B.**

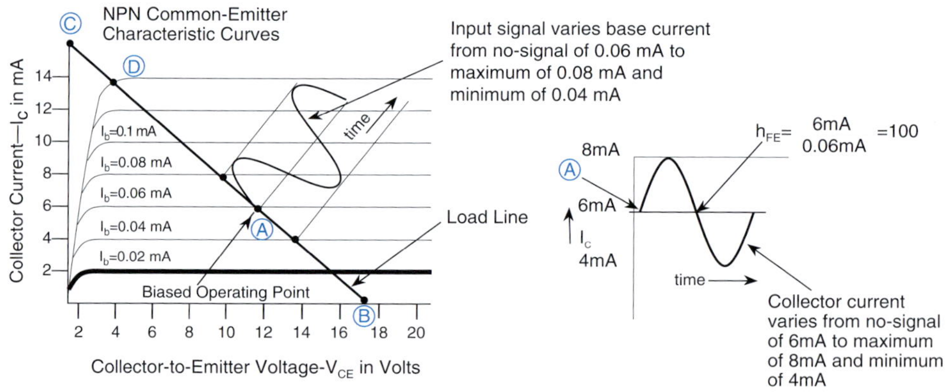

a. Characteristic Curves   b. Output Signal

Biasing sets "No-Signal" operating point

Bipolar Transistor Basics – PNP and NPN

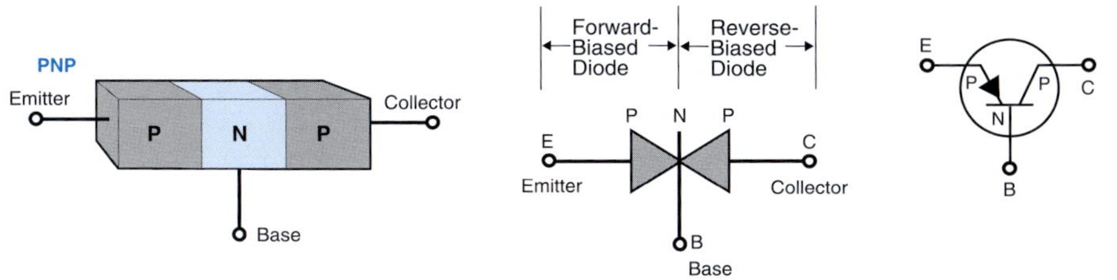

a. Junction Structure b. Diode Juction Equivalent c. PNP Symbol

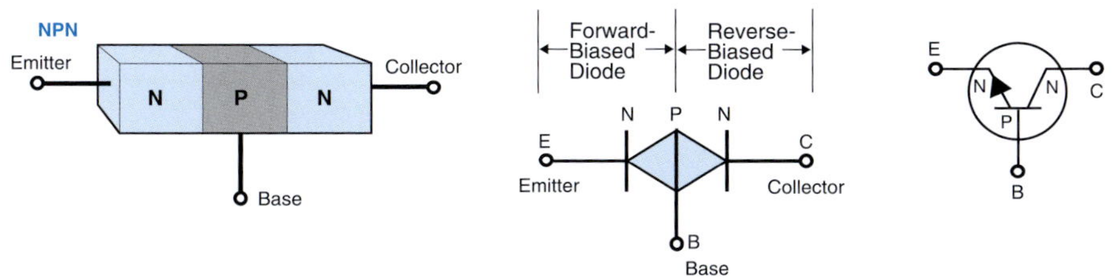

d. Junction Structure e. Diode Junction Equivalent f. NPN Symbol

There are two types of transistors, bipolar and field-effect. A bipolar transistor is a combination of two junctions of semiconductor material built into a semiconductor chip (usually silicon). There are two types – PNP and NPN. Their junction structures are shown in Figure a and d, respectively. For a transistor that produces gain, the emitter-base junction is a forward-biased diode, and the collector-base junction is a reverse-biased diode. The diode equivalents of PNP and NPN transistors are shown in Figure b and e, respectively. However, unlike a reverse-biased diode that does not conduct current (except for a small leakage current), the reverse-biased collector-base junction of a bipolar transistor conducts collector current that is controlled by the current into the base at the base-emitter junction. The normal active device operation is shown below in Figure g and h.

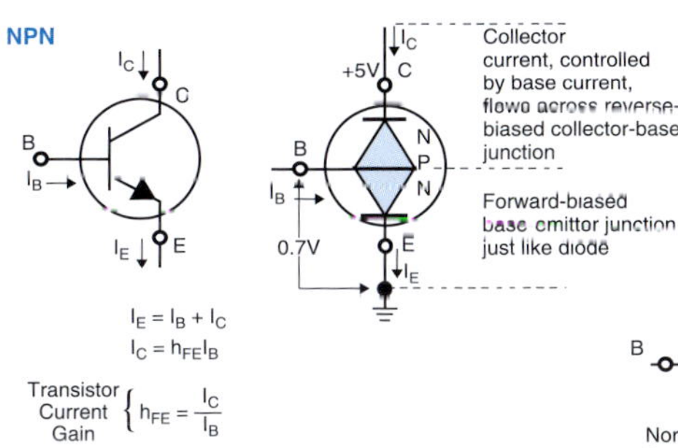

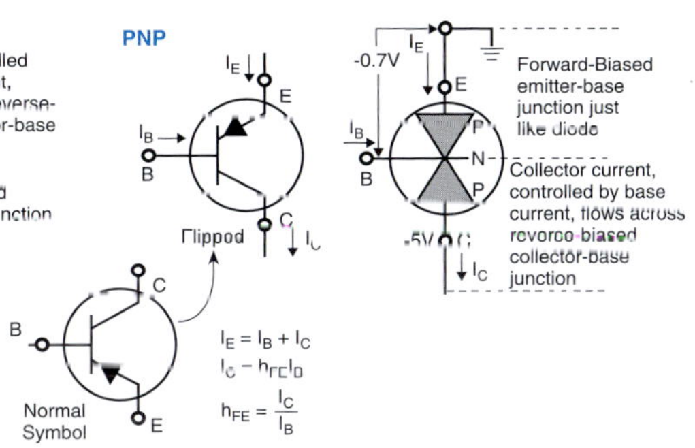

g. NPN Operation h. PNP Operation

An NPN silicon transistor P base is 0.7 V more positive that its N emitter, and the N collector is several more volts more positive than the emitter. The emitter current is the sum of the base and collector current. The current gain under any DC operating condition is hFE, the ratio of I_C to I_B, and current gains of 50 to 200 are common in modern-day silicon transistors. hFE is actually called "the common-emitter" current gain because the emitter is common in the circuit.

A PNP silicon transistor N base is 0.7 V negative with respect to its emitter in order to have the P emitter more positive than the N base. The P collector is several volts negative from the emitter to keep the collector-base junction reverse biased. As shown in Figure h, the same current equations apply, and the current gain, hFE, is the same. The major difference is in the polarity of the voltages for operation. For NPN common-emitter operation the base and collector voltages are positive with respect to the emitter; while for the PNP the voltages are negative.

Source: *Basic Communications Electronics*, Hudson & Luecke, © 1999 Master Publishing, Inc., Niles, IL

E6A07 Which of the following indicates that a silicon NPN junction transistor is biased on?

A. Base-to-emitter resistance of approximately 6 ohms to 7 ohms
B. Base-to-emitter resistance of approximately 0.6 ohms to 0.7 ohms
C. Base-to-emitter voltage of approximately 6 volts to 7 volts
D. Base-to-emitter voltage of approximately 0.6 volts to 0.7 volts

When an NPN transistor is biased "on," the voltage on the base is positive with respect to the emitter. Free electrons can flow from the emitter to the base. A characteristic of a ***forward-biased NPN junction*** is that the ***voltage drop*** across it is always approximately ***0.6 to 0.7V***. You don't need to remove the transistor from the circuit for this easy transistor check. **ANSWER D.**

E6A08 What is the term for the frequency at which the grounded-base current gain of a bipolar junction transistor has decreased to 0.7 of the gain obtainable at 1 kHz?

A. Corner frequency
B. Alpha rejection frequency
C. Beta cutoff frequency
D. Alpha cutoff frequency

Grounded-base is the same as a common base configuration. Common base means alpha. Since ***1 kHz*** is mentioned in this question, they are referring to the ***alpha cutoff frequency*** in a transistor. **ANSWER D.**

E6A09 What is a depletion-mode field-effect transistor (FET)?

A. An FET that exhibits a current flow between source and drain when no gate voltage is applied
B. An FET that has no current flow between source and drain when no gate voltage is applied
C. An FET that exhibits very high electron mobility due to a lack of holes in the N-type material
D. An FET for which holes are the majority carriers.

Your handheld battery may drain unexpectedly if ***depletion-mode FETs*** are used in many of the circuits. Even though there is no gate voltage applied, ***current still continues to flow*** from these components. **ANSWER A.**

E6A10 In Figure E6-1, which is the schematic symbol for an N-channel dual-gate MOSFET?

A. 2
B. 4
C. 5
D. 6

Note that the ***arrow is pointing IN***, and there are now ***two gates***. **ANSWER B.**

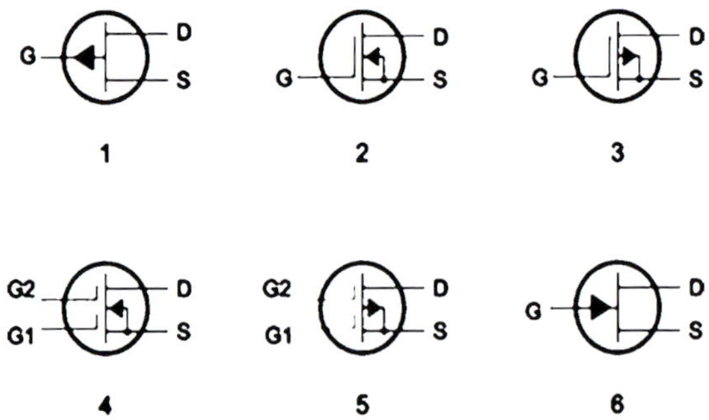

Figure E6-1

E6A11 In Figure E6-1, which is the schematic symbol for a P-channel junction FET?

A. 1
B. 2
C. 3
D. 6

In the ***P-channel junction FET***, the ***arrow is pointing out***. **ANSWER A.**

E6A12 What is the purpose of connecting Zener diodes between a MOSFET gate and its source or drain?

A. To provide a voltage reference for the correct amount of reverse-bias gate voltage
B. To protect the substrate from excessive voltages
C. To keep the gate voltage within specifications and prevent the device from overheating
D. To protect the gate from static damage

Most MOSFET devices are somewhat vulnerable to static discharge damage. The insulation between the GATE and the CHANNEL is extremely thin and doesn't take much voltage to arc across the insulator and destroy the device. A Zener diode (or sometimes two back-to-back Zeners) across the gate and source terminals will "fire" before a static voltage acquires a harmful level and shorts out the static charge. **ANSWER D.**

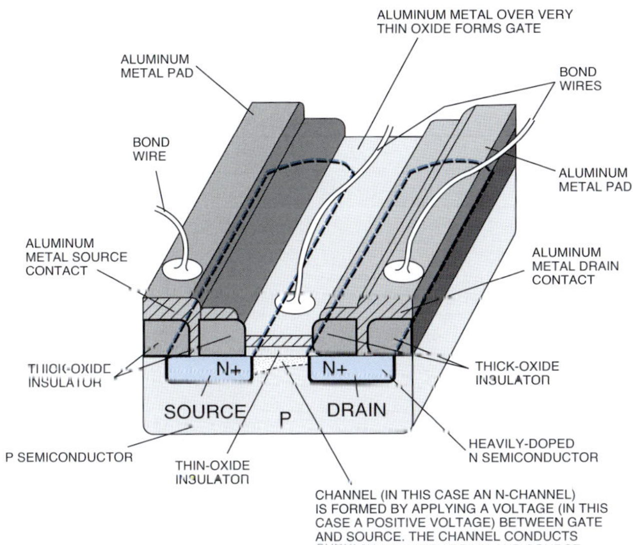

Pictorial of FET Construction (N-Channel Enhancement)

Unlike bipolar transistors that depend on current into the base to control collector current, field-effect transistor current between source and drain is controlled by a voltage on a gate. The basic structure of an N-channel MOSFET is shown here. Heavily-doped N semiconductor material forms source and drain regions in a P semiconductor material substrate. The region between the source and drain is the gate region, where a thin layer of oxide insulates the P semiconductor substrate underneath from a metal plate that is deposited over the thin oxide. A thick oxide layer over the source and drain regions insulates metal connection pads from the substrate. Holes in this thick oxide layer allow the metal pads to contact the source and drain. There are four common types of MOSFETs: P-channel depletion and enhancement mode devices; and N-channel depletion and enhancement mode devices.

Source: *Basic Communications Electronics*, Hudson & Luecke, © 1999 Master Publishing, Inc., Niles, IL

E6B01 What is the most useful characteristic of a Zener diode?

A. A constant current drop under conditions of varying voltage
B. A constant voltage drop under conditions of varying current
C. A negative resistance region
D. An internal capacitance that varies with the applied voltage

Always remember that a ***Zener diode*** is for voltage regulation, which gives you ***constant voltage*** even though current changes. **ANSWER B.**

Zener Diode

Here is the schematic symbol of a Zener diode. Since a diode only passes energy in one direction, look for that one-way arrow, plus a "Z" indicating it is a Zener diode. Doesn't that vertical line look like a tiny "Z?"

E6B02 Which characteristic of a Schottky diode makes it a better choice than a silicon junction diode for use as a power supply rectifier?

A. Much higher reverse voltage breakdown
B. More constant reverse avalanche voltage
C. Longer carrier retention time
D. Lower forward voltage drop

The Schottky diode is regularly found in solar power and power supply circuits because of its lower forward resistance and a slightly higher peak inverse voltage (PIV) rating. They are slightly more expensive than traditional power supply diodes, but they are used regularly in mobile and home solar systems because the ***Schottky diode offers minimal voltage drop*** while charging the sealed battery system, and they prevent current from flowing out of the battery system and back into the solar panel at night. **ANSWER D.**

E6B03 What property of an LED's semiconductor material determines its forward voltage drop?

A. Intrinsic resistance
B. Band gap
C. Junction capacitance
D. Junction depth

The ***BAND GAP*** of an atom is a fundamental quantum mechanical value. It is the energy necessary to move an electron into a higher energy orbit in an atom. Since in any practical device we're working with vast numbers of atoms, we normally can only observe the behavior of the bulk material. ***The forward voltage drop of an LED is the result of all the material's band gap voltages, averaged***. It's fortunate that we don't have to know VERY MUCH about the internal physics of a semiconductor in order to make good use of it! **ANSWER B.**

E6B04 What type of semiconductor device is designed for use as a voltage-controlled capacitor?

A. Varactor diode
B. Tunnel diode
C. Silicon-controlled rectifier
D. Zener diode

The ***varactor diode*** is used to tune VHF and UHF circuits by ***varying the voltage*** applied to it. **ANSWER A.**

E6B06 Which of the following is a common use of a Schottky diode?

A. As a rectifier in high current power supplies
B. In oscillator circuits as the negative resistance element
C. In power supplies as a constant voltage reference
D. As a VHF/UHF mixer or detector

The Schottky diode is also known as a hot carrier diode, with a PN junction. When you get involved with ***VHF and UHF*** equipment, you'll find ***Schottky diodes*** commonly used as ***mixers or detectors*** because of their low noise figure characteristics. Hot-carrier diodes also have a lower forward voltage drop, making them more sensitive in mixer circuits and even in crystal radio detectors. **ANSWER D.**

E6B10 In Figure E6-2, which is the schematic symbol for a Schottky diode?

A. 1 C. 2
B. 6 D. 3

A Schottky diode is known primarily for its very low ***forward voltage drop***, which makes it suitable as a weak signal detector. The ancient cat's whisker detector of the earliest days of radio was actually a primitive Schottky or ***hot carrier*** diode, long before physicists actually knew how it worked! Don't confuse a Schottky diode symbol for the very similar Zener Diode symbol. **ANSWER B.**

E6B07 What causes a junction diode to fail from excessive current?

A. Excessive inverse voltage C. Insufficient forward voltage
B. Excessive junction temperature D. Charge carrier depletion

Guess what will kill most electronic components? High temperature! The limit of the maximum forward current in a junction diode is the junction temperature. This is why you'll find these diodes mounted to the chassis of your equipment. The chassis acts as a heat sink to keep the ***junction temperature*** from ***exceeding its maximum limit***. **ANSWER B.**

E6B08 Which of the following is a Schottky barrier diode?

A. Metal-semiconductor junction C. PIN junction
B. Electrolytic rectifier D. Thermionic emission diode

A Schottky barrier diode is a semiconductor diode which consists of a ***metallic electrode in contact with a bar of semiconductor material***, rather than the standard doped semiconductor "PN" junction. The primitive "cat's whisker" detector used in primitive crystal radios was actually the first Schottky diode. Schottky diodes have much lower forward voltage drop than standard semiconductor diodes, making them more suitable for detector purposes. **ANSWER A.**

E6B09 What is a common use for point-contact diodes?

A. As a constant current source
B. As a constant voltage source
C. As an RF detector
D. As a high-voltage rectifier

Because it only conducts in one direction, the little ***point contact diode*** may be used as an ***RF detector*** in some VHF and UHF equipment. **ANSWER C.**

Figure E6-2

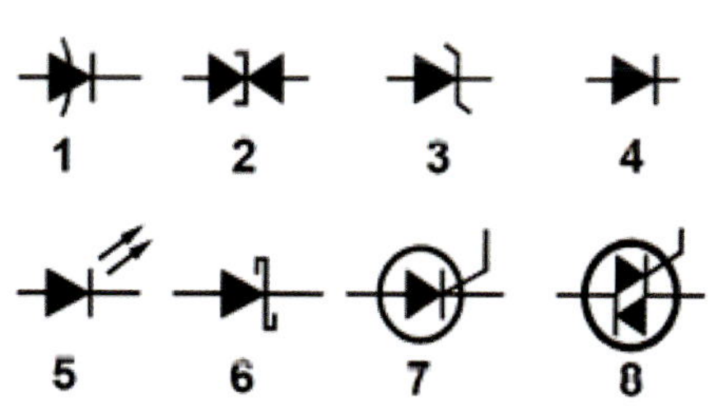

E6A03 Why does a PN-junction diode not conduct current when reverse biased?

A. Only P-type semiconductor material can conduct current
B. Only N-type semiconductor material can conduct current
C. Holes in P-type material and electrons in the N-type material are separated by the applied voltage, widening the depletion region
D. Excess holes in P-type material combine with the electrons in N-type material, converting the entire diode into an insulator

A depletion region is the area of a transistor junction where charge carriers are very few to non-existent. The ***greater the electric field*** applied to this junction, the ***greater the depletion region***. **ANSWER C.**

E6B05 What characteristic of a PIN diode makes it useful as an RF switch?

A. Extremely high reverse breakdown voltage
B. Ability to dissipate large amounts of power
C. Reverse bias controls its forward voltage drop
D. Low junction capacitance

The PIN diode, operating at forward bias, will act as a closed switch. This has replaced older style T/R (transmit/receive) relay switching in many new transceivers. The PIN diode offers ***low junction capacitance*** between the N and P layers, with a lightly doped intrinsic layer insulating between the two layers. **ANSWER D.**

E6B11 What is used to control the attenuation of RF signals by a PIN diode?

A. Forward DC bias current
B. A sub-harmonic pump signal
C. A variable RF reference voltage
D. Capacitance of an RF coupling capacitor

The ***PIN diode*** can not only work as an RF switch, but also as an ***RF attenuator*** if the ***forward DC bias current*** is reduced. With just the right amount of forward bias, the PIN diode will act as a variable resistance to RF signals. **ANSWER A.**

E6F07 What is a solid-state relay?

A. A relay that uses transistors to drive the relay coil
B. A device that uses semiconductors to implement the functions of an electromechanical relay
C. A mechanical relay that latches in the on or off state each time it is pulsed
D. A semiconductor switch that uses a monostable multivibrator circuit

The ***PIN diode,*** sometimes called a ***solid-state relay***, is capable of passing microwave and RF signals when forward biased. The PIN diode is not capable of replacing the high voltage and high current handling properties of mechanical relays in automatic antenna tuners. Electromechanical relays in tuners will stick around for some time to come. **ANSWER B.**

E6C05 Which of the following digital logic families has the lowest power consumption?

A. Schottky TTL
B. ECL
C. NMOS
D. CMOS

While a single digital logic gate consumes a minuscule amount of power, when you put thousands or millions of them together in a large integrated circuit, the power consumption becomes immense. Complementary Metal Oxide Semiconductor technology is the most economical and power wise of the common logic families. **ANSWER D.**

E6C06 Why do CMOS digital integrated circuits have high immunity to noise on the input signal or power supply?

A. Large bypass capacitance is inherent
B. The input switching threshold is about twice the power supply voltage
C. The input switching threshold is about half the power supply voltage
D. Bandwidth is very limited

The ***CMOS*** device is relatively ***immune to noise*** because its ***switching threshold is half the power supply voltage***. As a result, much larger power supply noise transients or input transients will not cause a transition in the input state when compared to other logic circuit types. **ANSWER C.**

E6E04 Which is the most common input and output impedance of MMICs?

A. 50 ohms
B. 300 ohms
C. 450 ohms
D. 75 ohms

The beauty of the ***MMIC*** (Monolithic Microwave Integrated Circuit) is that it has a ***characteristic impedance of 50 ohms***: just what the microwavers love. **ANSWER A.**

Monolithic Microwave Integrated Circuit
Photo Courtesy of Hewlett Packard Co.

E6E03 Which of the following materials supports the highest frequency of operation when used in MMICs?

A. Silicon
B. Silicon nitride
C. Silicon dioxide
D. Gallium nitride

When most people think of semiconductors they think of silicon. While silicon has a lot going for it – it's cheap, abundant, and can handle high temperatures – it is not the world's best semiconductor. Many other elements and compounds are much faster than silicon or have other highly desirable traits, such as exceptionally low noise. A monolithic microwave integrated circuit (***MMIC***) usually uses something other than silicon to do its job; one of the best performers being ***gallium nitride***. **ANSWER D.**

E6E06 What characteristics of MMICs make them a popular choice for VHF through microwave circuits?

A. The ability to retrieve information from a single signal even in the presence of other strong signals
B. Extremely high Q factor and high stability over a wide temperature range
C. Nearly infinite gain, very high input impedance, and very low output impedance
D. Controlled gain, low noise figure, and constant input and output impedance over the specified frequency range

A monolithic microwave integrated circuit (MMIC) is a radio frequency "chip" that is designed for high gain, wide bandwidth, and good stability. The MMIC usually only has an input terminal, an output terminal, and a power terminal, so it often resembles a simple "pill" style RF transistor. However, MMICs are often incorporated into blocks or modules with SMA or other convenient RF connectors. MMICs can be cascaded to provide extremely ***high values of RF gain and provide close to 50 ohm impedance matching at the input and output*** ports across their normal operating range. A complete receiving system can be built by plugging MMICs together with a minimum of external components. **ANSWER D.**

E6E08 How is power supplied to the most common type of MMIC?

A. Through a capacitor and RF choke connected to the amplifier input lead
B. MMICs require no operating bias
C. Through a resistor and/or RF choke connected to the amplifier output lead
D. Directly to the bias voltage (Vcc) lead

Unlike most "normal" Op-Amps, MMICs (Monolithic Microwave Integrated Circuits) do not have separated power pins. An MMIC most commonly has just three terminals: an input, an output, and a ground. DC power is supplied to the output terminal through an external "bias" network, consisting of an external resistor or inductor, or both. **ANSWER C.**

E6E07 What type of transmission line is often used for connections to MMICs?

A. Miniature coax
B. Circular waveguide
C. Parallel wire
D. Microstrip

Just about any passive electrical component (and some active ones, as well) can be constructed using microstrip methods; shaping the traces on a piece of printed circuit board (PCB) to form inductors, capacitors, and transmission lines. The ***MMIC*** (Monolithic Microwave Integrated Circuit) advances microstrip methods to a high art form, creating stable, high-performance, and inexpensive microwave components. **ANSWER D.**

E6E01 Why is gallium arsenide (GaAs) useful for semiconductor devices operating at UHF and higher frequencies?

A. Higher noise figures
B. Higher electron mobility
C. Lower junction voltage drop
D. Lower transconductance

While silicon is the most common element used for semiconductors of all kinds, it is ***not*** the best choice for every solid-state application. Electron movement through Silicon is relatively ***slow*** when compared to its movement in many newer compounds or even much older germanium. A number of ***compound*** semiconductors such as ***Gallium Arsenide*** or Indium Antimonide are ***much faster*** and more suitable for microwave applications. These tend to be a lot more expensive than silicon, however, so don't expect silicon to go away any time soon. **ANSWER B.**

E6C01 What is the function of hysteresis in a comparator?

A. To prevent input noise from causing unstable output signals
B. To allow the comparator to be used with AC input signals
C. To cause the output to continually change states
D. To increase the sensitivity

Hysteresis can be thought of as electrical "stickiness" or friction. It can be designed into a circuit to require a little extra "nudge" to make it change state, thus ***making the circuit less susceptible to random noise*** or other signals. **ANSWER A.**

E6C02 What happens when the level of a comparator's input signal crosses the threshold voltage?

A. The IC input can be damaged
B. The comparator changes its output state
C. The reference level appears at the output
D. The feedback loop becomes unstable

As the name suggests, a comparator compares the input at one terminal to the input at another terminal and makes a concrete decision as to which input is larger. ***When the level of a comparator's input signal crosses the threshold voltage, it changes its output state.*** A comparator is the simplest analog-to-digital converter, sometimes called a "1 bit" converter (which usually costs less than two bits). **ANSWER B.**

E6E10 What advantage does surface-mount technology offer at RF compared to using through-hole components?

A. Smaller circuit area
B. Shorter circuit board traces
C. Components have less parasitic inductance and capacitance
D. All these choices are correct

Admittedly, surface mount components are much more difficult to work with than "old school" through-hole components. Many surface mount components are almost microscopic. However, if you can convince yourself to learn the fine art of surface mount soldering, the benefits are great: ***smaller, higher performance, and more stable circuits,*** especially at higher radio frequencies. Even some of us old geezers have learned to work with the things. **ANSWER D.**

Lead-less components require surface mount soldering skills if you plan to remove and replace a defective component.

E6E09 Which of the following component package types have the least parasitic effects at frequencies above the HF range?

A. TO-220
B. Axial lead
C. Radial lead
D. Surface mount

Surface mount devices can be made extremely small, sometimes bordering on the microscopic. As such, they have very small parasitic inductance making them suitable for VHF and microwave applications. ***Surface mount*** soldering is a skill worth learning. It just takes some practice and a good magnifying glass. **ANSWER D.**

E6E02 Which of the following device packages is a through-hole type?

A. DIP
B. PLCC
C. BGA
D. SOT

DIP (dual in-line package) construction is the traditional form of integrated circuit. The DIP device is mounted on top of a printed circuit board (PCB) with its leads passing through holes to be soldered on the trace side of the board. Individual transistors and other "old school" components are generally ***through-hole construction***. Unfortunately, many parts are becoming more difficult to find as surface mount devices are becoming the standard. It's a good idea to learn how to do surface mount soldering if you plan on doing much home brewing or repairing modern radio gear. **ANSWER A.**

E6E12 Why are DIP through-hole package ICs not typically used at UHF and higher frequencies?

A. Excessive dielectric loss
B. Epoxy coating is conductive above 300 MHz
C. Excessive lead length
D. Unsuitable for combining analog and digital signals

One of the factors limiting the high frequency performance of radio circuits is the stray or "parasitic" inductance of component leads. ***Lead lengths*** were greatly reduced when radio circuits evolved from point-to-point wiring to printed circuit board construction, using standard "through hole" components with somewhat shorter (***DIP***) leads. Lead lengths were even further reduced with the advent of surface mount components, such as chip capacitors, chip resistors, and chip inductors, where the lead lengths are nearly non-existent, allowing even better performance at UHF and higher frequencies. **ANSWER C.**

E6E11 What is a characteristic of DIP packaging used for integrated circuits?

A. Extremely low stray capacitance (dielectrically isolated package)
B. Extremely high resistance between pins (doubly insulated package)
C. Two chips in each package (dual in package)
D. Two rows of connecting pins on opposite sides of package (dual in-line package)

Back before everything became microscopic, most of us semi-old-timers used SIP (Single In-Line Package) or ***DIP*** (Dual In-Line Package) through-hole components for most of our construction projects. These components were named not for the chips they contained, as that could be many, but rather for the ***number of rows of connecting pins*** radiating from the package. DIPs have a row of pins on each side (Dual). SIP and DIP chips are still available, but becoming less standard, as SMD takes over the landscape. (Eric's done everything from point-to-point, pre-circuit- board vacuum tube wiring, to the latest SMD construction. He says it's a good idea to learn it all!) **ANSWER D.**

Logically Speaking of Counters

E6C08 In Figure E6-3, which is the schematic symbol for a NAND gate?

A. 1 C. 3
B. 2 D. 4

Since the ***NAND*** gate has a ***logic "0"*** at its output, spot the little, ***tiny "o" on the nose of the symbol that looks like the letter D***. It looks the same as the AND gate, except for the small circle at its output. **ANSWER B.**

Figure E6-3

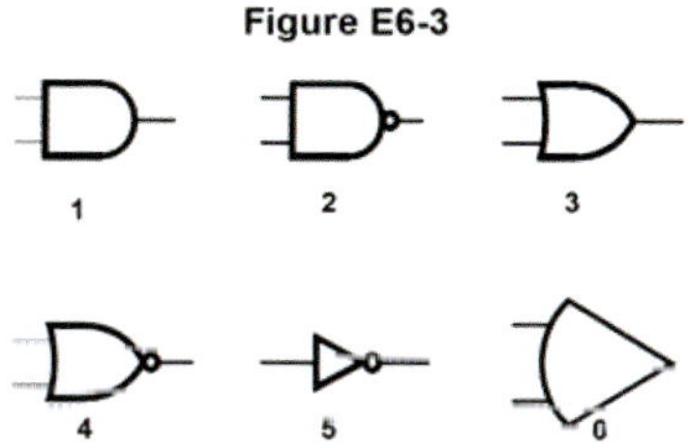

Figure E6-3

E6C10 In Figure E6-3, which is the schematic symbol for a NOR gate?

A. 1 C. 3
B. 2 D. 4

Since the ***NOR gate*** will give us a logic "0" at its output if any inputs are logic "1," look for the little "o" on the nose and a concave input. It looks just like an OR gate, except for the ***small circle on its nose***. **ANSWER D.**

E6C11 In Figure E6-3, which is the schematic symbol for the NOT operation (inversion)?

A. 2 C. 5
B. 4 D. 6

The ***NOT gate is triangular*** in shape and has a ***little "o" on its nose***. It looks altogether different from any of the other logic gates. You'll probably have one question based on gate symbols. Make up flashcards, and see if you can easily call out which is which. **ANSWER C.**

It's Only Logical!

Since in recent years "technology" and "digital" are nearly synonymous, it's easy to overlook just how recent these concepts are. Searching through the ARRL periodical archives, I discovered that the very first mention of digital logic was in the November, 1962 QST article, "Logic for Amateurs," which described some simple decision making circuits using vacuum tubes. Actually, the article never used the term "digital"...that was to appear about six years later in that same venerable publication. Shucks, I was already in high school and building my own stereo systems by that time, which shows us just how recent the concept is...at least in Amateur Radio circles!

How...or why....did digital technology (and thought processes) come to dominate our "electronics" thinking? After all, we'd been doing radio just fine, thank you, for nearly two-thirds of a century before digital and logic were familiar terms.

Well, not exactly. Things like light switches and telegraphs keys and rotary phone dials were around at least as long as radio, and these are all logical devices. If you have a room with a pair of "three way" light switches, you actually have an "exclusive OR" function. If Switch A is up, the light lights; if switch B is up, the light lights; and if BOTH are up, the light stays dark. This is simple combinational logic.

If you have ever used a dial telephone (gasp!) you have worked with sequential logic. The first digit you dial operates one set of 10 contact rotary relays at the phone office, which then transfers you to the next set of 10 contact rotary relays, which transfers you to the next set of 10 contact relays, and so on. What the system does with each digit you dial is dependent on the previous digit you dialed...in other words, you need some kind of memory...in this case, the physical position of each rotary relay is stored. (By the way, if you've never seen mechanical phone relays in action, it's something to behold. There are some great videos of this ancient technology on YouTube!)

Digital truth tables are nothing more than maps that allow you to follow through a maze of open or closed switches to get an electrical voltage to the final destination. Of course, we don't often use mechanical switches for digital logic, as these have been largely replaced by transistors, or other more exotic semiconductors.

Transistors are a lot smaller, quieter, and faster than mechanical switches, but they perform the same task. MOST digital logic gates are of the inverting type, NAND, NOR, because common emitter transistors invert the input and output voltages. This is no mystery, as we've already studied those, right? Transistor switches can be placed in series to form an AND (or NAND) function, while transistor switches in parallel form an OR (or NOR) function. It's that simple! Just plumbing and pipes.

If you have enough time, space, and patience, you can build most digital devices with discrete transistors, and it wasn't too long ago that most electronics students did just that. In high school, I built a 10:1 counter using a pile of "bare" transistors and it actually worked! Nowadays, of course, digital integrated circuits can have thousands or millions of transistors, but they're still nothing but collections of switches.

DIAGRAM SYMBOLS **TRUTH TABLES**

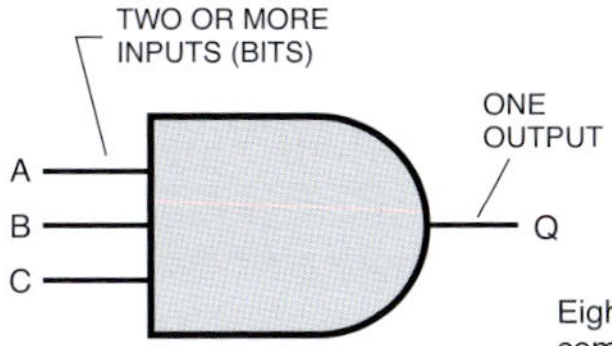

Eight possible combinations of three input bits.

INPUT BITS / RESULTING OUTPUT BIT

A	B	C	Q
0	0	0	0
0	0	1	0
0	1	0	0
0	1	1	0
1	0	0	0
1	0	1	0
1	1	0	0
1	1	1	1

Output of an AND gate is 1 only when *all* inputs are 1; that is, Input A AND Input B AND Input C must be 1 for Output Q to be 1.

a. A 3-Input AND Gate

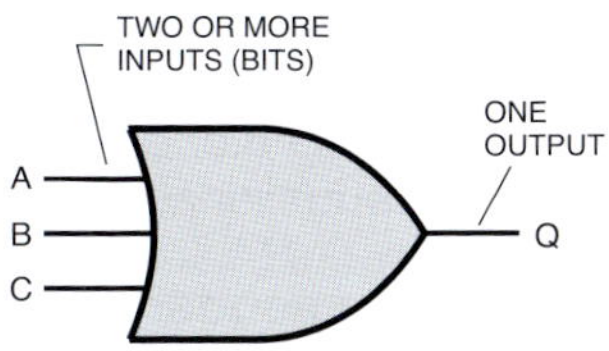

INPUT BITS / RESULTING OUTPUT BIT

A	B	C	Q
0	0	0	0
0	0	1	1
0	1	0	1
0	1	1	1
1	0	0	1
1	0	1	1
1	1	0	1
1	1	1	1

Output of an OR gate is 1 when *at least one* input is 1; that is, Input A OR Input B OR Input C, or any combination of A,B,C, must be 1 for Output Q to be 1.

b. A 3-Input OR Gate

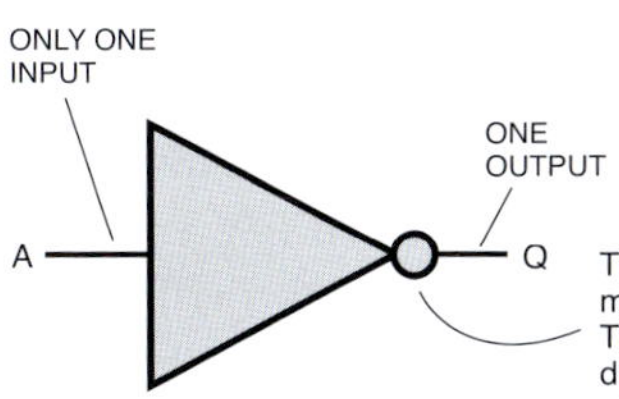

The little circle means *inversion*. This is an inverting digital amplifier.

INPUT BIT / RESULTING OUTPUT BIT

A	Q
0	1
1	0

An inverter's output is always opposite of the input.

c. A NOT Gate Inverter

AND, OR, and NOT Truth Tables

Source: *Basic Electronics* © 1994, Master Publishing, Inc., Arlington, Texas

E7A07 What logical operation does a NAND gate perform?

A. It produces a 0 at its output only if all inputs are 0
B. It produces a 1 at its output only if all inputs are 1
C. It produces a 0 at its output if some but not all inputs are 1
D. It produces a 0 at its output only if all inputs are 1

The word ***NAND*** reminds us of the word "naw", meaning no or nothing. The NAND gate ***produces a logic 0 at its output only if all inputs are logic 1***. **ANSWER D.**

E7A08 What logical operation does an OR gate perform?

A. It produces a 1 at its output if any input is 1
B. It produces a 0 at its output if all inputs are 1
C. It produces a 0 at its output if some but not all inputs are 1
D. It produces a 1 at its output if all inputs are 0

The ***OR*** gate will produce a ***logic 1*** at its output ***if any one or all inputs are a logic 1***. **ANSWER A.**

E7A09 What logical operation is performed by a two-input exclusive NOR gate?

A. It produces a 0 at its output only if all inputs are 0
B. It produces a 1 at its output only if all inputs are 1
C. It produces a 0 at its output if one and only one of its inputs is 1
D. It produces a 1 at its output if one and only one input is 1

Any digital logic gate can be converted to its inverse function by putting an inverter or "NOT" gate at its output. In other words, you can make an AND gate a NAND gate by putting an inverter on its output. You can make an OR gate a NOR gate by putting an inverter on its output. Likewise, you can make an exclusive NOR gate by putting an inverter on the output of an exclusive OR gate. Looking at truth tables, if you were to invert the output states (making all 0s 1s and vice versa) of the exclusive OR gate, you would have the truth table for an exclusive NOR gate. ***The exclusive NOR gate will produce a logic 0 if only one input is a logic 1.*** **ANSWER C.**

E7A10 What is a truth table?

A. A list of inputs and corresponding outputs for an op-amp
B. A list of inputs and corresponding outputs for a digital device
C. A diagram showing logic states when the digital gate output is true
D. A table of logic symbols that indicate the logic states of an op-amp

We use a ***truth table*** in digital circuitry to characterize a digital device's function. It ***lists the inputs and outputs*** for a digital device. If you buy the owner's technical manual for a piece of Amateur Radio gear, you will usually find a page describing the truth tables of the digital devices used in the equipment. **ANSWER B.**

E7A11 What does "positive logic" mean in reference to logic devices?

A. The logic devices have high noise immunity
B. High voltage represents a 1, low voltage a 0
C. The logic circuit is in the "true" condition
D. 1s and 0s are defined as different positive voltage levels

For practical reasons, most simple logic gates are negative logic, which means that a low voltage represents a 1 and a high voltage represents a zero. NAND, NOR, and EXCLUSIVE NOR gates are typical examples. POSITIVE logic, while rarer in in practice, is more intuitive, where a 1 is represented by a voltage presence, and a 0 represented by a voltage absence. Any positive logic device can be converted to a negative logic device by putting a simple inverter (or NOT gate) on the output of the device, or vice versa. **ANSWER B.**

E7A01 Which circuit is bistable?

A. An AND gate
B. An OR gate
C. A flip-flop
D. A bipolar amplifier

In electronics, a ***flip-flop*** circuit has two stable states and can be used to store state information. Bistable stands for ***two stable states***. **ANSWER C.**

E7A03 Which of the following can divide the frequency of a pulse train by 2?

A. An XOR gate
B. A flip-flop
C. An OR gate
D. A multiplexer

A ***flip-flop*** can be used to ***divide a pulse train*** (the frequency of an AC signal). **ANSWER B.**

E7A04 How many flip-flops are required to divide a signal frequency by 16?

A. 4 C. 8
B. 6 D. 16

A standard flip-flop, or ***bistable multivibrator***, is often used as a two-to-one frequency divider. Multiple flip-flops can be cascaded to form frequency division by 2, 4, 8, 16, or any other ***power of two*** (2^x where x is the power of 2 and is also the number of flip flops required). Division by any other factor requires some kind of "Texas Two-step" circuitry, which involves additional gating and recursion. In this example 2^4=16, so ***4*** is the number of flip-flops required. **ANSWER A.**

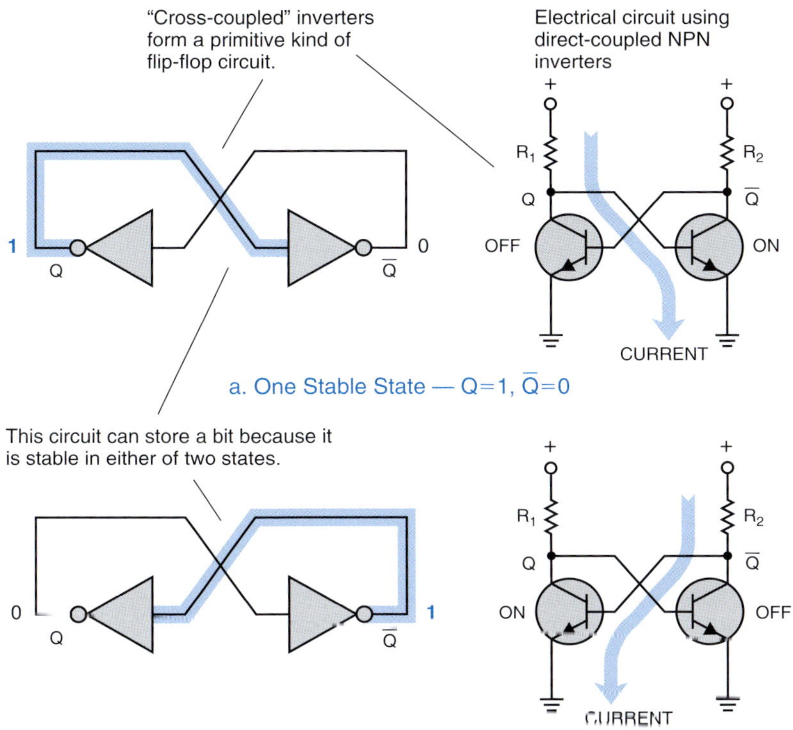

a. One Stable State — Q=1, $\overline{Q}$=0

b. Second Stable State — Q=0, $\overline{Q}$=1

Basic concept of the flip-flop, also called a bistable element or a static memory element.

Source: *Basic Electronics* © 1994, Master Publishing, Inc., Arlington, Texas

E7A05 Which of the following circuits continuously alternates between two states without an external clock signal?

A. Monostable multivibrator C. T flip-flop
B. J-K flip-flop D. Astable multivibrator

An ***astable multivibrator*** continuously switches back and forth between ***two unstable states without an external clock***. **ANSWER D.**

E7A06 What is a characteristic of a monostable multivibrator?

A. It switches temporarily to an alternate state for a set time
B. It produces a continuous square wave
C. It stores one bit of data
D. It maintains a constant output voltage, regardless of variations in the input voltage

This type of ***multivibrator may momentarily be monostable*** (MO MO): it stays in an original state until triggered to its other state, where it remains for a time usually determined by external components, after which it ***returns to the original state***. This is sometimes called a "one-shot" multivibrator. It is often used to create a precise time delay. The extremely popular 555 IC is capable of operating as a one-shot multivibrator. **ANSWER A.**

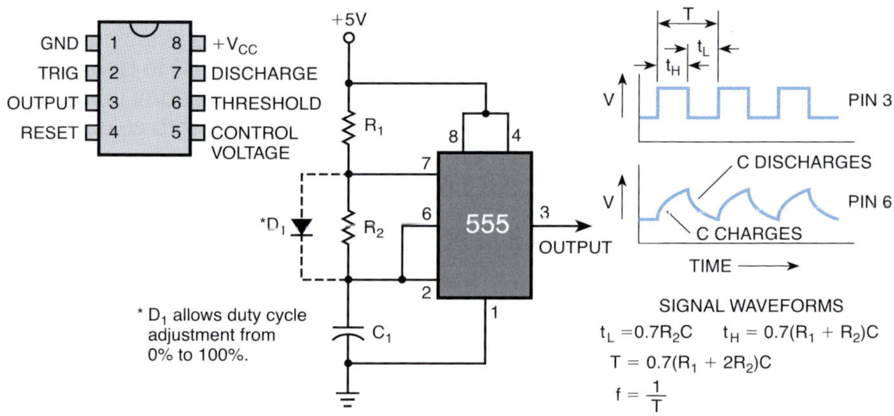

An IC 555 timer can be used as a bistable, astable, or monostable multivibrator. This schematic shows it wired as an astable multivibrator. The resulting astable voltage waveforms at pins 3 and 6 are shown.
Source: *Basic Digital Electronics,* Evans, ©1996, Master Publishing, Inc., Niles, IL

E8A09 How many different input levels can be encoded by an analog-to-digital converter with 8-bit resolution?

A. 8
B. 8 multiplied by the gain of the input amplifier
C. 256 divided by the gain of the input amplifier
D. 256

The number of values that any digital number can represent is calculated by 2^N, where N is the number of bits. ***2^8 is 256***. Keep in mind that the first number is 0! So the actual numbers represented are 0-255. **ANSWER D.**

E6C07 What best describes a pull-up or pull-down resistor?

A. A resistor in a keying circuit used to reduce key clicks
B. A resistor connected to the positive or negative supply used to establish a voltage when an input or output is an open circuit
C. A resistor that ensures that an oscillator frequency does not drift
D. A resistor connected to an op-amp output that prevents signals from exceeding the power supply voltage

Digital logic gates don't generally like inputs "flapping in the breeze." Such a condition can lead to some "interesting" effects, to say the least. Depending on the type of gate, a high resistance ***pull-up or pull-down resistor may be connected to an input gate*** to "tie" that gate to a specific voltage when not being toggled, thus ***preventing false outputs***. Sometimes the pull-up or pull-down resistor is built into the chip, sometimes not. Always check your data sheets! **ANSWER B.**

E6C09 What is used to design the configuration of a field-programmable gate array (FPGA)?

A. Karnaugh maps
B. Hardware description language (HDL)
C. An auto-router
D. Machine and assembly language

The Field Programmable Gate Array (FPGA) can be thought of as an integrated circuit that can be designed and rewired "on the fly." They are used in a large number of very fast processing applications. A special language designed just for programming these jewels is known as HDL for Hardware Description Language. Several Graphic input front ends to HDL exist, making programming an FPGA a snap. Well, a semi-snap, anyway. **ANSWER B.**

This is a UV erasable EEPROM, which makes it possible for a manufacturer to upgrade your rig's receiver to take in new bands, such as those proposed below 500 kHz.

E8A10 What is the purpose of a low-pass filter used at the output of a digital-to-analog converter?

A. Lower the input bandwidth to increase the effective resolution
B. Improve accuracy by removing out-of-sequence codes from the input
C. Remove spurious sampling artifacts from the output signal
D. All these choices are correct

The problem with the ***D-to-A*** converter is ***spurious sampling artifacts***, which are usually easier to deal with using a ***low pass filter***. You don't need the same "brick wall" anti-aliasing filter required in the A-to-D device. **ANSWER C.**

E6C03 What is tri-state logic?

A. Logic devices with 0, 1, and high-impedance output states
B. Logic devices that utilize ternary math
C. Logic with three output impedances which can be selected to better match the load impedance
D. A counter with eight states

Tri-state logic devices can cut down on the number of individual OR gates needed to route data in a bus wire where there may be many "listeners." The tri-state device is identified by one additional control lead. The truth table looks like this for the tri-gate device:

A	Control	B
0	on	0
1	on	1
0	off	high impedance
1	off	high impedance

Tri-state logic devices offer three levels of output: "0," "1," and high impedance. **ANSWER A.**

E6C04 Which of the following is an advantage of BiCMOS logic?

A. Its simplicity results in much less expensive devices than standard CMOS
B. It is immune to electrostatic damage
C. It has the high input impedance of CMOS and the low output impedance of bipolar transistors
D. All these choices are correct

The ***BiCMOS*** device offers a ***high input impedance*** associated with CMOS, yet a ***low output impedance*** that you would find with a bipolar transistor. It also provides increased gain with low power dissipation in heat. The BiCMOS inverter can be used as a buffer to drive large load capacitances offering increased speed in critical design networks. **ANSWER C.**

E4A05 What is the purpose of using a prescaler with a frequency counter?

A. Amplify low-level signals for more accurate counting
B. Multiply a higher frequency signal so a low-frequency counter can display the operating frequency
C. Prevent oscillation in a low-frequency counter circuit
D. Reduce the signal frequency to within the counter's operating range

You will find a prescaler in frequency counter, where it is used to ***reduce the signal frequency within the counter's operating range***. **ANSWER D.**

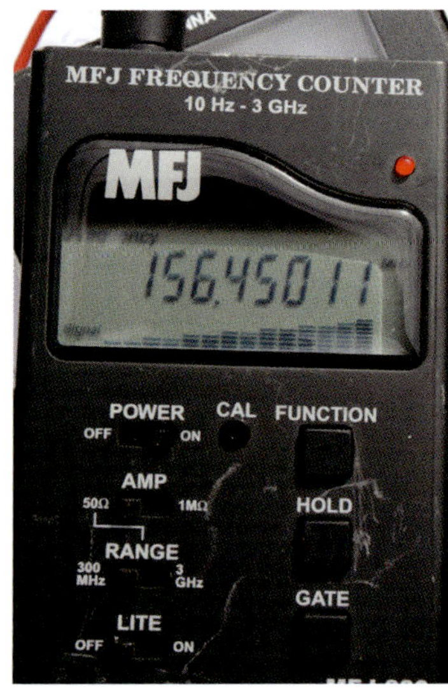

The portable frequency counter is a good way to see what frequency is found on an un-marked land mobile radio that you plan to convert to 2 meters. The counter is reading marine VHF channel 9.

E7A02 What is the function of a decade counter?

A. It produces one output pulse for every 10 input pulses
B. It decodes a decimal number for display on a seven-segment LED display
C. It produces 10 output pulses for every input pulse
D. It decodes a binary number for display on a seven-segment LED display

A ***decade counter*** digital IC gives ***one output pulse for every 10 input pulses***. **ANSWER A.**

E4B01 Which of the following factors most affects the accuracy of a frequency counter?

A. Input attenuator accuracy
B. Time base accuracy
C. Decade divider accuracy
D. Temperature coefficient of the logic

When shopping for a ***frequency counter***, check out the frequency ***time base accuracy***. The stability of the counter will depend on the internal crystal reference. You should regularly check your crystal reference when calibrating transmit equipment for digital operation with no frequency error greater than 5 Hz. **ANSWER B.**

Optos and OpAmps Plus Solar

E7G12 What is an operational amplifier?

A. A high-gain, direct-coupled differential amplifier with very high input impedance and very low output impedance
B. A digital audio amplifier whose characteristics are determined by components external to the amplifier
C. An amplifier used to increase the average output of frequency modulated amateur signals to the legal limit
D. A RF amplifier used in the UHF and microwave regions

The ***operational amplifier (op-amp)*** is a direct-coupled ***differential amplifier*** that offers high gain and high input impedance. The "differential" wording refers to the op-amp input design where the output is determined by the difference in voltage between the two inputs. Different op-amp circuits and characteristics are determined by the external components connected to the amplifier. This integrated circuit op-amp offers high-gain with very ***high input*** impedance and very ***low output impedance***. The gain and frequency response of an op-amp are almost entirely determined by external feedback components, primarily resistors and capacitors. **ANSWER A.**

E7G02 What is the frequency response of the circuit in E7-3 if a capacitor is added across the feedback resistor?

A. High-pass filter
B. Low-pass filter
C. Band-pass filter
D. Notch filter

In an operational amplifier, the gain is determined by the ratio of the feedback resistor to the series input resistor. If a capacitor is paralleled with the feedback resistor, it applies more feedback at higher frequencies as the capacitive reactance decreases. The ***increased feedback at higher frequencies results in a low pass filter*** response. (See p. 195 for full diagram.) **ANSWER B.**

E7G05 How can unwanted ringing and audio instability be prevented in an op-amp audio filter?

A. Restrict both gain and Q
B. Restrict gain but increase Q
C. Restrict Q but increase gain
D. Increase both gain and Q

In order to keep an ***op-amp*** from going into oscillation, ***both the gain and the Q are restricted*** and set by the feedback circuit. **ANSWER A.**

E7G06 What is the gain-bandwidth of an operational amplifier?

A. The maximum frequency for a filter circuit using that type of amplifier
B. The frequency at which the open-loop gain of the amplifier equals one
C. The gain of the amplifier at a filter's cutoff frequency
D. The frequency at which the amplifier's offset voltage is zero

Operational amplifiers (op amps) are almost never used without a significant amount of negative feedback. However, the open-loop gain (the gain without negative feedback) is a useful measure of how well the op-amp will work in the real world. The gain bandwidth product is ***the frequency at which the gain of an op amp drops to 1***, when no inverse feedback is applied. The actual useful bandwidth of the op-amp is increased by the use of inverse feedback, which decreases the *low frequency gain* and "flattening" the overall response. **ANSWER B.**

E7G08 How does the gain of an ideal operational amplifier vary with frequency?

A. It increases linearly with increasing frequency
B. It decreases linearly with increasing frequency
C. It decreases logarithmically with increasing frequency
D. It does not vary with frequency

The ***gain*** of an ideal op-amp ***does not vary with frequency***. Op-amp design is primarily based on the ideal op-amp, where the characteristics are determined only by the feedback components. However, all real world op-amps have an open loop gain bandwidth product (GBP) that limits the ultimate performance of an op-amp. **ANSWER D.**

E7G07 What voltage gain can be expected from the circuit in Figure E7-3 when R1 is 10 ohms and RF is 470 ohms?

A. 0.21
B. 4700
C. 47
D. 24

Take a look at the operational amplifier basics chart on page 192 and remember the gain formula of an inverting IC op-amp:

$$G = -R_F \div R_1 = R_F/R_1.$$

The minus sign means the output is out of phase with the input. R_F = 470 ohms; R_1 = 10 ohms. ***Dividing 10 into 470 yields 47***. On a calculator, the keystrokes are: Clear, 470 ÷ 10 = 47. Not really a brain-buster, was it? **ANSWER C.**

Op-Amp devices.

E7G09 What will be the output voltage of the circuit shown in Figure E7-3 if R1 is 1,000 ohms, RF is 10,000 ohms, and 0.23 volts DC is applied to the input?

A. 0.23 volts
B. 2.3 volts
C. -0.23 volts
D. -2.3 volts

This time we work the formula: $V_{OUT}/V_{IN} = -R_F/R_1$, when $R_1 = 1000$, and $R_F = 10{,}000$, the gain of the op-amp is $V_{OUT}/V_{IN} = -10$. As a result, the ***output = -10 X VIN = -10 X 0.23 = -2.3 volts***. The gain of an op-amp is easy to calculate because it is determined entirely by the external resistors: being the simple quotient of the feedback resistance over the input resistance. Since the output voltage is 180 degrees out of phase with the input voltage, op-amps will be operated in the ***inverting*** mode for stability. A major clue that you're using an inverting amplifier is the minus sign in the terminal into which you're putting the signal! In this case, the gain is 10,000/1000 or 10:1. So any voltage we apply will be amplified by 10. But since it's an ***inverting*** amplifier, we have to use the minus sign. So our signal is ultimately multiplied by -10X the input signal. Nothing to it! **ANSWER D.**

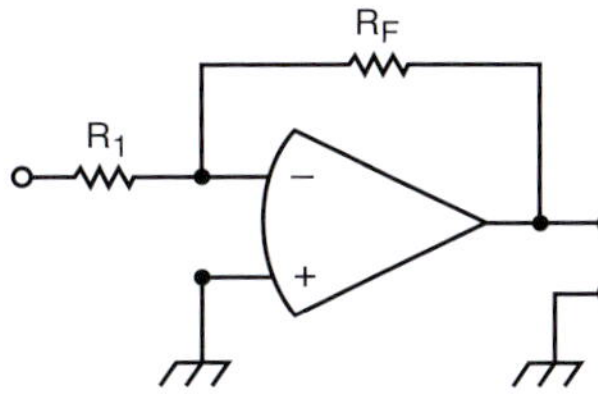

Figure E7-3

E7G10 What absolute voltage gain can be expected from the circuit in Figure E7-3 when R1 is 1,800 ohms and RF is 68 kilohms?

A. 1
B. 0.03
C. 38
D. 76

Now we're back to the gain of an inverting IC operational amplifier. ***Simply divide 1,800 ohms into 68,000 ohms***. Remember, the gain is the quotient of the feedback resistance over the input resistance. Your calculator keystrokes are: clear, 68000 ÷ 1800 = 37.777, rounded off to your ***correct answer of 38***. Simple, huh? **ANSWER C.**

E7G11 What absolute voltage gain can be expected from the circuit in Figure E7-3 when R1 is 3,300 ohms and RF is 47 kilohms?

A. 28
B. 14
C. 7
D. 0.07

One more time, an easy one – ***divide*** the small number ***(3,300) into*** the large number ***(47,000),*** and you end up with 14.24, rounded to the correct answer of ***14***. **ANSWER B.**

Operational Amplifier Basics

Ideal Operational Amplifier

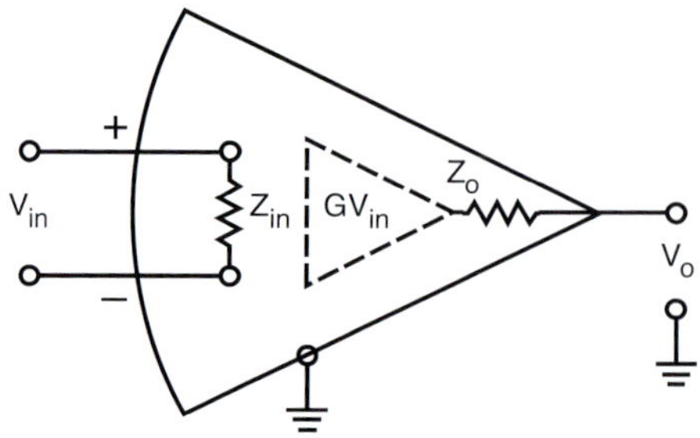

Z_{in} = Infinity

G = Gain = Infinity

Z_o = Zero

Bandwith = Infinity

V_o has no offset
(V_o= 0 when V_{in}= 0)

IC Operational Ampliifer

Even though IC operational amplifiers do not meet all the ideal specifications, G and Z_{in} are very large. Because G is very large, any small input voltage would drive the output into saturation (for practical supply voltages); therefore, normal operational amplifier operation is with feedback to set the gain. Here's an example for an inverting amplifier:

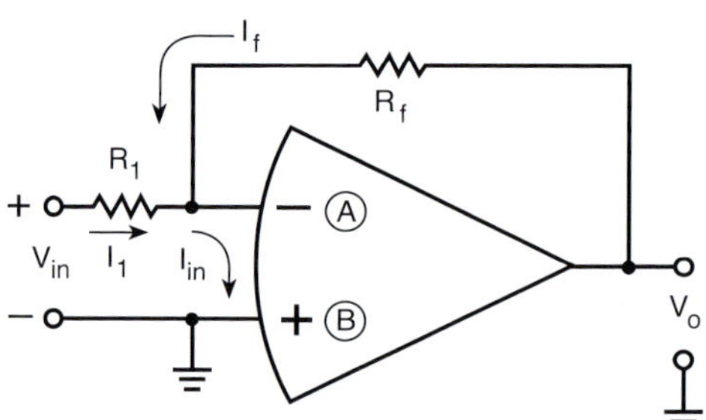

Since Z_{in} is very large, I_{in} = 0

$$\therefore I_1 + I_f = 0 \qquad \therefore \frac{V_{in}}{R_1} + \frac{V_o}{R_f} = 0$$

$$I_1 = \frac{V_{in}}{R_1} \qquad \frac{V_o}{R_f} = -\frac{V_{in}}{R_1}$$

$$I_f = \frac{V_o}{R_f} \qquad \therefore \frac{V_o}{V_{in}} = -\frac{R_f}{R_1}$$

(A) Inverting Input — An input voltage that is more positive on this input will cause the output voltage to be less positive.

(B) Non-Inverting Input — An input voltage that is more positive on this input will cause the output voltage to be more positive.

The gain of an *inverting IC operational amplifier* is:

$$G = -\frac{R_f}{R_1}$$

The minus sign means the output is out of phase with the input.

E7G04 What is meant by the term "op-amp input offset voltage"?

A. The output voltage of the op-amp minus its input voltage
B. The difference between the output voltage of the op-amp and the input voltage required in the immediately following stage
C. The differential input voltage needed to bring the open loop output voltage to zero
D. The potential between the amplifier input terminals of the op-amp in an open loop condition

If the op-amp has been constructed properly, you should see almost ***no voltage between the amplifier input terminals*** when the feedback loop is closed. **ANSWER C.**

E7G03 What is the typical input impedance of an op-amp?

A. 100 ohms
B. 10,000 ohms
C. Very low
D. Very high

Remember that the ***input impedance*** to the ideal ***op-amp*** is always ***Hi***. **ANSWER D.**

E7G01 What is the typical output impedance of an op-amp?

A. Very low
B. Very high
C. 100 ohms
D. 10,000 ohms

Remember that the ***output impedance*** of the ideal op-amp is always ***low out***. **ANSWER A.**

E6F01 What absorbs the energy from light falling on a photovoltaic cell?

A. Protons
B. Photons
C. Electrons
D. Holes

Photon energy from sunlight is absorbed by the electrons in the photovoltaic cell thus creating voltage output from sunlight. When sunlight photons strike the panel, an ***electron will absorb the energy*** and develop a new energy level, ***creating the photovoltaic process***. **ANSWER C.**

E6F02 What happens to photoconductive material when light shines on it?

A. Resistance decreases
B. Resistance increases
C. Reflectivity increases
D. Reflectivity decreases

When a solar panel's photoconductive material is exposed to sunlight, resistance decreases, allowing the solar panel's voltage and current to charge up your battery system. **ANSWER A.**

E6F03 What is the most common configuration of an optoisolator or optocoupler?

A. A lens and a photomultiplier
B. A frequency-modulated helium-neon laser
C. An amplitude-modulated helium-neon laser
D. An LED and a phototransistor

Optocouplers are found in many modern, high-frequency transceivers. When you spin the dial to tune, you are not turning a giant capacitor, but rather interrupting an ***LED*** light beam on a ***phototransistor***. **ANSWER D.**

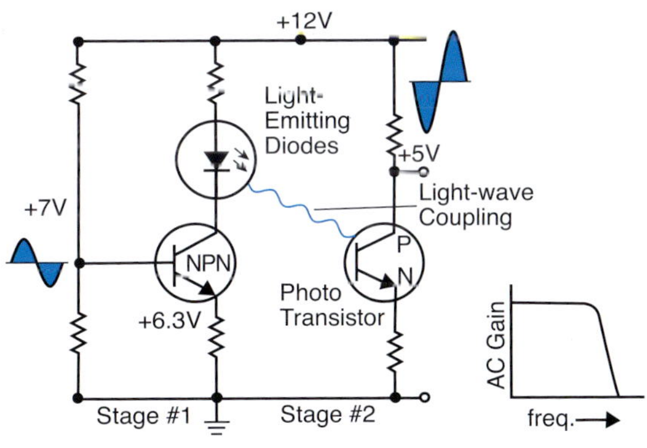

Optical Coupling

E6F08 Why are optoisolators often used in conjunction with solid-state circuits that control 120 VAC circuits?

A. Optoisolators provide a low-impedance link between a control circuit and a power circuit
B. Optoisolators provide impedance matching between the control circuit and power circuit
C. Optoisolators provide an electrical isolation between a control circuit and the circuit being switched
D. Optoisolators eliminate the effects of reflected light in the control circuit

The chief advantage of the ***optoisolator*** at household voltages is to provide a ***very high degree of electrical isolation between*** the control circuit and the house power circuit. Look for the answer *isolation* to agree with the optoisolator device. **ANSWER C.**

E6F05 Which of the following describes an optical shaft encoder?

A. A device that detects rotation by interrupting a light source with a patterned wheel
B. A device that measures the strength of a beam of light using analog-to-digital conversion
C. An optical computing device in which light is coupled between devices by fiber optics
D. A device for generating RTTY signals by means of a rotating light source

When you turn the big VFO knob of that new high-frequency transceiver you awarded to yourself for passing the Extra Class exam, an ***LED light source shining through*** clear ***patterns on a wheel*** will mimic rotation of a control while a logic circuit counts the number of light pulses making it through to the ***optical shaft encoder***. **ANSWER A.**

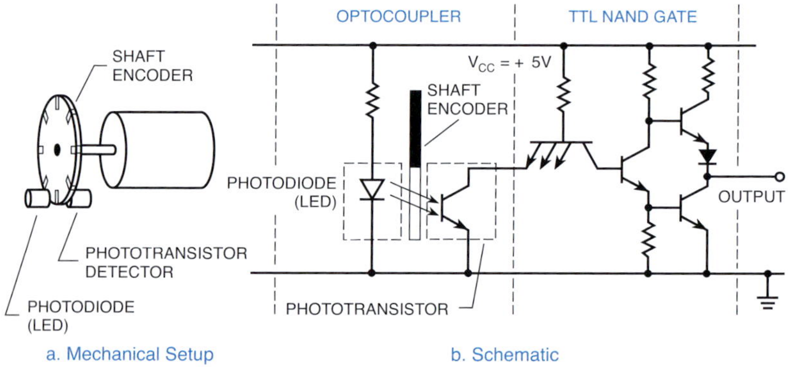

Optocoupler Used for Shaft Encoder

E6F04 What is the photovoltaic effect?

A. The conversion of voltage to current when exposed to light.
B. The conversion of light to electrical energy.
C. The conversion of electrical energy to mechanical energy.
D. The tendency of a battery to discharge when exposed to light.

The term ***photovoltaic*** refers to ***light photons being converted to electrical energy*** such as in solar panels. **ANSWER B.**

E6F09 What is the efficiency of a photovoltaic cell?

A. The output RF power divided by the input DC power
B. The output in lumens divided by the input power in watts
C. The open-circuit voltage divided by the short-circuit current under full illumination
D. The relative fraction of light that is converted to current

Lead sulfide offers good efficiency in a photovoltaic cell – lead takes in the infrared and cadmium takes in visible light. ***Solar cell efficiency is the relative fraction of light that ultimately gets converted to current***. **ANSWER D.**

In full sunlight these four solar panels, which are wired in parallel, can produce better than 10 amps 12 VDC charging power.

E7D09 How is battery operating time calculated?

A. Average current divided by capacity in amp-hours
B. Average current divided by internal resistance
C. Capacity in amp-hours divided by average current
D. Internal resistance divided by average current

Energy is equal to ***POWER*** times ***TIME*** and is measured in ***WATT-HOURS***. Since power is equal to volts times amps, as long as the voltage is constant, we can determine our power consumption by monitoring the amperage. Or simply stated, watt-hours are just ampere-hours times voltage. If you know what the ampere-hour rating of a battery is, it's easy to figure out how long it will last; simply divide that figure by the average current draw. Keep in mind that ***PEAK*** current during transmitting can be many times the average current, averaged over many hours. **ANSWER C.**

E7D11 What is the purpose of an inverter connected to a solar panel output?

A. Reduce AC ripple on the output
B. Maintain voltage with varying illumination levels
C. Prevent discharge when panel is not illuminated
D. Convert the panel's output from DC to AC

A solar panel generates DC power, which may or may not be directly usable by a ham station. (Direct panel-to-rig solar operation can be practical for QRP/ portable operation). In most cases however, we want to convert DC from a solar panel to 120 V AC 60 Hz for use with most common appliances. An ***inverter*** is the device that does this job by "chopping up" the DC voltage and then applying the chopped voltage to a ***step-up transformer***. Widely varying levels of sophistication are used in such inverters for different levels of AC purity and efficiency. **ANSWER D.**

E6F06 Which of these materials is most commonly used to create photoconductive devices?

A. Polyphenol acetate
B. Argon
C. Crystalline semiconductor
D. All these choices are correct

The ***crystalline semiconductor*** is found inside the optical shaft encoder mechanism inside your transceiver. It might also be found outside your ham shack as part of your solar-charging system. **ANSWER C.**

Resistance Variation of a Photodiode

Light Incidence	Current I	Resistance	Test Conditions
Dark	5 nA	2000 MΩ	$V_R = 10$ V $^*E_e = 0$
Light	15 μA	666.7 kΩ	$V_R = 10$ V $E_e = 250$ μW/cm^2 at 940 nm

*Irradiance (E_e) is the radiant power per unit area incident on surface

E6F11 What is the approximate open-circuit voltage produced by a fully illuminated silicon photovoltaic cell?

A. 0.5 volts
B. 0.7 volts
C. 1.1 volts
D. 1.5 volts

Photovoltaic cells, commonly referred to as solar cells, produce a voltage which is proportional to the amount of light that falls upon them...up to a point. A silicon photocell's ***output voltage tops out at about half a volt***. For most solar-powered applications, multiple photocells are wired in series to obtain practical voltages for charging storage batteries. **ANSWER A.**

E6F10 What is the most common material used in power-generating photovoltaic cells?

A. Selenium
B. Silicon
C. Cadmium sulfide
D. Indium arsenide

The photovoltaic material of a ***solar cell*** is made up of ***silicon wafers***, which offer relatively high voltage output and durability for the many years it will be out in the sunlight. **ANSWER B.**

Test Gear, Testing, Testing 1, 2, 3

E4A02 Which of the following parameters does a spectrum analyzer display on the vertical and horizontal axes?

A. Signal amplitude and time
B. Signal amplitude and frequency
C. SWR and frequency
D. SWR and time

A spectrum analyzer resembles an oscilloscope, but instead of the ***horizontal axis*** representing time, it ***represents frequency***, while the ***vertical axis represents signal amplitude***. A spectrum analyzer can take several forms, ranging from a simple "panadapter" used for checking band activity to a precise measuring instrument for alignment and troubleshooting of radio equipment of all kinds. The waterfall display of a sound card digital software package is an audio frequency spectrum analyzer. **ANSWER B.**

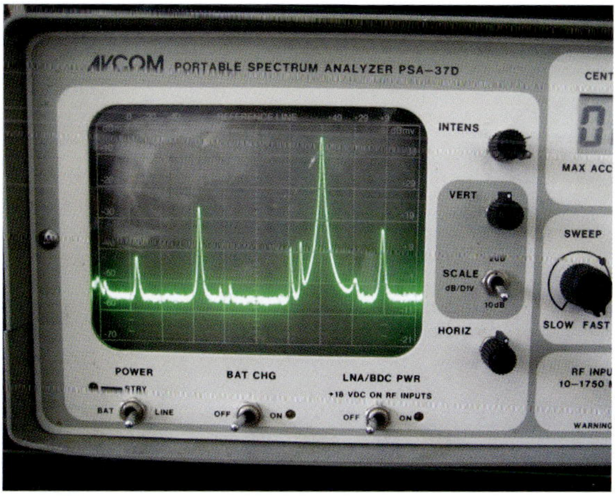

This analog CRT spectrum analyzer is battery operated, but the CRT really eats up the juice!

E4A03 Which of the following test instruments is used to display spurious signals and/or intermodulation distortion products generated by an SSB transmitter?

A. Differential resolver
B. Spectrum analyzer
C. Logic analyzer
D. Network analyzer

A ***spectrum analyzer*** is the most convenient way to check your transmitter's RF emissions. While expensive, a well-equipped radio shop will have a spectrum analyzer. But it's not the only way to check your on-the-air signal. A general coverage receiver can be a valuable and much less expensive piece of test equipment. No ham shack should be without one. **ANSWER B.**

E4B10 Which of the following methods measures intermodulation distortion in an SSB transmitter?

A. Modulate the transmitter using two RF signals having non-harmonically related frequencies and observe the RF output with a spectrum analyzer
B. Modulate the transmitter using two AF signals having non-harmonically related frequencies and observe the RF output with a spectrum analyzer
C. Modulate the transmitter using two AF signals having harmonically related frequencies and observe the RF output with a peak reading wattmeter
D. Modulate the transmitter using two RF signals having harmonically related frequencies and observe the RF output with a logic analyzer

When looking at your signal on a spectrum analyzer, a whistle into the microphone is not a good test for transmitter signal distortion and intermodulation. Using ***two non-harmonically related AF signals*** from a two tone generator is a much better way to observe the characteristics of the signal on your spectrum analyzer. **ANSWER B.**

E4B02 What is the significance of voltmeter sensitivity expressed in ohms per volt?

A. The full scale reading of the voltmeter multiplied by its ohms per volt rating is the input impedance of the voltmeter
B. The reading in volts multiplied by the ohms per volt rating will determine the power drawn by the device under test
C. The reading in ohms divided by the ohms per volt rating will determine the voltage applied to the circuit
D. The full scale reading in amps divided by ohms per volt rating will determine the size of shunt needed

When measuring anything in a circuit, it is important that the measuring instrument cause as little change to the circuit as possible. Since a voltmeter is placed across a circuit (connected between two points in that circuit), it acts as a resistance in parallel with that part of the circuit. In order to cause the least change, the meter should be as high a resistance as possible. Further, ***by multiplying the ohms per volt rating times the maximum voltage on a given scale of the meter, you can determine the equivalent resistance your meter will look like to the circuit***. **ANSWER A.**

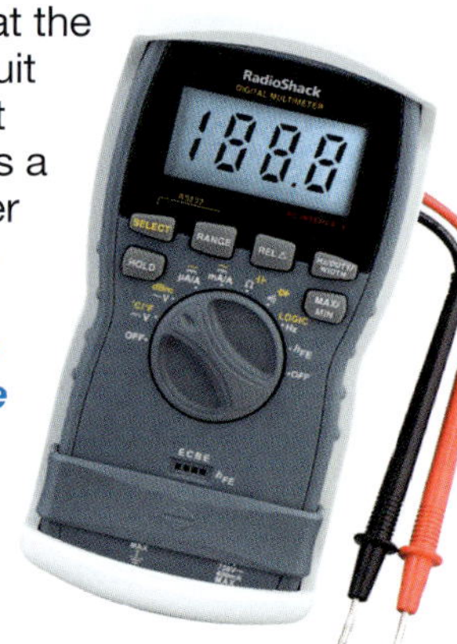

Digital Multitester

E4B06 How much power is being absorbed by the load when a directional power meter connected between a transmitter and a terminating load reads 100 watts forward power and 25 watts reflected power?

A. 100 watts
B. 125 watts
C. 112.5 watts
D. 75 watts

A directional wattmeter is a valuable instrument for every radio amateur. While functionally similar to an SWR meter, the directional wattmeter gives you actual forward and reflected power and is more revealing in many cases. The actual power delivered to a load (which may be an antenna or some intermediate device, such as an amplifier) is always the forward power minus the reflected power; in the case of this question, ***75 watts***. While it's usually best to adjust an antenna for zero reflected power, it's not always convenient or possible. So, one of the best investments you can make is in a good low-loss antenna tuner in conjunction with low-loss transmission line. By means of conjugate matching the antenna tuner causes any reflected power to be re-reflected and added back to the original forward power, assuring that all the power is eventually delivered to the antenna. Required reading for every radio amateur is Maxwell's *Reflections II*. **ANSWER D.**

E4A11 Which of the following can be measured with an antenna analyzer?

A. Velocity factor
B. Cable length
C. Resonant frequency of a tuned circuit
D. All these choices are correct

Most of us old timers wish we had something like a modern antenna analyzer when we were entering the hobby. The modern antenna analyzer has features we could only dream of a few years ago, replacing a whole bench full of instruments like grid dip oscillators, signal generators, and impedance bridges. One handy variation of the antenna analyzer is the ***Vector Network Analyzer (VNA)***, which nowadays is so inexpensive and powerful that there is no reason for any ham not to have one. **ANSWER D.**

E4A01 Which of the following limits the highest frequency signal that can be accurately displayed on a digital oscilloscope?

A. Sampling rate of the analog-to-digital converter
B. Analog-to-digital converter reference frequency
C. Q of the circuit
D. All these choices are correct

The well-known Nyquist Sampling Theorem tells us that in order to accurately sample and digitize an analog waveform, you need to take two samples per cycle of the ***highest*** frequency present in the analog signal. What this means is that if you want to sample a 100 MHz analog signal, ***the sampling rate of the A to D converter*** needs to be 200 Ms/s (200 million samples per second) at a *bare minimum.* It is preferable to have the sampling rate much greater than this, however. **ANSWER A.**

E4A09 Which of the following is good practice when using an oscilloscope probe?

A. Minimize the length of the probe's ground connection
B. Never use a high-impedance probe to measure a low-impedance circuit
C. Never use a DC-coupled probe to measure an AC circuit
D. All these choices are correct

When working on radio equipment using an ***oscilloscope probe***, the ***probe's*** signal ground lead should be ***kept as short as possible*** and clipped to the ground chassis near where you are testing. The short ground lead will minimize stray pickup from other stages of the radio. **ANSWER A.**

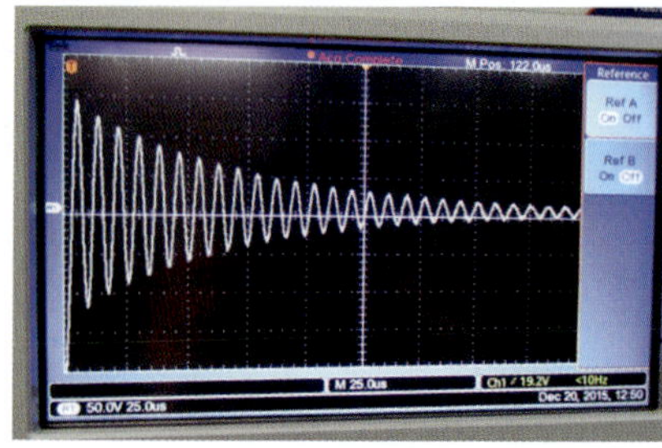

Damped oscillation seen on a digital oscilloscope.

When working on sensitive component equipment, make sure YOU have a grounded body connection, and that the scope ground lead is tied into a close-by ground bus, with the shortest lead length possible.

E4A04 How is the compensation of an oscilloscope probe performed?

A. A square wave is displayed, and the probe is adjusted until the horizontal portions of the displayed wave are as nearly flat as possible
B. A high frequency sine wave is displayed, and the probe is adjusted for maximum amplitude
C. A frequency standard is displayed, and the probe is adjusted until the deflection time is accurate
D. A DC voltage standard is displayed, and the probe is adjusted until the displayed voltage is accurate

If you just purchased a new probe for the oscilloscope you regularly use when working on your ham radio equipment, use your ***square wave*** generator and adjust the horizontal portion of the displayed wave ***to be as flat as possible***. **ANSWER A.**

E4A06 What is the effect of aliasing on a digital oscilloscope when displaying a waveform?

A. A false, jittery low-frequency version of the waveform is displayed
B. The waveform DC offset will be inaccurate
C. Calibration of the vertical scale is no longer valid
D. Excessive blanking occurs, which prevents display of the waveform

Any digital sampling system is capable of generating "alias" signals, and the Digital Storage Oscilloscope is no exception. You can blame Nyquist's Sampling Theorem for the inevitable existence of aliases. There are a number of ways to identify and avoid ***false, jittery, low-frequency alias waveforms***. The operator's manual for your DSO will show combinations of settings and frequencies to avoid to side-step aliasing. **ANSWER A.**

When upgrading to a digital scope, be aware that you will need to read the manual and learn to recognize tiny "alias" signals that are not part of the real signal you are measuring.

E4B08 Which of the following can be used to measure the Q of a series-tuned circuit?

A. The ratio of inductive reactance to capacitive reactance
B. The frequency shift
C. The bandwidth of the circuit's frequency response
D. The resonant frequency of the circuit

In an antenna just as in a series-tuned circuit, the higher the Q the narrower the bandwidth of frequencies it will pass. A high-frequency mobile whip antenna with a small helical-coil has a relatively low Q, offering good ***bandwidth***, but with less efficiency. **ANSWER C.**

E4B09 Which of the following can be measured by a two-port vector network analyzer?

A. Phase noise
B. Filter frequency response
C. Pulse rise time
D. Forward power

Probably 90% of the time, hams who use Vector Network Analyzers (VNAs) use them in the one-port mode, essentially as an antenna analyzer. But the VNA can do a lot more, and one of the more common 2-port functions of the VNA is measuring filter response. When used in this manner, the VNA can plot both the amplitude and the phase response of the filter inserted between the two ports. As with any VNA measurement, be sure you perform the complete calibration before you start; the ***2-port*** calibration has a couple of extra steps beyond the normal "three point" calibration needed for one-port measurements. **ANSWER B.**

The R.F. ammeter is a severely underutilized instrument in present day amateur radio. In the early days of ham radio, it was the only piece of test equipment most hams had at their disposal. It can tell you more about what's really happening in an antenna system than just about any other piece of test equipment even to this day.

E4A10 Which trigger mode is most effective when using an oscilloscope to measure a linear power supply's output ripple?

A. Single-shot
B. Edge
C. Level
D. Line

Learning how to properly trigger an oscilloscope to look at fleeting or complex electrical signals can be a bit of an art form. However, modern oscilloscopes are far simpler to use in this respect than those of just a few years back. There are many cases where a signal is some harmonic of the 60 Hz power line frequency, such as the inevitable ripple on linear power supply, no matter how small. Almost every oscilloscope has a means of synchronizing the sweep to 60 Hz line frequency (or 50 Hz in some parts of the world). Line triggering was also very commonly used to diagnose analog TV signals, such as NTSC, but of course, this is pretty rare these days. **ANSWER D.**

E4B07 What do the subscripts of S parameters represent?

A. The port or ports at which measurements are made
B. The relative time between measurements
C. Relative quality of the data
D. Frequency order of the measurements

S parameter notation can seem like Klingon if you're used to "normal" (lumped constant) circuit analysis, but is extremely revealing, as well as a whole lot of fun, once you get used to it. In S parameter notation, the first subscript ***identifies the port*** we're looking at for a response, and the second subscript is the port through which the "stimulus" is transmitted. Now that more hams are using network analyzers, S parameter notation is becoming more common. **ANSWER A.**

E4B03 Which S parameter is equivalent to forward gain?

A. S11
B. S12
C. S21
D. S22

In S parameter notation, the first subscript identifies the port we're looking at for some response, and the second subscript is the "stimulus" which causes the response. So to translate ***S21*** into English: Let's look at port 2 and see what happens here when we do something at port 1. This sounds like an ***amplifier***, doesn't it? What does the output of an amplifier (port 2) look like when we put a signal into the input (Port 1)? **ANSWER C.**

E4B04 Which S parameter represents input port return loss or reflection coefficient (equivalent to VSWR)?

A. S11
B. S12
C. S21
D. S22

In S parameter notation, ***S11 describes a one-port network***. We're looking at the same port for both a stimulus and a response. In a typical ***VSWR*** measurement, we are looking at a reflected wave at the same port to which we're applying a forward wave. A network analyzer is capable of dealing with both one-port and two-port networks. **ANSWER A.**

E4B05 What three test loads are used to calibrate an RF vector network analyzer?

A. 50 ohms, 75 ohms, and 90 ohms
B. Short circuit, open circuit, and 50 ohms
C. Short circuit, open circuit, and resonant circuit
D. 50 ohms through 1/8 wavelength, 1/4 wavelength, and 1/2 wavelength of coaxial cable

One of the neat things about a vector network analyzer (VNA) is that it will tell you the impedance at the far end of a transmission line of any arbitrary length, provided you perform the proper calibration procedure. This is extremely useful if you don't care to lug a network analyzer up the tower with you! You can make all the measurements on the ground. ***The VNA has three calibration points: shorted, open, and terminated (50 ohms)***. Perform these calibrations at the load end of the transmission line you intend to use. It will automatically compensate for the length and loss of the transmission line. Then connect the transmission line to your antenna and take the readings from your VNA in the comfort of your shack. They will be accurate! **ANSWER B.**

E4B11 Which of the following can be measured with a vector network analyzer?

A. Input impedance
B. Output impedance
C. Reflection coefficient
D. All these choices are correct

The vector network analyzer (VNA) used to be a very exotic and expensive instrument, usually costing many tens of thousands of dollars. Nowadays, high quality network analyzers are affordable for most radio amateurs. The ***VNA*** can be thought of as an "electronic Smith Chart" ***capable of analyzing and displaying anything*** having to do with transmission lines and complex impedances. **ANSWER D.**

E8C13 What is described by the constellation diagram of a QAM or QPSK signal?

A. How many carriers may be present at the same time
B. The possible phase and amplitude states for each symbol
C. Frequency response of the signal stream
D. The number of bits used for error correction in the protocol

Many digital modes employ both amplitude and phase modulation of one or more baseband carriers or "tones." The health of such a digital signal is best determined by looking at its ***constellation*** on an X-Y display oscilloscope, usually fitted with a proper ***graticule*** designed for this purpose. The digital bits will appear as dots properly centered in their quadrants when all is well. Defects in the transmission path can result in "smearing" of the dots in the display, or appearing in the wrong locations. **ANSWER B.**

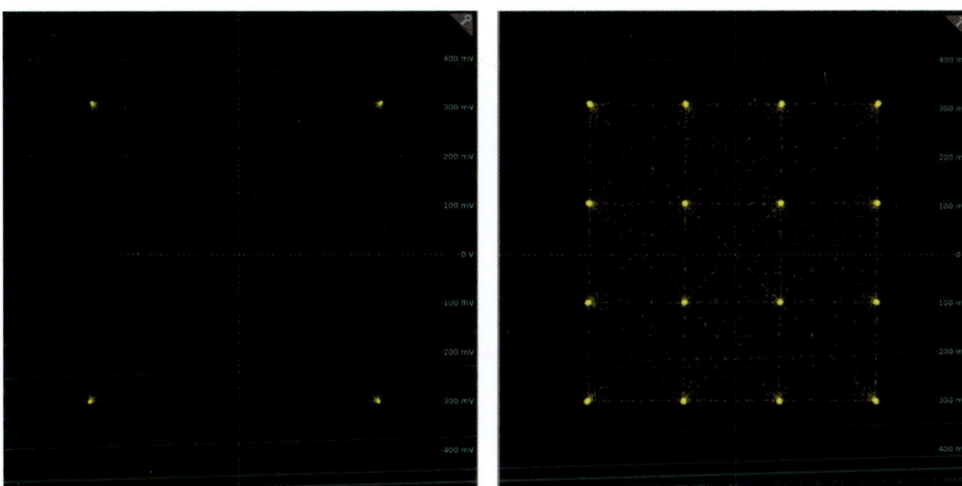

Examples of quadrature phase shift keying or QPSK (four dots) and 16 quadrature amplitude modulation or 16QAM (16 dots), on an oscilloscope. Courtesy of Alan Wolke, W2AEW.

Antennas

E9A03 What term describing total radiated power takes into account all gains and losses?

A. Power factor
B. Half-power bandwidth
C. Effective radiated power
D. Apparent power

The actual radiated field from an antenna is a function of antenna ***gain***, as well as accumulated ***losses***, such as that incurred in long runs of transmission line. An antenna achieves gain by concentrating the radiated field in one (or perhaps a couple) of specific directions. Antenna gain is compared to that of an isotropic radiator, which radiates equally in all directions. It is the ***Effective Radiated Power (ERP)*** that determines the actual effectiveness (field strength) of an amateur radio transmission in a desired direction. **ANSWER C.**

E9A04 Which of the following factors affect the feed point impedance of an antenna?

A. Transmission line length
B. Antenna height
C. The settings of an antenna tuner at the transmitter
D. The input power level

You're thinking about building a new beam. Be sure to consider the three largest factors that will affect the ***feed point impedance*** of that antenna. This includes getting it more than a half wavelength in ***height above the ground***. Element length to diameter ratio, and how close it is to a nearby conductive object (like another band beam below it) also affect feed point impedance. **ANSWER B.**

Multiple antennas on the same mast can influence the impedance of all the antennas in the stack.

E9A09 What is antenna efficiency?

A. Radiation resistance divided by transmission resistance
B. Radiation resistance divided by total resistance
C. Total resistance divided by radiation resistance
D. Effective radiated power divided by transmitter output

The efficiency of your ***antenna*** is an important consideration when you plan to upgrade your antenna system. ***Efficiency equals the radiation resistance divided by the total resistance.*** You want to keep your total (ohmic) resistance as low as possible to make your antenna more efficient. Generally, bigger antenna elements have greater radiation resistance, lower ohmic resistance, higher efficiency, and increased bandwidth. Bigger is always better! **ANSWER B.**

E9A10 Which of the following improves the efficiency of a ground-mounted quarter-wave vertical antenna?

A. Installing a ground radial system
B. Isolating the coax shield from ground
C. Shortening the radiating element
D. All these choices are correct

Most ***vertical antennas*** are one-quarter wavelength long and use a mirror image counterpoise to compare to that of a halfwave antenna. A good ***ground radial system*** will give your signal more punch, lower the noise floor, lower your take-off angle, and give you good DX. **ANSWER A.**

As soon as Jim Ford, N6JF, ran 4 radials per band, ERP was much improved for some fun QRP contacts. Yes, a wet lawn will help, too.

E9A11 Which of the following determines ground losses for a ground-mounted vertical antenna operating on HF?

A. The standing wave ratio
B. Distance from the transmitter
C. Soil conductivity
D. Take-off angle

Dirt is ground, right? Likely not – dry ***soil*** is not nearly as effective a ***conductor*** as 4 quarter-wavelength radials per band on that new HF vertical antenna in your back yard. Looking to really increase your 5-band trap vertical performance on HF? Replace those quarter wavelength ground wires with 3-inch-wide copper foil strips, 4 per band, each a quarter wavelength long. **ANSWER C.**

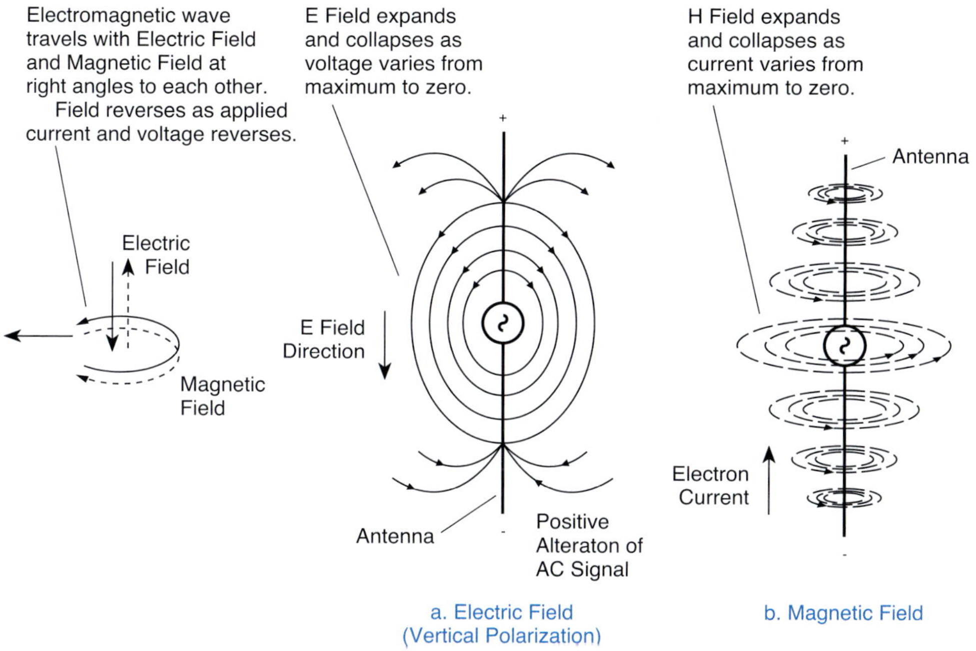

Propagating Electromagnetic Waves

Source: *Antennas – Selection and Installation*, A.J. Evans, Copyright ©1986 Master Publishing, Inc., Arlington, Texas

E9C11 How is the far-field elevation pattern of a vertically polarized antenna affected by being mounted over seawater versus soil?

A. Radiation at low angles decreases
B. Additional lobes appear at higher elevation angle
C. Separate elevation lobes will combine into a single lobe
D. Radiation at low angles increases

Here we go – mount it over ***salt water*** and achieve a desired increase in radiation at low angles, resulting in longer range skywave signals. **ANSWER D.**

E0A01 What is the primary function of an external earth connection or ground rod?

A. Prevent static build up on power lines
B. Lightning charge dissipation
C. Reduce RF current flow between pieces of equipment
D. Protect breaker panel from power surges

While there isn't much you can do if a lightning bolt has your number on it, you can reduce the possibility of it doing too much damage if you direct the charge to ground outside your radio shack. There is no silver bullet when it comes to effectively grounding your station; you may have to decide which is more important for your situation, ***lightning charge dissipation***, RF suppression, or electrostatic discharge protection. **ANSWER B.**

E9C01 What type of radiation pattern is created by two 1/4-wavelength vertical antennas spaced 1/2-wavelength apart and fed 180 degrees out of phase?

A. Cardioid
B. Omni-directional
C. A figure-eight broadside to the axis of the array
D. A figure-eight oriented along the axis of the array

An array of two quarter-wave vertical antennas spaced one-half wavelength apart and fed 180 degrees out of phase results in a ***figure-8 radiation pattern*** with best radiation ***in line with the antennas***. **ANSWER D.**

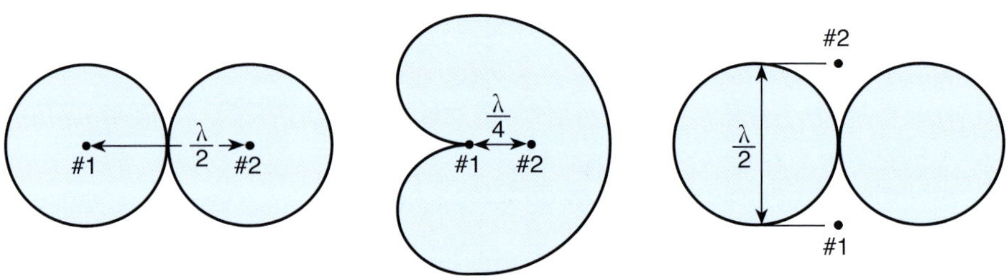

a. Two λ/4 Verticals Fed 180° Out of Phase (Figure 8 End-Fire In-Line)

b. Two λ/4 Verticals Fed 90° Out of Phase (Cardioid)

c. Two λ/4 Verticals Fed In Phase (Figure 8 Broadside)

Antenna Radiation Patterns

E9C02 What type of radiation pattern is created by two 1/4 wavelength vertical antennas spaced 1/4 wavelength apart and fed 90 degrees out of phase?

A. Cardioid
B. A figure-eight end-fire along the axis of the array
C. A figure-eight broadside to the axis of the array
D. Omni-directional

An array of ***two quarter-wave vertical antennas*** spaced one-quarter wavelength apart and ***fed 90 degrees out of phase*** results in a ***cardioid pattern***. This cardioid or unidirectional pattern is needed for direction finding. **ANSWER A.**

E9C03 What type of radiation pattern is created by two 1/4-wavelength vertical antennas spaced 1/2 wavelength apart and fed in phase?

A. Omni-directional
B. Cardioid
C. A figure-eight broadside to the axis of the array
D. A figure-eight end-fire along the axis of the array

An array of two quarter-wave vertical antennas spaced ***one-half wavelength apart*** and ***fed in phase*** will also result in a ***figure-8 pattern***, but maximum gain will be ***broadside to the antennas***. **ANSWER C.**

E9C04 What happens to the radiation pattern of an unterminated long wire antenna as the wire length is increased?

A. Fewer lobes form with the major lobes increasing closer to broadside to the wire
B. Additional lobes form with major lobes increasingly aligned with the axis of the antenna
C. The elevation angle increases, and the front-to-rear ratio decreases
D. The elevation angle increases, while the front-to-rear ratio is unaffected

If you have an infinite amount of property, a long wire antenna can give you an infinite amount of gain! However, the point of diminishing returns is usually reached once a long wire antenna exceeds six or seven wavelengths. The major lobe of the radiation pattern of a long wire antenna tends to follow the direction of the wire. The longer the wire is (in terms of wavelength), the closer the ***major lobes are increasingly aligned with the axis of the antenna wire*** itself. Combinations of long-wire antennas in the form of Rhombics or V-beams achieve very high gain by aligning the major lobes of the individual long-wire portions. **ANSWER B.**

E9C08 What is a folded dipole antenna?

A. A dipole one-quarter wavelength long
B. A center-fed dipole with the ends folded down 90 degrees at the midpoint of each side
C. A half-wave dipole with an additional parallel wire connecting its two ends
D. A dipole configured to provide forward gain

The folded dipole is a fun one to build out of television twin-lead cable. The twin-lead ***folded dipole is constructed by using a half-wavelength piece of twin lead to form a loop that is one full wavelength long***. Scrape the insulation from the wire ends and twist the wires together on each end. In the center of one of the leads in the twin-lead, cut the lead and scrape some of the insulation off to expose bare wires on the end of each quarter-wave piece. Feed these wires with standard 300-ohm twin-lead or connect a balun transformer to the bare wires to permit 50 ohm or 75-ohm coax cable to feed the folded dipole. Just remember, a folded dipole is still one-half wavelength long with an additional parallel wire connecting its two ends, as we described when explaining how to build it. **ANSWER C.**

E9C07 What is the approximate feed point impedance at the center of a two-wire half-wave folded dipole antenna?

A. 300 ohms
B. 72 ohms
C. 50 ohms
D. 450 ohms

The ***impedance at the center of a half-wave folded dipole*** may be a whopping ***300 ohms***. You'll need an impedance-matching transformer to feed it with 50-ohm coax; however, ordinary TV twin-lead will match very well. **ANSWER A.**

E9C05 What is the purpose of feeding an off-center-fed dipole (OCFD) between the center and one end instead of at the midpoint?

A. To create a similar feed point impedance on multiple bands
B. To suppress off-center lobes at higher frequencies
C. To resonate the antenna across a wider range of frequencies
D. To reduce common-mode current coupling on the feed line shield

The off-center fed dipole is a relatively recent resurrection of the ancient TRUE WINDOM antenna, which was popular a century ago. The Windom was noted for its ability to operate on both Even and Odd harmonics with a reasonably good match. The feedpoint is approximately 14 percent off center, which yields an impedance of about 600 ohms on several bands. **ANSWER A.**

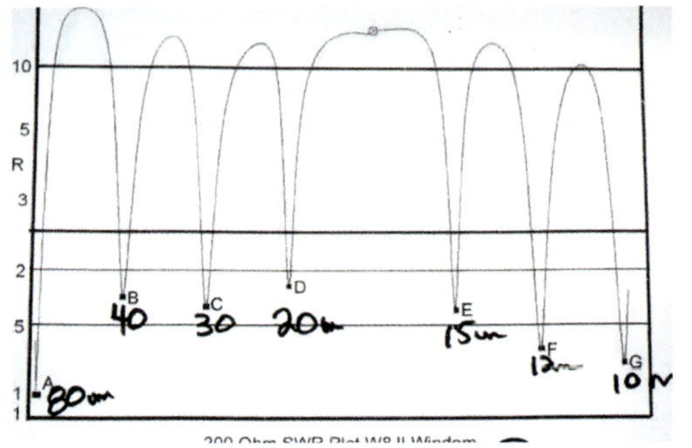

This off center fed antenna has a 4:1 balun at the feed point, spaced away from the metal mast. The dipole offers resonance on all bands provided you work with it long enough when pruning each end.

E9C09 Which of the following describes a G5RV antenna?

A. A wire antenna center-fed through a specific length of open-wire line connected to a balun and coaxial feed line
B. A multi-band trap antenna
C. A phased array antenna consisting of multiple loops
D. A wide band dipole using shorted coaxial cable for the radiating elements and fed with a 4:1 balun

There are times when having a high SWR isn't such a bad thing; in fact, it might be absolutely necessary. The specific length ***open wire transmission line section of a G5RV*** antenna is just such a case. This line section acts as an impedance matching transformer and tuning "stub." While there is a high SWR on the open wire line, the bottom end of the line is close to 50 ohms, providing a reasonable match to the 50-ohm coax preceding it. A balun at the bottom end assures that the line radiation on the coaxial segment is low. **ANSWER A.**

E9C10 Which of the following describes a Zepp antenna?

A. A horizontal array capable of quickly changing the direction of maximum radiation by changing phasing lines
B. An end-fed half-wavelength dipole
C. An omni-directional antenna commonly used for satellite communications
D. A vertical array capable of quickly changing the direction of maximum radiation by changing phasing lines

The ***Zepp*** antenna, so named for the Zeppelin airships on which it was commonly deployed, ***is an end fed halfwave dipole***. A quarter-wave ladder line

"Q-section" matches the transmitter's low output impedance to the extremely high impedance at the end of the Zepp dipole. Incidentally, the popular J-pole vertical antenna is, for all practical purposes, a "stretched out" Zepp. **ANSWER B.**

E9C12 Which of the following describes an extended double Zepp antenna?

A. An end-fed full-wave dipole antenna
B. A center-fed 1.5-wavelength dipole antenna
C. A center-fed 1.25-wavelength dipole antenna
D. An end-fed 2-wavelength dipole antenna

The ***Extended Double Zepp (EDZ) antenna*** has the greatest possible broadside gain for a single wire antenna, about 3dB of gain over a dipole at typical heights. (The same is true for its half-size vertical variation, the ***5/8 wave*** vertical or ground plane). The EDZ, like the 5/8 wave vertical, is not self-resonant. Its feed point consists of some capacitive reactance, so it must have inductive loading of some kind. In addition, the resistive component is somewhat higher than 50 ohms, so some form of matching network is usually necessary. **ANSWER C.**

E9A01 What is an isotropic radiator?

A. A calibrated, unidirectional antenna used to make precise antenna gain measurements
B. An omnidirectional, horizontally polarized, precisely calibrated antenna used to make field measurements of antenna gain
C. A hypothetical, lossless antenna having equal radiation intensity in all directions used as a reference for antenna gain
D. A spacecraft antenna used to direct signals toward Earth

An ***isotropic radiator*** is simply a theoretical antenna; and in every April 1st issue of many magazines, advertisers try to sell "isotropic radiators" to the unsuspecting ham! We use an isotropic antenna as a ***theoretical lossless antenna, having equal radiation intensity in all directions, to calculate the gain*** or loss of real antenna systems. **ANSWER C.**

E9A02 What is the effective radiated power (ERP) of a repeater station with 150 watts transmitter power output, 2 dB feed line loss, 2.2 dB duplexer loss, and 7 dBd antenna gain?

A. 469 watts
B. 78.7 watts
C. 420 watts
D. 286 watts

For quick calculations combine gains as positive and losses as negative. Our total loss is about 4.2 dB (just add the losses together), and our total gain is 7 dB. This is an ***approximate 3 dB overall gain which*** almost ***doubles your 150 watts to an approximately 286 watts***. The rule of thumb for antennas is that for every 3 dB of gain, radiated power almost doubles. But don't get too excited just yet because you need to ***double*** the size of an antenna for every additional 3 dB of gain! The number one rule of thumb in ham radio is NFL… No Free Lunch! **ANSWER D.**

The Isotropic Radiator

An isotropic radiator is an imaginary antenna that radiates perfectly equally in all directions. Although you will never encounter a true isotropic radiator, it is a very useful reference for comparing any other antenna.

The FCC specifies maximum Effective Isotropic Radiated Power (EIRP) on our two newest bands on 2200 and 630 meters, as well as on 60 meters. But what does that mean?

EIRP is a measure of radiated FIELD STRENGTH. The radiated field strength of any antenna is the product of transmitter power and antenna gain. And the gain of an antenna is relative to our imaginary isotropic radiator. To put it very simply, you can achieve a high EIRP by putting a lot of transmitter power into low gain antenna, like an isotropic radiator, or by putting a small amount of transmitter power into an antenna with a lot of gain, like a Yagi.

On our low frequency bands, because of the extremely long wavelengths involved, transmitting antennas will have gains far less than an isotropic radiator, sometimes by a factor of 100 or more! For this reason, on 2200 meters, it may take every morsel of 1500 watts of legal transmitter power to achieve 1 watt EIRP... the same power radiated by an isotropic radiator with just 1 watt of transmitter power!

E9A06 What is the effective radiated power (ERP) of a repeater station with 200 watts transmitter power output, 4 dB feed line loss, 3.2 dB duplexer loss, 0.8 dB circulator loss, and 10 dBd antenna gain?

A. 317 watts
B. 2,000 watts
C. 126 watts
D. 300 watts

Our total loss is 8 dB, and our total gain is 10 dB. This gives us a net gain of 2 dB; therefore, from Decibels chart in the appendix, we have a gain of 1.585 times. With 200 watts input, ***the ERP is 317 watts (200 x 1.585)***. Now let's check our answer by working backwards through the formula for calculating gain or loss when we know input power and ERP.

$$\text{dB} = 10 \log_{10}\left(\frac{P_1}{P_2}\right)$$

where: P_1 = Power Output
P_2 = Power Input

Therefore,

$$\text{dB} = 10 \log_{10}\left(\frac{317}{200}\right)$$

$$\text{dB} = 10 \log_{10} 1.585$$

The $\log_{10}$ of 1.585 is 0.2

$$\text{dB} = 10 \times 0.2 = 2$$

Since the output power is 317 watts and the input power is 200 watts, there is a net gain of 2 dB, which was the total gain that we calculated by combining the gains and losses of the system. **ANSWER A.**

Although you can't build an isotropic radiator, you can build a reference antenna that is just as useful, and almost as accurate – the half-wave dipole. A dipole in free space has a gain of 2.15 dB over an isotropic radiator, so it is also a very useful "yardstick."

At the other end of the scale from our low frequency bands, a typical UHF Moonbounce setup may have an EIRP of tens of thousands of watts... all perfectly legal, because on UHF frequencies there is no restriction on EIRP, just transmitter power, which is a very good thing for experimentally-minded hams. (A cheap hand held laser pointer, by the way, might have quadrillions of watts of EIRP!)

E9A07 What is the effective isotropic radiated power (EIRP) of a repeater station with 200 watts transmitter power output, 2 dB feed line loss, 2.8 dB duplexer loss, 1.2 dB circulator loss, and 7 dBi antenna gain?

A. 159 watts
B. 252 watts
C. 632 watts
D. 63.2 watts

Our total loss adds up to a 6 dB loss, and our total gain is 7 dB, for an actual gain of 1 dB. This would be a slight increase to our 200 watts output in effective radiated power. So, from the answers given, ***252 watts ERP*** is the most reasonable selection. **ANSWER B.**

E9B07 What is the difference in radiated power between a lossless antenna with gain and an isotropic radiator driven by the same power?

A. The power radiated from the directional antenna is increased by the gain of the antenna
B. The power radiated from the directional antenna is stronger by its front-to-back ratio
C. They are the same
D. The power radiated from the isotropic radiator is 2.15 dB greater than that from the directional antenna

This is a good question to get you thinking about the isotropic antenna – an antenna that radiates equally well in all directions. If you were to pump 100 watts into this theoretical antenna, you would get 100 watts out, unity gain, in every direction, including up and down. Now you switch to the beam antenna and presto, those 100 watts are focused in one direction, with minimum up and minimum down radiation. By taking energy in all directions and concentrating it in just one direction with a beam, you have improved your signal on transmit and receive. But the total amount of radiation emitted is that same 100 watts at the feed point, so there is no magical new power created in the beam. Thus, ***there is no difference between the two antennas***. A 100-watt light bulb hanging by a wire and a 100 watt bulb with a polished reflector is still 100 watts, right? Sure, but one bulb looks brighter, with the same amount of power, thanks to the reflector! **ANSWER C.**

E9H03 What is receiving directivity factor (RDF)?

A. Forward gain compared to the gain in the reverse direction
B. Relative directivity compared to isotropic
C. Relative directivity compared to a dipole
D. Peak antenna gain compared to average gain over the hemisphere around and above the antenna

When dealing with receiving antennas, overall antenna *gain* isn't always the most important factor. Sometimes you can greatly improve a receiver's signal-to-noise ratio by sacrificing antenna gain in the direction (or directions) from which unwanted noise is arriving. The RDF of an antenna is generally a better indicator of a receiving antenna's effectiveness than the gain alone. The receiving directivity factor, or ***RDF, is defined as peak antenna gain compared to average gain over the entire hemisphere around and above the antenna***. **ANSWER D.**

E9A12 How much gain does an antenna have compared to a half-wavelength dipole if it has 6 dB gain over an isotropic radiator?

A. 3.85 dB
B. 6.0 dB
C. 8.15 dB
D. 2.79 dB

The halfwave dipole has about 2.15 dB gain over an isotropic radiator. If an antenna under measurement has 6 dB gain over an isotropic radiator, it has ***6 dB - 2.15 dB = about 3.85 dB gain over the dipole***. **ANSWER A.**

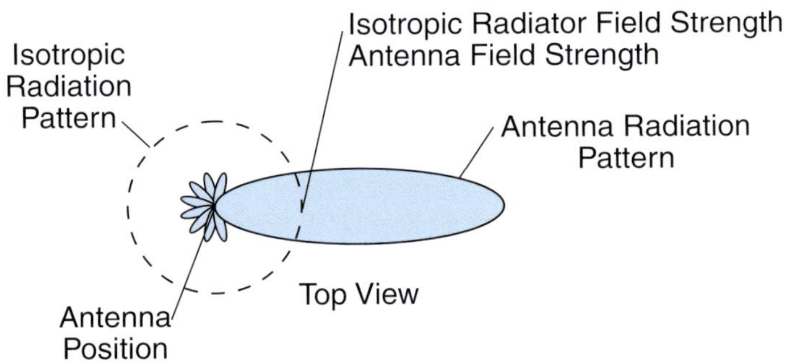

"Gain" of an Antenna

E9D08 What happens as the Q of an antenna increases?

A. SWR bandwidth increases
B. SWR bandwidth decreases
C. Gain is reduced
D. More common-mode current is present on the feed line

As with any other resonant circuit, the Q of an antenna is inversely proportional to the bandwidth (***as Q increases, bandwidth decreases***). In the calculations for an antenna, the R used in the equation Q = X/R is *radiation resistance*. Antennas with higher radiation resistance tend to be more broad banded and have a lower Q. **ANSWER B.**

E9D09 What is the function of a loading coil in an electrically short antenna?

A. To increase the SWR bandwidth by increasing net reactance
B. To lower the losses
C. To lower the Q
D. To resonate the antenna by cancelling the capacitive reactance

Since X_L (inductive reactance) must equal X_C (capacitive reactance) in order for a whip antenna to achieve resonance, we need to ***add a loading coil*** on an antenna for lower frequency operations to ***tune out capacitive reactance***. **ANSWER D.**

E9A08 Which frequency band has the smallest first Fresnel zone?

A. 5.8 GHz
B. 3.4 GHz
C. 2.4 GHz
D. 900 MHz

A Fresnel Zone is a volume of space that needs to be clear above and beyond the line-of-sight path between a transmitter and receiver antenna. Freznel zones are a series of concentric watermelon-shaped region that define where ground reflections can be destructive, due to phase cancellation. The higher the frequency (or shorter the wavelength) the smaller the Fresnel zone that needs to be accommodated. 5.8 GHZ has the shortest wavelength of all the examples, so it has the smallest Fresnel zone requirement. **ANSWER A.**

E9D06 What happens to SWR bandwidth when one or more loading coils are used to resonate an electrically short antenna?

A. It is increased
B. It is decreased
C. It is unchanged if the loading coil is located at the feed point
D. It is unchanged if the loading coil is located at a voltage maximum point

Loading coils are used when an antenna is shorter than necessary for self-resonance. As the length of an antenna is decreased from its natural resonant length, the radiation resistance decreases, resulting in an increase of the Q of the antenna. For a series resonant circuit, the Q is the ratio of reactance to resistance. Decreasing the resistance will increase the Q, ***decreasing the working SWR bandwidth***. **ANSWER B.**

A high Q mobile coil (left) with the contactor disc seen in the upper 2/3 of the antenna coil. These motorized high Q antennas really improve performance on 40 and 75 meters.

E9D04 Why should antenna loading coils have a high ratio of reactance to resistance?

A. To swamp out harmonics
B. To lower the radiation angle
C. To maximize efficiency
D. To minimize the Q

Those giant open-air coils halfway up the mast on a mobile whip antenna are ***high Q coils that maximize efficiency and minimize losses***. The bigger the coil, the lower the losses. **ANSWER C.**

E9D03 What is the most efficient location for a loading coil on an electrically short whip?

A. Near the center of the vertical radiator
B. As low as possible on the vertical radiator
C. At a voltage maximum
D. At a voltage null

To minimize coil losses and increase performance of a loaded vertical antenna, the ***loading coil*** should be near the ***center of the vertical radiator***, not at the base. **ANSWER A.**

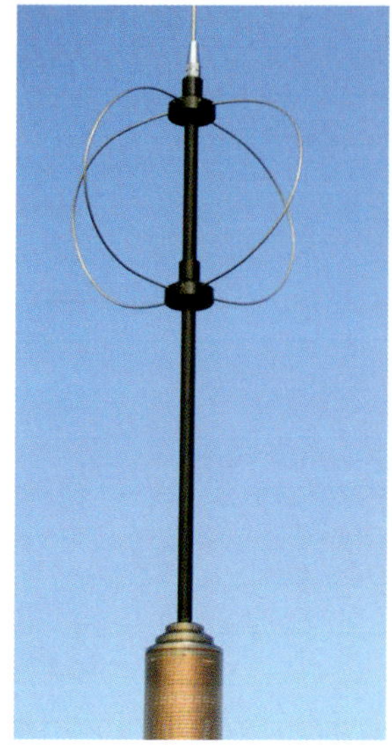

See the bird cage capacity loading hat above the coil? It increases performance on the lower bands, like 40 meters.

E9D07 What is an advantage of top loading an electrically short HF vertical antenna?

A. Lower Q
B. Greater structural strength
C. Higher losses
D. Improved radiation efficiency

Top loading and helical loading are used on a shortened HF vertical antenna to ***improve radiation efficiency***. With this in mind, it's easy to understand why a little 96-inch CB whip antenna with an automatic antenna tuner will not perform as well as a resonant center-loaded, helical-loaded, or top-loaded vertical antenna. It doesn't matter how much power you put into an antenna by fooling the radio with the tuner; if the antenna is not resonant great amounts of power are lost. **ANSWER D.**

E9D05 Approximately how long is a Yagi's driven element?

A. 234 divided by frequency in MHz
B. 1005 divided by frequency in MHz
C. 1/4 wavelength
D. 1/2 wavelength

The driven element of a Yagi antenna is essentially a dipole, and its dimensions are determined accordingly. Like any dipole, the driven element of a dipole is ½ wavelength. However, the other parasitic elements in proximity adjust this length slightly, but more profoundly, the feedpoint impedance is substantially lowered by the parasitic elements. The feedpoint impedance of a typical three element Yagi is between around 25 and 35 ohms, so some form of matching device is usually desired. **ANSWER D.**

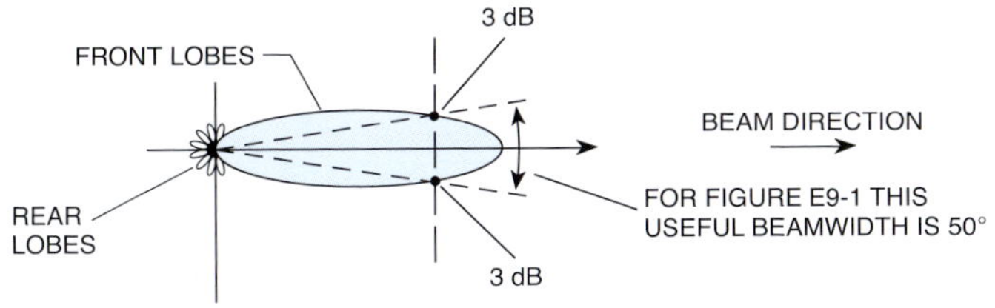

Directional Radiation Pattern of a Yagi Beam

Source: *Antennas – Selection and Installation*, A.J. Evans, Copyright ©1986 Master Publishing, Inc., Arlington, Texas

E9D11 Why do most two-element Yagis with normal spacing have a reflector instead of a director?

A. Lower SWR
B. Higher receiving directivity factor
C. Greater front-to-side
D. Higher gain

In a Yagi antenna with typical element spacing of .12 wavelengths or so, more forward gain can be achieved with a single reflector than a single director. With extremely close spacing, a single director can actually yield higher gain than a single reflector, but this system is ***extremely*** touchy and narrow-banded and is almost never used. Since the question specifies "normal spacing" we can eliminate the single director configuration. **ANSWER D.**

E9D12 What is the purpose of making a Yagi's parasitic elements either longer or shorter than resonance?

A. Wind torque cancellation
B. Mechanical balance
C. Control of phase shift
D. Minimize losses

All directional antennas rely on wave ***interference*** to achieve their directional properties, and the Yagi is no exception. The length and spacing of the parasitic elements determine in which direction wave reinforcement occurs, and in which direction wave cancellation (or partial cancellation) occurs. A reflector is generally about 5% longer than the driven element and the first director, about 5% shorter. **ANSWER C.**

E9B08 What is the far field of an antenna?

A. The region of the ionosphere where radiated power is not refracted
B. The region where radiated power dissipates over a specified time period
C. The region where radiated field strengths are constant
D. The region where the shape of the radiation pattern no longer varies with distance

The ***far field zone*** of an antenna is where the ***antenna's radiation pattern remains constant, independent of distance***. This is generally a safe area when calculating MPE. **ANSWER D.**

E9A05 What does the term "ground gain" mean?

A. The change in signal strength caused by grounding the antenna
B. The gain of the antenna with respect to a dipole at ground level
C. To force net gain to 0 dB by grounding part of the antenna
D. An increase in signal strength from ground reflections in the environment of the antenna

Very seldom does an amateur radio antenna function as if it's in free space. The far field pattern of an antenna is the product of both the direct rays from the antenna and reflected rays from any nearby objects. In many cases, ground reflections can reinforce the direct rays (while invariably creating nulls as well). When a ground reflected ray is in phase with the direct ray signal reinforcement occurs, which it can be substantial, at least in certain directions or elevation angles. **ANSWER D.**

E9B09 What type of analysis is commonly used for modeling antennas?

A. Graphical analysis
B. Method of Moments
C. Mutual impedance analysis
D. Calculus differentiation with respect to physical properties

Method of Moments is a computer program ***for modeling a proposed antenna*** on the computer monitor. **ANSWER B.**

E9B10 What is the principle of a Method of Moments analysis?

A. A wire is modeled as a series of segments, each having a uniform value of current
B. A wire is modeled as a single sine-wave current generator
C. A wire is modeled as a single sine-wave voltage source
D. A wire is modeled as a series of segments, each having a distinct value of voltage across it

Method of Moments (MOM) software models a wire as a series of segments where ***each segment has a uniform value of antenna current***. **ANSWER A.**

E9B11 What is a disadvantage of decreasing the number of wire segments in an antenna model below 10 segments per half-wavelength?

A. Ground conductivity will not be accurately modeled
B. The resulting design will favor radiation of harmonic energy
C. The computed feed point impedance may be incorrect
D. The antenna will become mechanically unstable

When modeling antennas on a computer, we always stay above 10 segments per half wavelength, so we don't cause the ***feed point impedance to be miscalculated using less than*** the recommended ***10 segments per half wavelength***. **ANSWER C.**

E9H01 When constructing a Beverage antenna, which of the following factors should be included in the design to achieve good performance at the desired frequency?

A. Its overall length must not exceed 1/4 wavelength
B. It must be mounted more than 1 wavelength above ground
C. It should be configured as a four-sided loop
D. It should be at least one wavelength long

The ***Beverage antenna*** is a tremendous signal-grabber. For best performance, it really ***needs to be L O N G!*** The longer the better for a Beverage antenna. **ANSWER D.**

E9H10 How can the output voltage of a multiple-turn receiving loop antenna be increased?

A. By reducing the permeability of the loop shield
B. By utilizing high impedance wire for the coupling loop
C. By increasing the number of turns and/or the area enclosed by the loop
D. All these choices are correct

To make a loop antenna more powerful, ***either increase the number of wire turns*** in the loop ***or make the loop physically larger***. **ANSWER C.**

The more turns you add to a loop the greater the receive output voltage.

E9H11 What feature of a cardioid pattern antenna makes it useful for direction-finding antennas?

A. A very sharp peak
B. A single null
C. Broadband response
D. High radiation angle

A ***cardioid pattern*** is shaped like a heart. There is a broad reception pattern in one direction and a ***deep null*** in the other. When the button is pushed to "sense" the direction of the incoming signal on a direction-finding system that has a cardioid antenna pattern, the deep null in the pattern is normally 180 degrees away from the incoming signal. It's easier to use the null point to sense the direction of the incoming signal than it is the broad pattern – look for the minimum signal, not the maximum. **ANSWER B.**

E9H04 What is the purpose of placing an electrostatic shield around a small-loop direction-finding antenna?

A. It adds capacitive loading, increasing the bandwidth of the antenna
B. It eliminates unbalanced capacitive coupling to the antenna's surroundings, improving the depth of its nulls
C. It eliminates tracking errors caused by strong out-of-band signals
D. It increases signal strength by providing a better match to the feed line

Transmitter hunting down on 75 meters is an international sport. Most operators use a shielded loop antenna for T-hunting because the ***shielding improves the nulls and offering better directivity***. **ANSWER B.**

E9H05 What challenge is presented by a small wire-loop antenna for direction finding?

A. It has a bidirectional null pattern
B. It does not have a clearly defined null
C. It is practical for use only on VHF and higher bands
D. All these choices are correct

Wire loop direction finding antennas will pick up maximum signal strength when the loop is broadside to the signal source. It is bi-directional – it will ***pick up signal sources on either side of the loop***. **ANSWER A.**

E9H08 What is the function of a sense antenna?

A. It modifies the pattern of a DF antenna to provide a null in only one direction
B. It increases the sensitivity of a DF antenna array
C. It allows DF antennas to receive signals at different vertical angles
D. It provides diversity reception that cancels multipath signals

On high frequency automatic radio direction finder systems, the double loop may employ an additional rigid whip to act as a ***sense antenna***. Push the ADF sense button and the fixed double loop DF (direction finding) antenna ***will null out*** ambiguity as to which direction the incoming signal is arriving from at that double loop. The sense antenna must be rigid, never a flexing whip. **ANSWER A.**

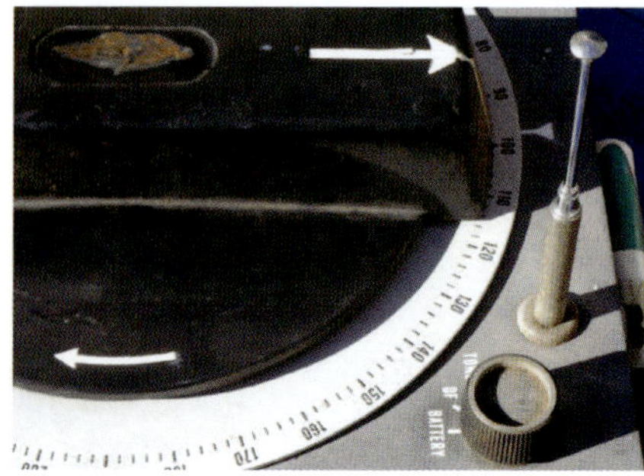

The telescopic whip on this direction finding loop extends up to provide a sense capability. It lets the operator know which end of the loop is pointing toward the transmitting station as the loop is turned slightly.

E9H06 What indicates the correct value of terminating resistance for a Beverage antenna?

A. Maximum feed point DC resistance at the center of the desired frequency range
B. Minimum low-angle front-to-back ratio at the design frequency
C. Maximum DC current in the terminating resistor
D. Minimum variation in SWR over the desired frequency range

The Beverage Antenna is unusual in that it is partly an antenna and partly a transmission line. Like any transmission line, it works best when it is properly terminated, which results in a "flat" line, one in which the variations in voltage (or current) along its length are minimized. In addition, a properly terminated Beverage will result in the greatest front to back ratio, which gives it its legendary low-noise performance. Incidentally, the Beverage antenna is not a great transmitting antenna, although a number of hams do use it as such, with variable results. **ANSWER D.**

E9H07 What is the function of a Beverage antenna's termination resistor?

A. Increase the front-to-side ratio
B. Absorb signals from the reverse direction
C. Decrease SWR bandwidth
D. Eliminate harmonic reception

The Beverage antenna is normally a non-resonant antenna, more closely akin to a transmission line than a traditional antenna. It also behaves partially like a directional coupler, which shows the most directionality when it is properly terminated. A termination resistor, usually on the order of 800 ohms, absorbs power from the "back end" making the antenna unidirectional instead of bidirectional. **ANSWER B.**

E9H09 What type of radiation pattern is created by a single-turn, terminated loop such as a pennant antenna?

A. Cardioid
B. Bidirectional
C. Omnidirectional
D. Hyperbolic

Although resistors in a transmitting antenna are almost universally a bad idea, they can do wonders for receiving antennas. A terminated small loop antenna, of nearly any shape, exhibits a cardioid pattern, which has a deep null in one direction. This is ideal for eliminating noise from that direction. The K9AY antenna is a very popular and effective terminated loop, with the added bonus of being electrically steerable. **ANSWER A.**

E9B01 What is the 3 dB beamwidth of the antenna radiation pattern shown in Figure E9-1?

A. 75 degrees
B. 50 degrees
C. 25 degrees
D. 30 degrees

Looking carefully at the diagram in Figure E9-1, notice that the main lobe of the pattern intersects the 3 dB points at approximately plus and minus 25 degrees. Hence, the ***3-dB beamwidth is approximately 50 degrees***. **ANSWER B.**

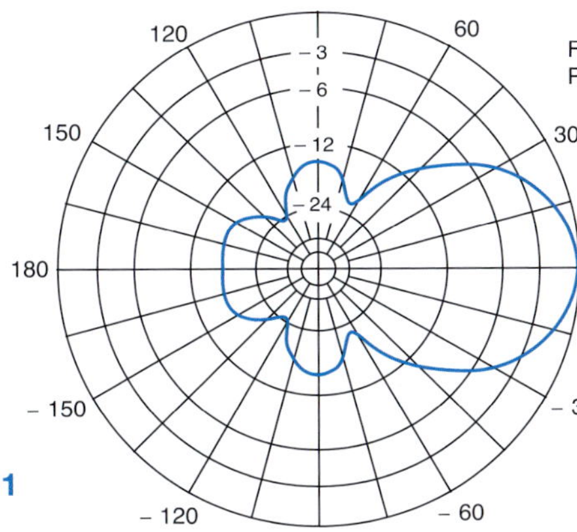

Figure E9-1

E9B02 What is the front-to-back ratio of the antenna radiation pattern shown in Figure E9-1?

A. 36 dB
B. 14 dB
C. 24 dB
D. 18 dB

If you look carefully at Figure E9-1, ***the rear lobe is about 18 dB down from the front lobe***. **ANSWER D.**

E9B03 What is the front-to-side ratio of the antenna radiation pattern shown in Figure E9-1?

A. 12 dB
B. 24 dB
C. 18 dB
D. 14 dB

Study Figure E9-1 carefully. The front-to-side ratio is measured where the lobes perpendicular to the main lobe cross the vertical axis. Notice the scale decreases as you travel out this axis. Since the lobe crosses just a little before 12 but closer to 12 than to 24, ***14 is your best choice*** from the answer's choices given here. **ANSWER D.**

E9H02 Which is generally true for 160- and 80-meter receiving antennas?

A. Atmospheric noise is so high that directivity is much more important than losses
B. They must be erected at least 1/2 wavelength above the ground to attain good directivity
C. Low loss coax transmission line is essential for good performance
D. All these choices are correct

Many hams forget that the function of a transmitting antenna can be very different from that of a receiving antenna. Further, the performance priorities of each can be very different as well. On low bands, especially, it can be a great advantage to use a separate receiving antenna. Small, tuned receiving loops can provide a large ***advantage in signal-to-noise ratio over a full-sized antenna, even if the overall gain is reduced***. **ANSWER A.**

E9B04 What is the front-to-back ratio of the radiation pattern shown in Figure E9-2?

A. 15 dB
B. 28 dB
C. 3 dB
D. 38 dB

The antenna ***front-to-back ratio*** in Figure E9-2 looks to be ***about 28 to 30 dB***. This is the ratio of forward "front" radiation to the slight amount of unwanted radiation in the reverse direction from the back of the antenna. This ratio is the same for transmit as well as receive. **ANSWER B.**

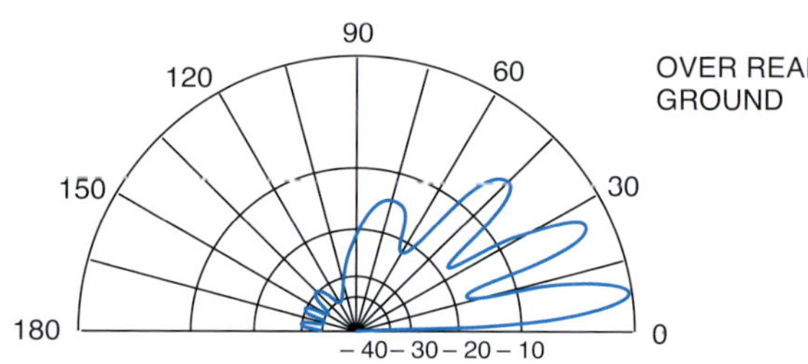

Figure E9-2

E9C14 How does the radiation pattern of a horizontally-polarized antenna mounted above a long slope compare with the same antenna mounted above flat ground?

A. The main lobe takeoff angle increases in the downhill direction
B. The main lobe takeoff angle decreases in the downhill direction
C. The horizontal beamwidth decreases in the downhill direction
D. The horizontal beamwidth increases in the uphill direction

If you live on the East coast, build your home and radio shack on a hill that slopes down to the ocean. Your beam will now have a main lobe whose takeoff angle decreases toward the ocean, giving you a dynamite signal to Europe. Out on the West coast, Gordo is up on some headlands with the ocean just down the way, giving him a good shot to Asia with the ***lower takeoff angle in the downhill direction***. **ANSWER B.**

E3B07 What effect does lowering a signal's transmitted elevation angle have on ionospheric HF skip propagation?

A. Faraday rotation becomes stronger
B. The MUF decreases
C. The distance covered by each hop increases
D. The critical frequency increases

A little simple geometry and some logic will tell you that, ***all things being equal, the lower your "launch angle" the greater the distance for each hop***. The same geometry will also explain why a higher ionosphere will yield longer skip distances. At night, the ionosphere recedes from the bottom up, which is why signals tend to "go long" after nightfall. **ANSWER C.**

E5D10 As a conductor's diameter increases, what is the effect on its electrical length?

A. Thickness has no effect on electrical length
B. It varies randomly
C. It decreases
D. It increases

If you've played with antennas much, you are aware that "fatter" conductors appear longer than thin ones. Antennas built with fat elements need to be shortened slightly to compensate for the diameter of the conductors. This is caused by the lowering of the velocity factor, due to the increase of the effective capacitance across the feed point. The decrease in velocity factor with "fatness" seems counterintuitive but can be explained with a little somewhat complex geometry. **ANSWER D.**

E9C13 How does the radiation pattern of a horizontally polarized antenna vary with increasing height above ground?

A. The takeoff angle of the lowest elevation lobe increases
B. The takeoff angle of the lowest elevation lobe decreases
C. The horizontal beamwidth increases
D. The horizontal beamwidth decreases

When you put up a 3-element Yagi, the high frequency antenna is usually mounted horizontally at a height above ground of at least one-half wavelength of the lowest band of operation. Most 3-element Yagis feature tri-band operation for 10, 15 and 20 meters. The lowest band, 20 meter, would like to see the antenna at one-half wavelength off the earth, or about 32 feet in the air. Anything lower than one-half wavelength from the operating frequency dramatically raises the main lobe takeoff angle. The higher the takeoff angle, the shorter the skip distance. You want to ***get your beam up at least one-half wavelength above the earth for a nice low takeoff angle*** for long-range DX. Increase the antenna height and the takeoff angle of the lowest lobe decreases. **ANSWER B.**

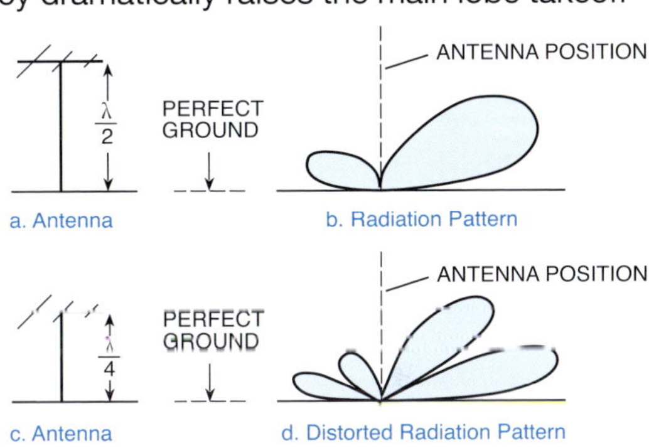

Radiation of 3-Element Yagi for Different Antenna Heights

E9B05 What type of antenna pattern is shown in Figure E9-2?

A. Elevation
B. Azimuth
C. Near field
D. Polarization

Figure E9-2 shows ***the takeoff elevation*** patterns of an antenna over real ground. **ANSWER A.**

E9B06 What is the elevation angle of peak response in the antenna radiation pattern shown in Figure E9-2?

A. 45 degrees
B. 75 degrees
C. 7.5 degrees
D. 25 degrees

From the pattern in the figure, we see that ***the main zero dB lobe is about 7.5 degrees up***, with two other significant lobes at higher elevations. **ANSWER C.**

E9D02 How can two linearly polarized Yagi antennas be used to produce circular polarization?

A. Stack two Yagis to form an array with the respective elements in parallel planes fed 90 degrees out of phase
B. Stack two Yagis to form an array with the respective elements in parallel planes fed in phase
C. Arrange two Yagis on the same axis and perpendicular to each other with the driven elements at the same point on the boom and fed 90 degrees out of phase
D. Arrange two Yagis collinear to each other with the driven elements fed 180 degrees out of phase

To get circular polarization out of a pair of Yagis, arrange the ***Yagis on the same axis perpendicular to each other*** with the driven elements in the same plane and ***fed 90 degrees out of phase***. **ANSWER C.**

E9C06 What is the effect of adding a terminating resistor to a rhombic or long-wire antenna?

A. It reflects the standing waves on the antenna elements back to the transmitter
B. It changes the radiation pattern from bidirectional to unidirectional
C. It changes the radiation pattern from horizontal to vertical polarization
D. It decreases the ground loss

A ***terminating resistor*** at the end of the rhombic opposite the feed line ***changes the radiation pattern from bidirectional to unidirectional***. **ANSWER B.**

E9D01 How much does the gain of an ideal parabolic reflector antenna increase when the operating frequency is doubled?

A. 2 dB
B. 3 dB
C. 4 dB
D. 6 dB

From 1270 MHz on up, you may wish to ***use a parabolic reflector dish antenna*** for microwave work. If you ***double the frequency, the gain of the dish increases by 6 dB***, a 4 times increase. **ANSWER D.**

Like a magnifying glass, the bigger the dish mechanism the smaller and more powerful the resulting radiated pattern. This 10,368.1 MHz X band microwave dish can work stations up to 500 miles away under good tropo conditions.

Feedlines & Safety

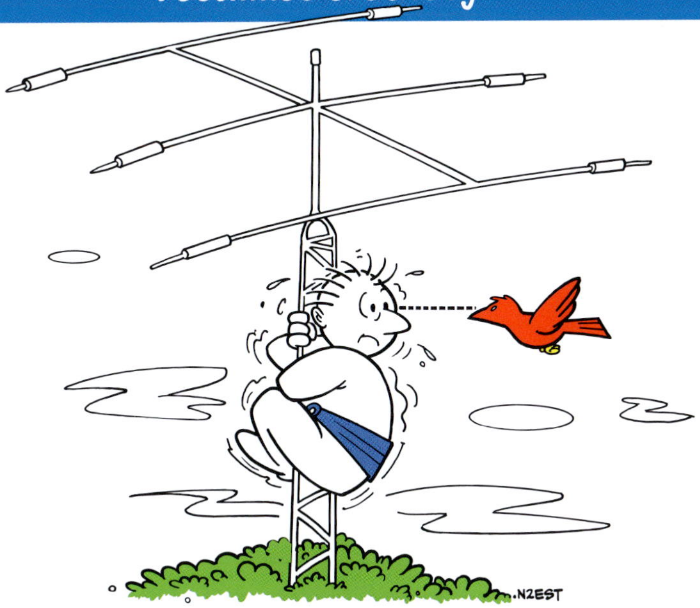

E9E01 Which matching system for Yagi antennas requires the driven element to be insulated from the boom?

A. Gamma
B. Beta or hairpin
C. Shunt-fed
D. T-match

There are a number of ways to match the impedance of a Yagi's feedpoint to a coaxial cable and achieve a good balance of antenna currents. While the gamma match is the most popular at HF frequencies, it is somewhat asymmetrical, both physically and electrically. The Beta or Hairpin match is nearly perfectly symmetrical but requires a few more components and insulation from the boom, which can make things a little weaker mechanically. **ANSWER B.**

E9E02 What antenna matching system matches coaxial cable to an antenna by connecting the shield to the center of the antenna and the conductor a fraction of a wavelength to one side?

A. Gamma match
B. Delta match
C. T-match
D. Stub match

A ***gamma match*** is used with coax cable that forms an ***unbalanced feed system*** going to a tube that acts as a capacitor in series with the connection point to change the impedance of the antenna to match that of the feed line. **ANSWER A.**

Close up of a Gamma match on a beam feed point. Look carefully behind the coaxial connector for the center insulated conductor going up and into the radiating element.

Why is Characteristic Impedance So Important?

Let's look at this Figure *to help us understand why the characteristic impedance, Z_0, of a transmission line is so important for the efficient transfer of energy.* **Figure a** *shows a transmitter coupled to an antenna with a transmission line that has a characteristic impedance of Z_0. The transmitter's output impedance is Z_T and the antenna input impedance is Z_A, the load impedance that the antenna presents to the transmission line.*

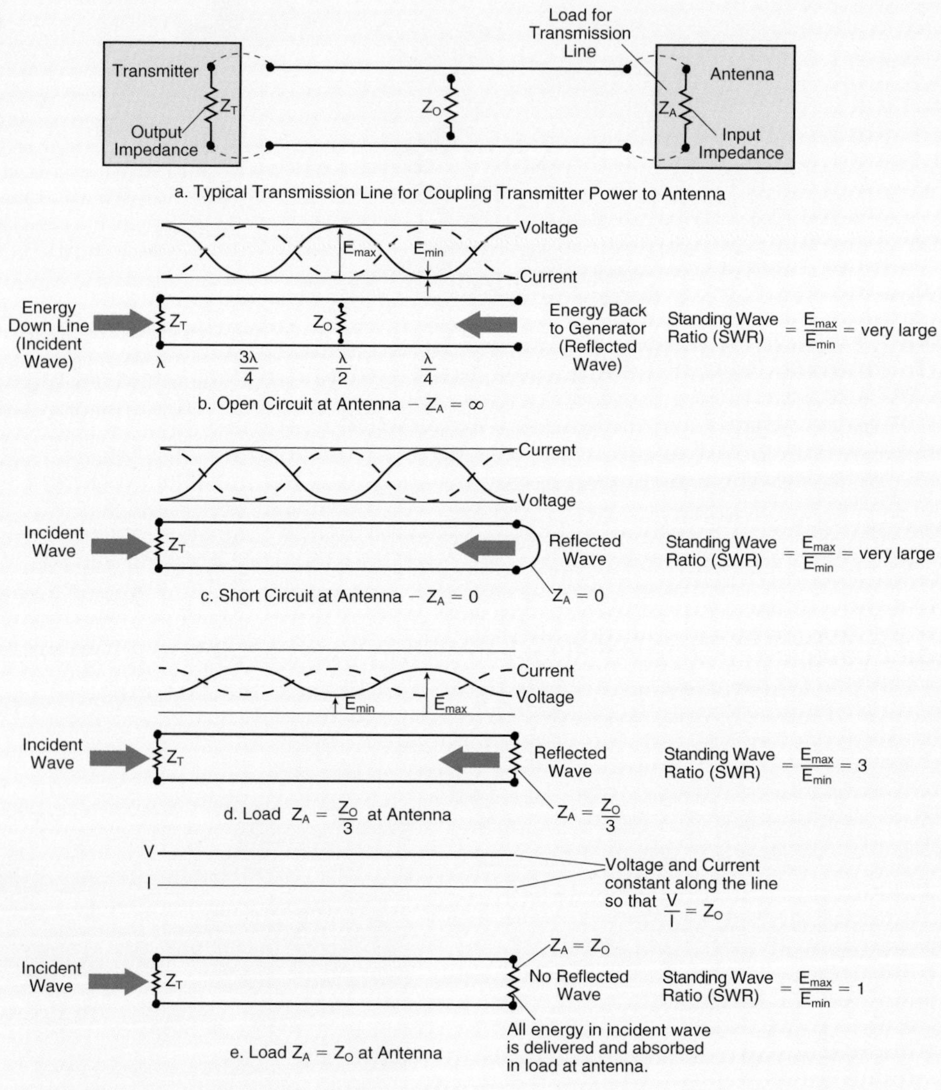

Let's assume that Z_A is infinity — in other words, the load on the transmission line at the antenna is an open circuit. Figure b *is a diagram of the interconnection and shows the voltages and currents that exist on the transmission line. Since Z_A is an open circuit, the current at the end of the line is zero and the voltage is maximum. Since the wave of energy sent from the transmitter has no load to absorb the energy, it is reflected back to the transmitter. The voltage and current waveforms shown are the combined resultant of the wave from the transmitter (called the incident wave) and the reflected wave from the load end. In this case, the total wave is reflected back and no signal is transmitted.*

Standing-Wave Ratio

The ratio of the maximum amplitude, E_{max}, of the resultant voltage waveform on the line to the minimum amplitude, E_{min}, is called the Standing-Wave Ratio, SWR, or more correctly the Voltage Standing-Wave Ratio, VSWR. In the case of an open load, since there is no energy absorbed at the load and all energy delivered is absorbed as losses in the line or absorbed by the transmitter's output resistance, the VSWR is very high (in the order of 20 or greater). VSWR is a measure of how well energy is being transferred over a transmission line to the load. VSWRs below 1.5 indicate quite efficient transfer of energy (VSWR = 1 is perfect), and VSWRs above 2 begin to indicate poor matching.

Now let's look at Figure c. *Here the Z_A is a short circuit. That means the voltage is zero at the end of the line and the current is maximum. Again, no energy is absorbed by the load. It is all reflected back down the line. The voltage and current waveforms vary in minimums and maximums every quarter wavelength, as in the open-circuit case, but are 180° out of phase from the open-circuit case. The standing-wave ratio again is very large and no energy is transferred.*

A Matched Case — $Z_A = Z_0$

Figure e is the case where all the energy in the incident wave is absorbed by the load. There is no reflected wave because $Z_A = Z_0$. The antenna load is matched to the characteristic impedance of the transmission line and maximum transfer of power from the transmitter to the antenna occurs. The standing-wave ratio is 1 because the voltage and current in the line are constant and are such that they equal Z_0 everywhere along the line. Thus, we see how important it is to match the load impedance to the transmission line Z_0. With $Z_A = Z_0$, there will be some small losses in the transmission line, but if there is a mismatch, energy will be reflected from the load back to the transmitter and large losses on the line will be dissipated as heat.

Source: *Basic Communications Electronics*, Hudson & Luecke, © 1999 Master Publishing, Inc., Niles, IL

E9E03 What matching system uses a short length of transmission line connected in parallel with the feed line at or near the feed point?

A. Gamma match
B. Delta match
C. T-match
D. Stub match

When ***a short section of transmission line*** is connected to the antenna feed line near the antenna, it is called ***stub matching***. This one is easy… think short = stub. Perhaps you once used a stub to cancel out interference in your TV receiver. **ANSWER D.**

Impedance Matching Stubs

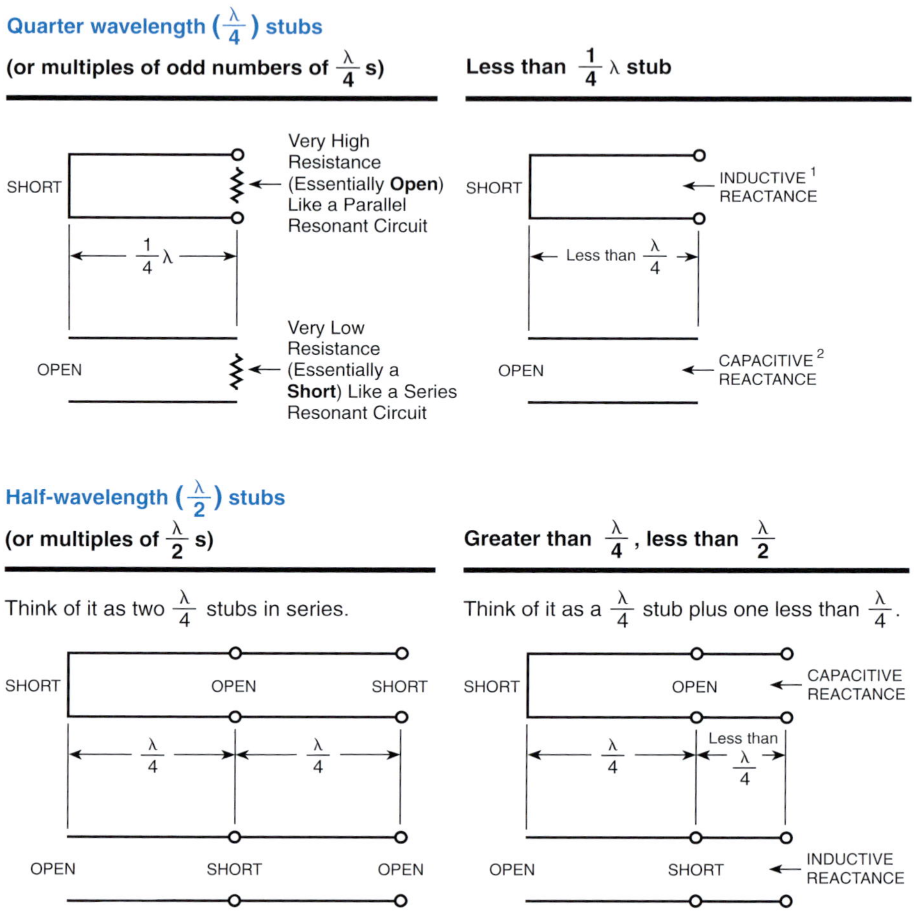

Notes

1. Since a short at end makes the diagram look like a long wire, this should remind the reader that impedance is an inductive reactance.
2. Since an open at end makes the diagram look like two parallel plates, this should remind the reader that impedance is a capacitive reactance.

E9E04 What is the purpose of the series capacitor in a gamma match?

A. To provide DC isolation between the feed line and the antenna
B. To cancel unwanted inductive reactance
C. To provide a rejection notch that prevents the radiation of harmonics
D. To transform the antenna impedance to a higher value

In some ***gamma***-type antenna ***matching*** networks, the center conductor and PVC insulation on the center conductor of coaxial cable are used inside a tube as a series capacitance. ***Series capacitance will cancel the unwanted inductive reactance*** of the matching network and result in resonance. **ANSWER B.**

E9E05 What Yagi driven element feed point impedance is required to use a beta or hairpin matching system?

A. Capacitive (driven element electrically shorter than 1/2 wavelength)
B. Inductive (driven element electrically longer than 1/2 wavelength)
C. Purely resistive
D. Purely reactive

For the ***hairpin matching system*** to work well, the ***driven element reactance must be capacitive, XC***. **ANSWER A.**

It's always a good idea to develop a coax choke at any HF antenna feed point. This will help trap any SWR feeding back out of the antenna and prevent any SWR signal from going back down on the coax outside shield. Keep your 10 turns nice and neat!

E9E07 What parameter describes the interactions of a load and transmission line?

A. Characteristic impedance
B. Reflection coefficient
C. Velocity factor
D. Dielectric constant

A ***mismatched*** transmission line will give you ***reflections***. **ANSWER B.**

E9E08 What is a use for a Wilkinson divider?

A. To divide the operating frequency of a transmitter signal so it can be used on a lower frequency band
B. To feed high-impedance antennas from a low-impedance source
C. To divide power equally between two 50-ohm loads while maintaining 50-ohm input impedance
D. To divide the frequency of the input to a counter to increase its frequency range

With many high frequency antennas now incorporating multiple-band driven elements all off one feed line, the ***Wilkinson divider distributes the antenna input power equally*** to the multiple driven elements ***while maintaining the 50-ohm input impedance to both loads***. **ANSWER C.**

E9E09 Which of the following is used to shunt-feed a grounded tower at its base?

A. Double-bazooka match
B. Beta or hairpin match
C. Gamma match
D. All these choices are correct

Gordo has a grounded tower that he can use as a halfwave antenna on 160 meters! The transmission line cable runs about halfway up the tower into a ***gamma match***. The outer braid taps the metal tower leg, and the center conductor is extended up and runs parallel to the same tower leg, and then makes a sharp turn and is connect to the same leg allowing for the impedance to match the transmission line. If you were to look at it with an ohm meter, it would look like a direct short. But your gamma match and a manual tuner can get you down to the band of choice. **ANSWER C.**

E9E06 Which of these transmission line impedances would be suitable for constructing a quarter-wave Q-section for matching a 100-ohm feed point impedance to a 50-ohm transmission line?

A. 50 ohms
B. 62 ohms
C. 75 ohms
D. 90 ohms

The quarter wave transmission line presents a special (and highly useful) case. For a quarter wave transmission line or "Q-section," the characteristic impedance of the transmission line is the ***geometric mean*** between the load resistance and the source resistance. (This only works when the load resistance is purely resistive). The geometric mean can be expressed simply: the characteristic impedance of the line squared is equal to the source impedance times the load resistance. In this example, we're looking for the characteristic impedance of the line (not often something you can select at will). But in this example, we have a 100-ohm load impedance, and we want a 50-ohm source impedance. 100 x 50 = 5000. Taking the square root of 5000, we get 70.7 ohms. It's not exactly ***75 ohms***, but that's the closest answer. **ANSWER C.**

E9E11 What is the purpose of using multiple driven elements connected through phasing lines?

A. To control the antenna's radiation pattern
B. To prevent harmonic radiation from the transmitter
C. To allow single-band antennas to operate on other bands
D. To create a low-angle radiation pattern

As an Extra Class operator, you likely will choose an exotic, multiband beam antenna to exercise your new privileges. Multiband antennas have a variety of methods to phase most elements for the desired beam pattern. "Trapless" designs may use rigid ***phasing lines to ensure each driven element operates in concert with all of the other elements*** on the antenna for the desired antenna radiation pattern. **ANSWER A.**

E9F08 Which of the following is a significant difference between foam dielectric coaxial cable and solid dielectric coaxial cable, assuming all other parameters are the same?

A. Foam dielectric coaxial cable has lower safe maximum operating voltage
B. Foam dielectric coaxial cable has lower loss per unit of length
C. Foam dielectric coaxial cable has higher velocity factor
D. All these choices are correct

What you look for in coax cable is a little different now that you are on HF. On HF bands, coax cable losses are significantly less than those experienced when operating on VHF and UHF. Popular RG-8 foam coax, on the 15 meter band, represents only 0.7 dB loss per 100 feet. Although lighter in weight, foam dielectric has lower safe operating voltage limits and a slightly higher velocity factor than solid dielectric. Some land mobile rated (LMR) coax cable that is about the same size as RG-8 exhibits wonderful qualities of being lightweight, having a non-contaminating jacket to resist sunlight cracks, a common 0.66 velocity factor, and slightly less attenuation on high frequency bands. It handles a little like a cold garden hose but, nonetheless, LMR type coax has plenty of benefits. ***All of the answer choices are correct for the difference between foam dielectric coax and solid dielectric coax***. **ANSWER D.**

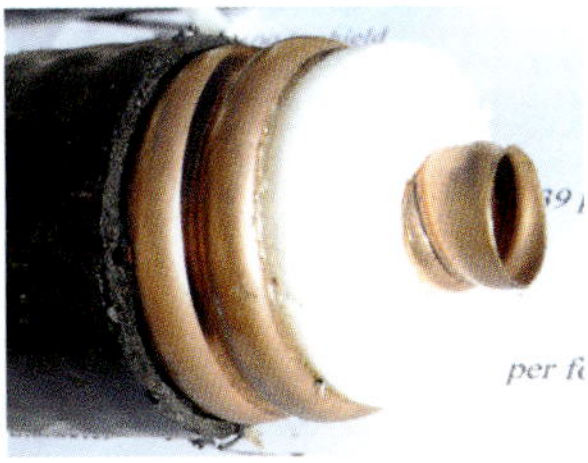

Skin effect saves copper! Notice that there is no solid copper in the center conductor. This is high-power foam dielectric hard line for microwave use.

Coax Cable Type, Size and Loss per 100 Feet

Coax Type	Size	Loss at HF 100 MHz	Loss at UHF 400 MHz
RG-6	Large	2.3 dB	4.7 dB
RG-59	Medium	2.9 dB	5.9 dB
RG-58U	Small	4.3 dB	9.4 dB
RG-8X	Medium	3.7 dB	8.0 dB
RG-8U	Large	1.9 dB	4.1 dB
RG-213	Large	1.9 dB	4.5 dB
Hardline	Large, Rigid	0.5 dB	1.5 dB

E9F01 What is the velocity factor of a transmission line?

A. The ratio of its characteristic impedance to its termination impedance

B. The ratio of its termination impedance to its characteristic impedance

C. The velocity of a wave in the transmission line multiplied by the velocity of light in a vacuum

D. The velocity of a wave in the transmission line divided by the velocity of light in a vacuum

In free space, radio waves travel at 300 million meters per second. But in coaxial cable transmission lines and other types of feed lines, the radio waves move more slowly. ***Velocity factor*** is the ***velocity of the radio wave on the transmission line divided by the velocity of the wave in free space***. It's a ratio which always has division. **ANSWER D.**

Remember, when you are cutting coax to a specific wavelength most amateur radio coax has a velocity factor of 0.66. Always seal your feed point coax connections against rain and snow.

E9F02 Which of the following has the biggest effect on the velocity factor of a transmission line?

A. The characteristic impedance
B. The transmission line length
C. The insulating dielectric material
D. The center conductor resistivity

There are some great low-loss coaxial cable types with different dielectrics separating the center conductor and the outside shield. It is the ***type of dielectric used that determines the velocity factor*** of the transmission line. **ANSWER C.**

E4E07 Which of the following can cause shielded cables to radiate or receive interference?

A. Low inductance ground connections at both ends of the shield
B. Common-mode currents on the shield and conductors
C. Use of braided shielding material
D. Tying all ground connections to a common point resulting in differential-mode currents in the shield

The best, most expensive coaxial cable in the world won't do its job properly if ***RF currents are induced or coupled on the outside shield***. A proper balun *and* transmission line orientation are usually necessary to prevent the outside of your transmission line from becoming a radiator; that is, an antenna. The degree to which feed line radiation is a problem varies from nothing to huge, depending on the application. There is no "silver balun" to solve all your transmission line problems, but understanding the principles and applying the appropriate ones will make you a happy ham. **ANSWER B.**

E9F04 What impedance does a 1/2-wavelength transmission line present to an RF generator when the line is shorted at the far end?

A. Very high impedance
B. Very low impedance
C. The same as the characteristic impedance of the line
D. The same as the output impedance of the RF generator

If a ***1/2 wavelength transmission line is shorted*** at the far end, it will have ***very low impedance***, essentially a short. **ANSWER B.**

E9F03 Why is the electrical length of a coaxial cable longer than its physical length?

A. Skin effect is less pronounced in the coaxial cable
B. Skin effect is more pronounced in the coaxial cable
C. Electromagnetic waves move faster in coaxial cable than in air
D. Electromagnetic waves move more slowly in a coaxial cable than in air

Electromagnetic radio waves move slightly ***slower inside coaxial cable*** than it does in free space. **ANSWER D.**

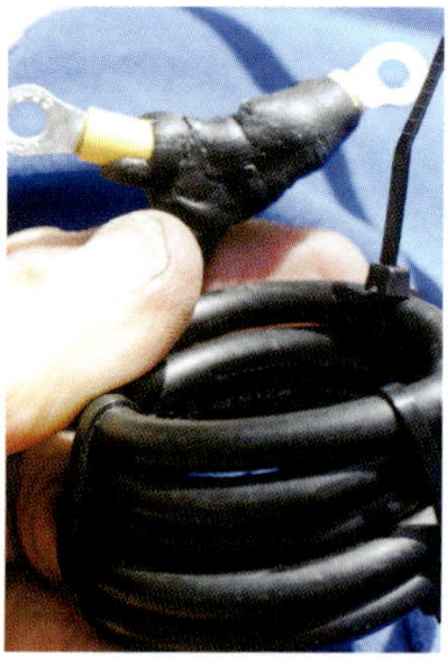

Here is where coax goes bad! In wet weather, water can enter the feed point connection and get into the dielectric. Treat the connection point with flexible sealant and then wrap it with self-vulcanizing waterproof tape.

E9F05 What is microstrip?

A. Special shielding material designed for microwave frequencies
B. Miniature coax used for low power applications
C. Short lengths of coax mounted on printed circuit boards to minimize time delay between microwave circuits
D. Precision printed circuit conductors above a ground plane that provide constant impedance interconnects at microwave frequencies

In VHF radio circuits, it's not always possible to reduce the length of wires so that they don't act as inductors (See question E5D02). But sometimes you can do the next best thing: make them look like transmission lines. If you place a wire close to a ground plane, such as a copper clad circuit board, it will act as a short length of transmission line or ***"microstrip,"*** which will cancel out the inductance. This construction practice is fairly standard in most VHF and UHF radio equipment nowadays. **ANSWER D.**

E9F06 What is the approximate physical length of an air-insulated, parallel conductor transmission line that is electrically 1/2 wavelength long at 14.10 MHz?

A. 7.0 meters
B. 8.5 meters
C. 10.6 meters
D. 13.3 meters

In this problem, the length of the parallel conductor feed line that will be one-half wavelength long at 14.10 MHz with a velocity factor of 0.95 is calculated by multiplying ***984 x 0.5 x 0.95*** and ***dividing by 14.10***, the frequency in MHz. ***Convert feet to meters***, and you end up with ***10.6 meters***. **ANSWER C.**

E9F07 How does parallel conductor transmission line compare to coaxial cable with a plastic dielectric?

A. Lower loss
B. Higher SWR
C. Smaller reflection coefficient
D. Lower velocity factor

Going to 450-ohm ladder line on the 6 meter band is one way to reduce losses over typical RG-58 coax cable. They both cost about the same, which keeps your expenses to a minimum. RG-58 should be relegated to test leads, never used for antenna feed line runs. The loss is simply too high. ***450-ohm ladder line will have much lower loss at 50 MHz than small RG-58 coax***. **ANSWER A.**

E9F09 What impedance does a 1/4-wavelength transmission line present to an RF generator when the line is shorted at the far end?

A. Very high impedance
B. Very low impedance
C. The same as the characteristic impedance of the transmission line
D. The same as the generator output impedance

If the far end a ***1/4 wavelength transmission line is shorted***, it at will look like a ***very high impedance*** at the generator. **ANSWER A.**

TRANSMISSION LINE MNEMONIC

If you move along a short circuited transmission line that is several wavelengths long from the load end toward the source the impedance will "cycle through" the values of (S)hort, (I)nductive, (O)pen, and (C)apacitive at 1/8th wave intervals. One way to remember the order is: Sit In Our Chair. Also remember that a ½ wave transmission line REPEATS the impedance, and a quarter wave transmission line INVERTS the impedance. If you have a1/4 wave transmission line shorted at the far end, it will look like an open circuit at the input end. If it is open at the far end, it will look like a short at the input. Just remember that a ¼ wavelength line is only ¼ wavelength at one frequency!

E9F10 What impedance does a 1/8-wavelength transmission line present to an RF generator when the line is shorted at the far end?

A. A capacitive reactance
B. The same as the characteristic impedance of the line
C. An inductive reactance
D. Zero

If a 1/8 wavelength transmission line is shorted out at the far end, it will look like an ***inductive reactance*** at the generator. **ANSWER C.**

E9F11 What impedance does a 1/8-wavelength transmission line present to an RF generator when the line is open at the far end?

A. The same as the characteristic impedance of the line
B. An inductive reactance
C. A capacitive reactance
D. Infinite

If a ***1/8 wavelength transmission line is open*** at the far end, it will look like a ***capacitive reactance*** at the generator. **ANSWER C.**

E9F12 What impedance does a 1/4-wavelength transmission line present to an RF generator when the line is open at the far end?

A. The same as the characteristic impedance of the line
B. The same as the input impedance to the generator
C. Very high impedance
D. Very low impedance

If the far end of a ***1/4 wavelength transmission line is open***, it will look like a ***very low impedance*** at the generator. **ANSWER D.**

Smith Chart—Finding Antenna Impedance

Finding Antenna Impedance, Z_A, when Fed with Known Length of Transmission Line (λ in Wavelengths) of Given Impedance, Z_o.
(Impedance and wavelength of transmission line are at the frequency of transmission)

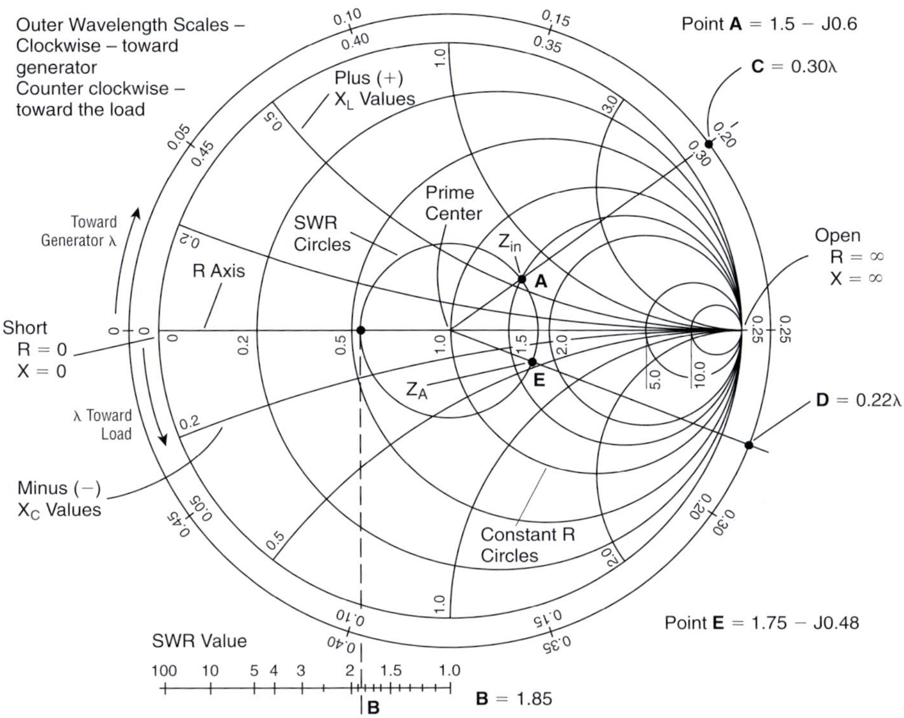

Known Values: $Z_o = 50$, $\lambda_Z = 0.42\lambda$

Steps

1. Connect transmission line to antenna.
2. Measure the input impedance Z_{in} to transmission line. (Example 75 + J30)
3. Normalize Z_{in} by dividing by Z_o.

$$Z_{in} = \frac{75}{50} + J\frac{30}{50} = 1.5 + J0.6$$

4. Plot 1.5 + J0.6 on chart (point A is at intersection of a resistance circle of 1.5 and a positive reactance curve of 0.6)
5. Draw SWR circle with prime center as center and radius through A. Project tangent to SWR circle at point where circle intersects the resistance axis down to linear scale and read value of SWR (Point B = 1.85).
6. Draw a line (radial line) from prime center through Z_{in} at A and project to wavelength scale. Use "toward load" scale because you are going from Z_{in} at generator to the load presented by Antenna. (Point C = 0.30λ)
7. To find the load impedance line on the wavelength (towards load) scale, add the transmission line length to Point C value (Point D = 0.42 + .30 = 0.72λ)
8. Since the wavelength scales are only plotted for 0.5λ, the values repeat every 0.5λ. Therefore, subtract 0.5 for 0.72 to arrive at 0.22. (Point D = 0.22 toward load)
9. Draw another radial line from prime center to Point D. The intersection with the SWR circle is the load impedance of the antenna. (Point E = 1.75 – J0.48). You may have to interpolate between lines.
10. This is still a normalized value so multiply by $Z_o = 50$ to obtain true value or antenna impedance. $Z_A = 50\,(1.75 - J0.48) = 87.5 - J24$. Which is an impedance of 87.5 ohms of resistance and 24 ohms of capacitive reactance.

The Importance of Knowing the Smith Chart

The Smith Chart is specifically designed to tell you what is happening within a transmission line. It won't actually tell you anything about the current distribution on an antenna itself, but if we apply a few "smarts" we can often make an educated guess as to what's happening "out there" at the far end of the coax based on the Smith Chart. Let's summarize the issue with a few principles we can rely on:

1. *If there is any reactance on the antenna, we will not have a 1:1 SWR. It is a common "newbie" misconception to conclude that if you have 25 ohms of resistance and 25 ohms of reactance the total "impedance" is 50 ohms, so the SWR will be flat. Wrong... dead wrong. The only time you will have no reflected power is if there is no reactance and the resistive component is matched.*

2. *Almost always, as you move off the resonant frequency of an antenna, the REACTANCE changes much faster than the RADIATION RESISTANCE. This is why SWR is a sensitive indicator of resonance, and a much poorer indication of "antenna impedance" (a bit of a misnomer).*

3. *The Smith chart can be normalized to any transmission line impedance you like. This is often forgotten. A 600 ohm transmission line with a 600 ohm load resistance has an SWR of 1:1... just like a 50 ohm transmission line with a 50 ohm load. This also tells you that you can feed a 50 ohm antenna DIRECTLY with 1/2 wavelength (or any multiple thereof) of 600 ohm ladder line and it will present a perfect 50 ohm impedance on the input end! This is in spite of the fact that the ladder line itself has a 12:1 SWR.*

4. *Impedances near the left side of the Smith Chart will always be more stable than those near the right side. A center-fed full wave antenna will have an extremely high impedance (between 5-10 thousand ohms) at the feed point, with a reactance that changes drastically (flopping between nearly infinite capacitive reactance to nearly infinite inductive reactance) with a minuscule change of frequency. This manic depressive impedance can be totally calmed down with a quarter wave transmission line or "Q-section" which will present a low, broadband impedance at the input side of the line. This is how the end fed Zepp antenna (and its red-headed stepchild, the J-pole) works.*

5. *Here's an interesting fact closely related to item #2: Reactance has zero effect... zip, zilch, nada... on the ability of an antenna to radiate! There is absolutely nothing sacred about a self-resonant antenna, nor does it give us any advantage whatsoever. The reason we need the system (not the antenna itself) to be resonant is so that we can effectively transfer power. See all the questions on power factor. The truth of the matter is that the power radiated by an antenna is determined by the current and the radiation resistance only! Reactance, no matter how much, does not even enter the equation! Need proof? Let's talk about AM broadcast antennas which must, by law, be nearly 100% efficient. There isn't one AM broadcast antenna out of 1,000 that's anywhere near self-resonant! AM towers are generally made of 60 foot sections of tower, so if you're within 60 feet of resonance, you're usually doing okay. They rely on highly efficient Antenna Tuning Units (ATUs) at the base of the towers, basically transmatches on steroids, to cancel out any reactance and match the feed point impedance (which can also be far different from 50 ohms). The amateur radio community has finally, if reluctantly, caught on to this truth with the rising popularity of the 43 foot non-resonant vertical, used across the entire HF band very effectively. It is important to have an efficient matching network for these to perform as advertised.*

E4A07 Which of the following is an advantage of using an antenna analyzer compared to an SWR bridge?

A. Antenna analyzers automatically tune your antenna for resonance
B. Antenna analyzers compute SWR and impedance automatically
C. Antenna analyzers display a time-varying representation of the modulation envelope
D. All these choices are correct

It is likely that no other amateur radio accessory has offered ham radio experimenters a bigger breakthrough in measuring antenna and feed line SWR than the ***antenna analyzer***. This allows you to check the antenna feed point impedance and resonance without putting a big test signal out on the air. Since the automatic ***micro transmitter is built into the analyzer, you don't need an actual radio to test your antenna system***. **ANSWER B.**

E4A08 Which of the following is used to measure SWR?

A. Directional wattmeter
B. Vector network analyzer
C. Antenna analyzer
D. All these choices are correct

Gordo always takes his ***portable antenna analyzer, directional wattmeter, and portable vector network analyzer*** whenever he goes out for any ham radio station checkout. Whether he's at the top of a sailboat mast, up a tower, or down inside a boat's lazarette, his portable analyzer can handle everything from automatic antenna tuners to marine VHF antennas to a big 3-element beam up 90 feet in the air. **ANSWER D.**

E9G01 Which of the following can be calculated using a Smith chart?

A. Impedance along transmission lines
B. Radiation resistance
C. Antenna radiation pattern
D. Radio propagation

The ***Smith Chart*** is an invaluable graph for ***calculating the impedance along Amateur Radio transmission lines***. Many times, you will see a completed, computer- generated Smith Chart illustrating the resonance of a commercial-quality VHF or UHF collinear base station or repeater antenna. **ANSWER A.**

E9G05 Which of the following is a common use for a Smith chart?

A. Determine the length and position of an impedance matching stub
B. Determine the impedance of a transmission line, given the physical dimensions
C. Determine the gain of an antenna given the physical and electrical parameters
D. Determine the loss/100 feet of a transmission line, given the velocity factor and conductor materials

The Smith Chart was designed to make life easier, contrary to popular opinion. The Smith Chart will show you the impedance at any point along a transmission line with a known load impedance, among other things. ***It will also reveal what you need to do to match any impedance to any other impedance***. A vector network analyzer provides a Smith Chart display of an actual antenna or other complex load. **ANSWER A.**

E9G02 What type of coordinate system is used in a Smith chart?

A. Voltage circles and current arcs
B. Resistance circles and reactance arcs
C. Voltage chords and current chords
D. Resistance lines and reactance chords

Looking at a ***Smith Chart***, you see ***two circles*** – ***resistance circles*** centered on a resistance axis – all tangent to an open circuit point (infinity resistance and infinity reactance), ***and*** portions of ***reactance arcs*** fanning out from the resistance axis and passing through the same open circuit point. Circles for positive inductive reactance are to the right of the resistance axis; circles for negative capacitance reactance are to the left of the resistance axis. **ANSWER B.**

E9G03 Which of the following is often determined using a Smith chart?

A. Beam headings and radiation patterns
B. Satellite azimuth and elevation bearings
C. Impedance and SWR values in transmission lines
D. Point-to-point propagation reliability as a function of frequency

Commercial-quality coaxial cable, coiled on a 500-foot spool, may also contain a computer-generated ***Smith Chart*** that shows ***impedance and SWR values***. **ANSWER C.**

E9G04 What are the two families of circles and arcs that make up a Smith chart?

A. Inductance and capacitance
B. Reactance and voltage
C. Resistance and reactance
D. Voltage and impedance

Resistance and reactance values comprise the "two families of circles" on the Smith Chart, as shown in the figure on the next page. Circles for positive inductive reactance are to the right of the resistance axis; circles for negative capacitance reactance are to the left of the resistance axis. **ANSWER C.**

E9G06 On the Smith chart shown in Figure E9-3, what is the name for the large outer circle on which the reactance arcs terminate?

A. Prime axis
B. Reactance axis
C. Impedance axis
D. Polar axis

The large outer circle is the reactance axis. Within the reactance axis are curved reactance circles that are tangent to the resistance axis. Where the reactance circle intersects with the reactance axis, you will see an assigned value of reactance. Inductive reactance is to the right; capacitive reactance is to the left. **ANSWER B.**

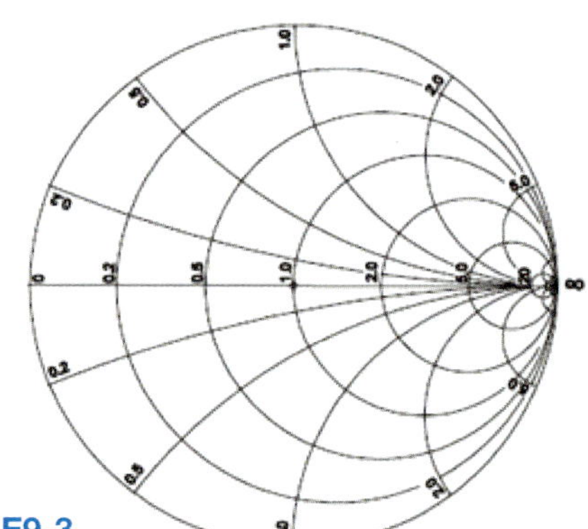

Figure E9-3

E9G07 On the Smith chart shown in Figure E9-3, what is the only straight line shown?

A. The reactance axis
B. The current axis
C. The voltage axis
D. The resistance axis

On a ***Smith chart***, the ***straight line is*** always the ***resistance axis***. **ANSWER D.**

E9G08 How is a Smith chart normalized?

A. Reassign the reactance axis with resistance values
B. Reassign the resistance axis with reactance values
C. Reassign the prime center's impedance value
D. Reassign the prime center to the reactance axis

For the Smith chart coordinate system to apply to all types of transmission lines, the coordinates are "customized" to a particular transmission line by dividing test measurements by the characteristic impedance of the transmission line. This ***process of reassigning resistance and reactance values with regard to the prime center is called "normalizing."*** **ANSWER C.**

E9G09 What third family of circles is often added to a Smith chart during the process of designing impedance matching networks?

A. Constant-SWR circles
B. Transmission line length circles
C. Coaxial-length circles
D. Radiation-pattern circles

When a computer is used to analyze an antenna design, the software will draw a ***third family of circles*** on the Smith chart ***showing constant standing wave ratios***. The SWR circle is one of several ***radially scaled parameters*** available on most commercially printed Smith Chart forms. **ANSWER A.**

E9G10 What do the arcs on a Smith chart represent?

A. Frequency
B. SWR
C. Points with constant resistance
D. Points with constant reactance

The arcs (curved lines) on a Smith chart ***are portions of reactance circles***. Remember that the large outer circle bounding the coordinate portion of the chart is the reactance axis. **ANSWER D.**

E9G11 In what units are the wavelength scales on a Smith chart calibrated?

A. In fractions of transmission line electrical frequency
B. In fractions of transmission line electrical wavelength
C. In fractions of antenna electrical wavelength
D. In fractions of antenna electrical frequency

The ***wavelength scales*** on the Smith chart are calibrated in ***portions of transmission line electrical wavelength***. **ANSWER B.**

E0A04 When evaluating a site with multiple transmitters operating at the same time, the operators and licensees of which transmitters are responsible for mitigating over-exposure situations?

A. Each transmitter that produces 20 percent or more of its MPE limit in areas where the total MPE limit is exceeded
B. Each transmitter operating with a duty cycle greater than 25 percent
C. Each transmitter that produces 5 percent or more of its MPE limit in areas where the total MPE limit is exceeded
D. Each transmitter operating with a duty cycle greater than 50 percent

If you are setting up a new repeater station on a mountaintop with many other transmitters, and ***if your station produces 5% or more of the maximum permissible*** exposure at that location, it will be ***your responsibility to mitigate*** this over-exposure MPE situation. **ANSWER C.**

E0A09 Which of the following types of equipment are exempt from RF exposure evaluations?

A. Transceivers with less than 7 watts of RF output
B. Antennas that radiate only in the near field
C. Hand-held transceivers sold before May 3, 2021
D. Dish antennas less than one meter in diameter

Every radio amateur needs to be aware of radio frequency (RF) exposure guidelines. Although it is unlikely that you will exceed RF exposure limits while operating at legal RF power, you need to be able to confirm that you are not subjecting yourself (or your neighbors) to excessive RF radiation. Some simple tools are available to readily determine your compliance, like the ARRL's RF exposure calculator: **www.arrl.org/rf-exposure-calculator**. Handheld radios sold before May 3, 2021, are given a blanket exemption from RF exposure evaluations. **ANSWER C.**

E0A10 When must an RF exposure evaluation be performed on an amateur station operating on 80 meters?

A. An evaluation must always be performed
B. When the ERP of the station is less than 10 watts
C. When the station's operating mode is CW
D. When the output power from the transmitter is less than 100 watts

Every radio amateur must perform an RF exposure evaluation to determine operating safety. At legal power levels, it is nearly impossible to exceed exposure levels, especially at HF frequencies. Fortunately, actual measurements aren't necessary except under unusual circumstances. Simple, readily available charts will allow you quickly to determine if your RF exposure levels are safe...and they almost always will be. The ARRL has a handy RF exposure calculator here: **http://www.arrl.org/rf-exposure-calculator** **ANSWER A.**

E0A05 What hazard is created by operating at microwave frequencies?

A. Microwaves are ionizing radiation
B. The high gain antennas commonly used can result in high exposure levels
C. Microwaves are in the frequency range where wave velocity is higher
D. The extremely high frequency energy can damage the joints of antenna structures

Gordo runs 10 GHz microwave with a relatively large 20 dB dish antenna. He's always cautious to never let anyone stand in the antenna main lobe when transmitting. He doesn't want to ***"cook" his neighbors with his microwave transmissions!*** **ANSWER B.**

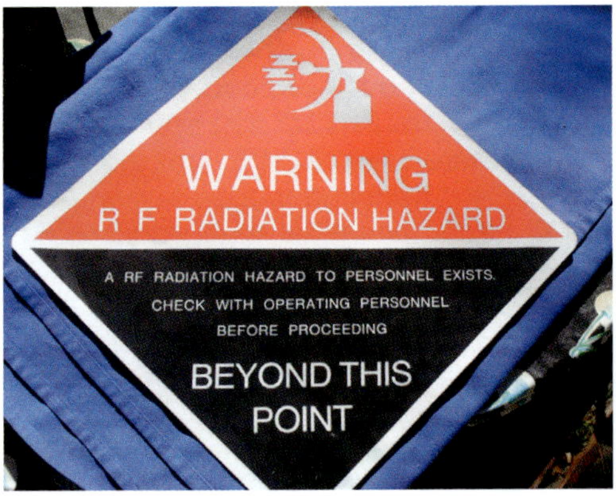

This sign says all!

E0A03 Over what range of frequencies are the FCC human body RF exposure limits most restrictive?

A. 300 kHz - 3 MHz
B. 3 - 30 MHz
C. 30 - 300 MHz
D. 300 - 3000 MHz

While operating within legally licensed amateur power limits will present minimal hazard to the human body at any frequency, the human body will absorb the most RF energy in the frequency range of ***30 – 300 MHz***, for a number of reasons, one being that the human body is about the length of a dipole at 100 MHz or so. At HF and lower frequencies, as well, most RF passes unrestricted through the body, while at UHF frequencies, RF is mostly reflected before penetrating the body significantly. Therefore, the recommended maximum ANSI exposure threshold is a minimum in the VHF range. **ANSWER C.**

E0A08 What does SAR measure?

A. Signal attenuation ratio
B. Signal amplification rating
C. The rate at which RF energy is absorbed by the body
D. The rate of RF energy reflected from stationary terrain

SAR stands for ***Specific Absorption Rate, the rate at which RF energy is absorbed by the human body***. Just like cooking a wiener in the microwave oven, the more time it gets microwaved, the more cooked it gets, not good for human tissue and innards! Be careful around Radio Frequency Exposure. **ANSWER C.**

E0A02 When evaluating RF exposure levels from your station at a neighbor's home, what must you do?

A. Ensure signals from your station are less than the controlled maximum permitted exposure (MPE) limits
B. Ensure signals from your station are less than the uncontrolled maximum permitted exposure (MPE) limits
C. Ensure signals from your station are less than the controlled maximum permitted emission (MPE) limits
D. Ensure signals from your station are less than the uncontrolled maximum permitted emission (MPE) limits

Your ***neighbor's house is considered "uncontrolled"*** when calculating maximum permissible exposure limits. Always make sure you do the calculations before putting up your new antenna system and make absolutely sure you are always ***less than the uncontrolled MPE limit***. **ANSWER B.**

E0A06 Why are there separate electric (E) and magnetic (H) MPE limits at frequencies below 300 MHz?

A. The body reacts to electromagnetic radiation from both the E and H fields
B. Ground reflections and scattering cause the field strength to vary with location
C. E field and H field radiation intensity peaks can occur at different locations
D. All these choices are correct

Radio transmissions contain both an electric field (E) and a magnetic field (H). These field radiations may peak at different distances from the antenna. In addition, ground reflections and scattering will vary the field intensity. The human body reacts differently to E and H fields. So, take ***all of these factors*** into account when do your calculations for ***both magnetic and electric fields***. **ANSWER D.**

E1B06 Which of the following additional rules apply if you are erecting an amateur station antenna structure at a site at or near a public use airport?

A. You may have to notify the Federal Aviation Administration and register it with the FCC as required by Part 17 of the FCC rules
B. You may have to enter the height above ground in meters, and the latitude and longitude in degrees, minutes, and seconds on the FAA website
C. You must file an Environmental Impact Statement with the EPA before construction begins
D. You must obtain a construction permit from the airport zoning authority per Part 119 of the FAA regulations

Live near an airport? If you do, you will need to ***consult Part 17 of the FCC Rules*** regarding the installation of that antenna tower. Likely, you will be ***working with the Federal Aviation Administration*** to get that tower sanctioned, as well as getting it documented, under Part 17 Rules. [97.15] **ANSWER A.**

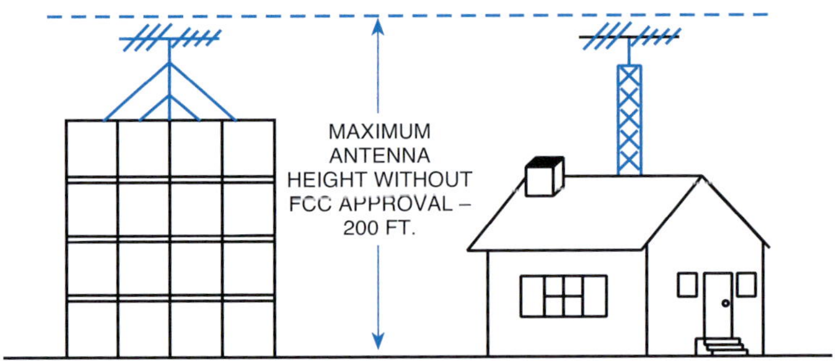

Antenna Height near a public use airport is governed by both the FCC and the FAA.

E0A07 What is meant by "100% tie-off" regarding tower safety?

A. All loose ropes and guys secured to a fixed structure
B. At least one lanyard attached to the tower at all times
C. All tools secured to the climber's harness
D. All circuit breakers feeding power to the tower must be tied closed with tape, cable, or ties

By having ***at least one lanyard attached between your body and a tower you're climbing at all times***, you will be fairly safe from falling (at least very far). Always use an energy absorbing lanyard; the sudden stop at the end of a "normal" rope can be a bit rough, if you do happen to fall. **ANSWER B.**

It takes two lanyards to maintain 100% tie-off. When moving one, the other remains attached.

E0A11 To what should lanyards be attached while climbing?

A. Antenna mast
B. Guy brackets
C. Tower rungs
D. Tower legs

The safest place to attach a climbing lanyard is around an actual tower leg (or two). Welded-on tower rungs have been known to fail, especially on older or poorly maintained towers. Be careful where you step as well, always be alert for loose or bent rungs. **ANSWER D.**

E0A12 Where should a shock-absorbing lanyard be attached to a tower when working above ground?

A. Above the climber's head level
B. To the belt of the fall-arrest harness
C. Even with the climber's waist
D. To the next lowest set of guys

Shock absorbing lanyards should always be attached to a tower leg above your head. As you ascend a tower, never climb above a previously attached anchor point (except momentarily). It may seem tedious to continually be "scooting" the lanyard up, especially when climbing a really tall tower, but it's the only way the lanyard can offer its full protection. **ANSWER A.**

E4E14 What is the purpose of a single point ground panel?

A. Remove AC power in case of a short-circuit
B. Prevent common-mode transients in multi-wire systems
C. Eliminate air gaps between protected and non-protected circuits
D. Ensure all lightning protectors activate at the same time

Most lightning damage is caused by a ***difference of potential*** between two or more items of equipment. By the way, it doesn't take a direct hit by lightning to cause tremendous damage; a large nearby lightning stroke can induce phenomenal electric currents in nearby conductors! In either case, by having everything in the shack properly ***bonded***, that is, at the same electrical potential, lighting damage can usually be minimized. The most convenient way of doing this is with a single point ground panel, which should also be the ground point for any lightning protection devices like Polyphasers. Polyphasers should also be inspected after any nearby lightning strokes, as they do have a finite number of "firings" they can withstand. ***Grounding and Bonding for the Radio Amateur*** is probably the best book on the topic. Every ham should have a copy. **https://home.arrl.org/action/Store/Product-Details/productId/133989**. **ANSWER D.**

5

Taking the Extra Class Examination

Get set to pass the Element 4 examination and gain additional band privileges as an Extra Class operator! If you are upgrading from General to Extra Class, you will gain a whopping 500 kHz of *additional* frequencies on the 75/80, 40, 20, and 15 meter worldwide bands. If you are upgrading from an Advanced Class license, you will gain an *additional* 250 kHz of band width on these same worldwide bands. Your Extra Class license gives you full privileges and full frequency coverage on every amateur band assigned by the Federal Communications Commission.

EXAMINATION ADMINISTRATION

Three Extra Class licensees who are also accredited Volunteer Examiners are required to administer your Element 4 exam.

Volunteer Examiner teams offer examinations on a regular basis at local sites to serve their communities. You can obtain information about VECs and exam sessions in your area by checking with your local radio club, ham radio store, or local packet bulletin boards. A list of VECs that was current at the time of publication is given in the Appendix (see page 255).

Once you have found your local VE team, contact them to select a test date and location and pre-register for your examination.

The Volunteer Examiners are permitted to charge you a fee for certain reimbursable expenses incurred in preparing, administering, and processing the examination. The fee is currently is about $15.00.

EXAM CONTENT

The questions, answers, and distractors for each question of the Extra Class written examination are public information. The question pool included in this book contains all 603 possible questions that can be used to make up your 50-question Element 4 written examination. The VEC is not permitted to change any wording, punctuation or numerical values included in the questions, answers, or distractors. The VEC can change the A-B-C-D order of the answer choices if it wishes.

Also, the VEC is required to select one question from each syllabus topic under each subelement so you should expect an exam that does just that. Look again at the question pool syllabus on page 258 to see how the test will be constructed. Remember, if there is a particular topic within a subelement that you just can't seem to grasp, there will be only one question from that topic. You will still pass the exam if you miss that one question.

Want to find a test site fast?
Call 1-860-594-0200, or visit www.arrl.org/exam.

WHAT TO BRING TO THE EXAM SESSION

Here's what you'll need to bring with you for your Extra Class examination:

1. The fee of $15.00.
2. The original plus a copy of your current General or Advanced Class license. If your new license has not yet been granted, make sure you bring the original plus two copies of your Certificate of Successful Completion of Examination (CSCE) indicating your most current license status.
3. A photo identification card – your driver's license is ideal for this purpose.
4. Some sharp pencils and fine-tip pens. It's good to have a backup.
5. Calculators may be used. However, the examiners may erase the memory before your exam begins.
6. Any other items that the VEC asks you to bring.

If you have a physical disability that requires additional assistance by your examination team, let them know ahead of time so they can prepare accommodations for your special requirements. This could be reading equipment, a volunteer examiner reader, access for a wheelchair, or any other special requirements that they may be able to accommodate for you. It is best for everyone if your requirements are known in advance. Be sure to contact the volunteer examiner team leader before your test session.

Arrive at the testing site with plenty of time to check in and complete the necessary forms that are required before you can begin your exam.

TAKING THE EXAMINATION

Get a good night's sleep before the exam day. Continue to study the Extra Class formulas right up to the moment you go into the examination room. Don't believe that old saying that too much study will cause you to forget the subject material. If you purchased the audio course, continue to play them up to the last minute.

Listen carefully for the instructions of your Volunteer Examiners. Most examination sessions will conduct all levels of testing, so listen carefully to the announcements so you get the correct testing materials.

Make no marks on the examination sheet. If you don't know an answer, lightly shade in one of the boxes on the answer sheet for that particular question and make a big note on your scratch paper to go back and work out the correct answer. Using this technique will keep you from accidentally misaligning your answer boxes on the answer sheet. If you finish the exam at question #49, chances are you skipped a question and all of your work is out of sync with your answer sheet. Double check that all 50 answer boxes indeed have a mark.

When you are taking the written examination, *take your time!* Some answers start out looking correct, but end up wrong at the end. Don't speed read the exam – getting careless after all this preparation could undo all the hard work you have put in on preparing for the Extra Class exam.

If you don't know an answer, eliminate the obviously wrong answers, and then take your best guess. NEVER LEAVE AN ANSWER BLANK! Leaving an answer blank is counted as a wrong answer, so you might as well make your best guess. We also find that going back and second-guessing an answer will sometimes cause you to change to the wrong answer. Usually your first impression is the correct one. Another trick is to read your answer first and then see if it agrees with what the question asks. Start this technique at Answer 50, and go backwards.

AND THE RESULTS ARE...

When you are satisfied that you have completed the Extra Class exam, turn in all of your examination papers – including scratch paper and the original test paper. Thank your examiners for their courtesy and follow their instructions while they prepare your score. *Don't stand over the examiners* as they grade your exam.

When you pass, the examiners may or may not indicate how many questions you missed. DON'T ASK – the examiners may tell you it is against their rules to divulge which questions you got wrong. Although we're not aware of any rule that states they cannot tell you what you got wrong, it's usually just a case of not enough time to spend with any one applicant pouring over their exam performance. They usually will tell you your score, and passing the exam is exactly what you want to hear!

When you pass your Element 4 exam, the VEs will issue you an official CSCE – Certificate of Successful Completion of Examination. This important document is official proof that you passed the exam. Check to see that it is signed by all three volunteer examiners. Be sure to sign your CSCE. Double-check that the date is correct. If the computer system should gobble up all of your test results, this is LEGAL PROOF that you indeed passed a particular examination element.

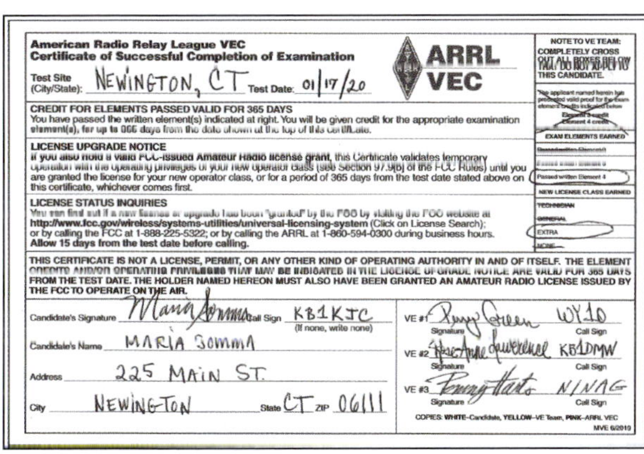

American Radio Relay League VEC
Certificate of Successful Completion of Examination

ARRL VEC

Test Site (City/State): NEWINGTON, CT Test Date: 01/17/20

NOTE TO VE TEAM: COMPLETELY CROSS OUT ALL BOXES BELOW THAT DO NOT APPLY TO THIS CANDIDATE.

CREDIT FOR ELEMENTS PASSED VALID FOR 365 DAYS
You have passed the written element(s) indicated at right. You will be given credit for the appropriate examination element(s), for up to 365 days from the date shown at the top of this certificate.

EXAM ELEMENTS EARNED

Passed written Element 4

LICENSE UPGRADE NOTICE
If you also hold a valid FCC-issued Amateur Radio license grant, this Certificate validates temporary operation with the operating privileges of your new operator class (see Section 97.9[b] of the FCC Rules) until you are granted the license for your new operator class, or for a period of 365 days from the test date stated above on this certificate, whichever comes first.

NEW LICENSE CLASS EARNED

TECHNICIAN

GENERAL

EXTRA

NONE

LICENSE STATUS INQUIRIES
You can find out if a new license or upgrade has been "granted" by the FCC by visiting the FCC website at http://www.fcc.gov/wireless/systems-utilities/universal-licensing-system (Click on License Search); or by calling the FCC at 1-888-225-5322; or by calling the ARRL at 1-860-594-0300 during business hours. Allow 15 days from the test date before calling.

THIS CERTIFICATE IS NOT A LICENSE, PERMIT, OR ANY OTHER KIND OF OPERATING AUTHORITY IN AND OF ITSELF. THE ELEMENT CREDITS AND/OR OPERATING PRIVILEGES THAT MAY BE INDICATED IN THE LICENSE UPGRADE NOTICE ARE VALID FOR 365 DAYS FROM THE TEST DATE. THE HOLDER NAMED HEREON MUST ALSO HAVE BEEN GRANTED AN AMATEUR RADIO LICENSE ISSUED BY THE FCC TO OPERATE ON THE AIR.

Candidate's Signature Maria Somma Call Sign KB1KJC (If none, write none)

Candidate's Name MARIA SOMMA

Address 225 MAIN ST.

City NEWINGTON State CT ZIP 06111

VE #1 Signature Call Sign WY1O

VE #2 Signature Call Sign KB1DMW

VE #3 Signature Call Sign N1NAG

COPIES: WHITE–Candidate, YELLOW–VE Team, PINK–ARRL VEC

MVE 6/2010

Congratulations! You Passed!

> **NOTE:** The renewal period on your new Extra Class license is NOT 10 years. It remains the SAME renewal date that is now on your General or Advanced Class license. Make sure your address is up-to-date, too! An upgrade to your license class does NOT change the original grant date of your previous license class.

CHANGING CALL SIGNS

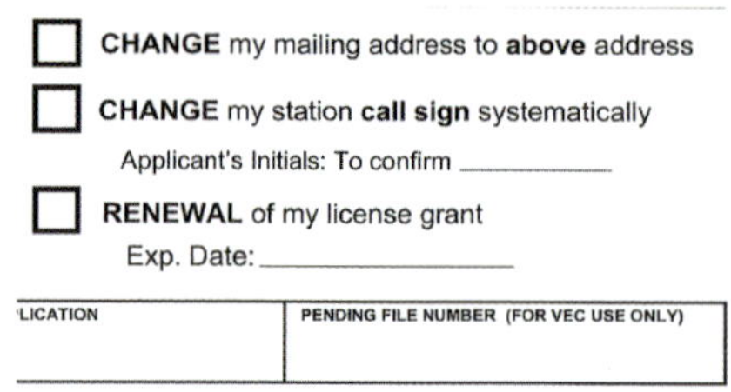

☐ **CHANGE** my mailing address to **above** address

☐ **CHANGE** my station **call sign** systematically

Applicant's Initials: To confirm ___________

☐ **RENEWAL** of my license grant

Exp. Date: ___________

LICATION	PENDING FILE NUMBER (FOR VEC USE ONLY)

Only check the "CHANGE" my station call sign box if you want the FCC computer to give you a new call sign. If you want to choose a call sign, apply for a vanity call.

If you check the box "CHANGE my station call sign systematically" on the NCVEC Form 605, the FCC computer will assign you the next Group "A" call sign. A Group "A" call sign will be in a 2 by 2 format with "A" as the first letter. The complete 2 by 2 format has two letters followed by a number indicating the part of the country where you live, then two more letters, assigned alphabetically. Once all the Group "A" call signs are assigned, the FCC will begin to assign Group "B" 2 by 2 call signs, which begin with "K", "N", or "W".

You must initial the request if you decide to systematically change your call sign. Once changed, you cannot easily change back to your previous call sign. The only way you can go back to a previously-held call sign is under the Vanity call sign program. So be thinking about it right now – do you want to stay with your present call sign, or do you want to change over to a new Extra Class 2-by-2 call sign?

VANITY CALL SIGNS

The ability to select a call sign of your choice began in 1996 and it has been immensely popular, especially with Extra Class amateurs who historically like to have short call signs. While only a little more than 10% of all radio amateurs ever make it to the top Extra Class level, they hold nearly half of all Vanity call signs issued!

Extra Class radio amateurs can choose an available call sign containing any Group, A, B, C or D Amateur format. Most try to get a 1-by-2 or a 2-by-1 format, which are only available to an Extra Class licensee. A 1-by-2 Group "A" call sign begins with either K, N or W (but not "A") followed by a numeral and two letters. For example: W5YI is a Group "A" call sign.

An example of a 2-by-1 call sign is AA1A. The two letter prefix must be from the AA to AL, KA to KZ, NA to NZ or the WA to WZ prefix block followed by a numeral and a single letter. To make matters more confusing, certain prefixes (AH, AL, KH, KL, KP, NH, NL, NP, WH, WL and WP) are reserved for amateurs who have a mailing address outside of the 48 contiguous U.S. states.

You can find out all the rules (and there are many!) surrounding obtaining a Vanity call sign by visiting the ARRL's website at **www.arrl.org/applying-for-a-vanity-call**.

You can file the application yourself online. Additional information on Vanity call signs is available from the FCC's Amateur Radio website located at:

https://www.fcc.gov/wireless/bureau-divisions/mobility-division/amateur-radio-service/vanity-faq

BECOMING A VOLUNTEER EXAMINER

When you successfully complete your Extra Class exam, ask your VE team for a volunteer examiner sign-up application. After only a short processing time, you'll join a cadre of other Extra Class hams who give something back to ham radio by helping with testing. With just a few hours a month, and some paper and computer processing, you will enjoy this rewarding "extra credit" as an Extra Class operator.

Accredited Volunteer Examiners also help develop new questions when the amateur radio question pools undergo periodic updating. Each pool gets updated once every four years with a public notice for radio experts to submit new or improved examination questions along with four suggested right and wrong answers. Your comments and suggestions are filed with the NCVEC Question Pool Committee. Join in – as an Extra Class operator. It's fun to see that you have contributed to the new question pool and the growth of the hobby.

Here's how to become a volunteer examiner:

If you are interested in becoming an ARRL Volunteer Examiner (VE) and serving the Amateur Radio community, it's easy and free!

Visit **www.arrl.org/become-an-arrl-ve** for resources and information.

FREE ACHIEVEMENT CERTIFICATE AND MORE!

You made it through to Extra Class! Congratulations! Allow me to send you a free certificate of achievement. Send a large envelope, self-addressed, with 12 first class stamps on the inside to me at Gordon West Radio School, 2414 College Drive, Costa Mesa, California 92626. It will take about 15 days to get the certificate back to you with a copy of the League's *On The Air* magazine, a color band chart and grid square maps, a list of country prefixes, a host of training aids, and a hearty congratulations on making it to the top.

Gordon West, WB6NOA, comes with the book and courses! Call him if you need help with your training at 714-549-5000, Monday – Thursday, 10am – 4pm, California time!

Eric and I hope to hear you soon on the worldwide airwaves or see you at an upcoming hamfest.

Welcome to the top!

Gordon West,
WB6NOA

I know congratulations will be in order very soon when you pass your Extra Class exam with flying colors!

Appendix

U.S. VOLUNTEER EXAMINER COORDINATORS IN THE AMATEUR SERVICE

American Radio Relay League (ARRL)
225 Main Street
Newington, CT 06111-1494
860-594-0200
860-594-0339
vec@arrl.org
www.arrl.org

Anchorage ARC VEC
P.O. Box 190192
Anchorage, AK 99519
907-632-3963
vec@kl7aa.org
https://kl7aa.org/vec/

Central America CAVEC, Inc.
P.O. Box 1984
Madison, AL 35758
938-653-5155
http://www.cavechamexam.com/

Golden Empire Amateur Radio Society (GEARS)
P.O. Box 508
Chico, CA 95927-0508
530-893-9211
www.gearsw6rhc.org/

Greater L.A. Amateur Radio Group
23371 Mulholland Drive, Unit 186
Woodland Hills, CA 91364
805-410-4169
www.glaarg.org

Jefferson Amateur Radio Club
P.O. Box 73665
Metairie, LA 70033
w5gad.vec@w5gad.org
504-636-8809
https://w5gad.org/

Laurel Amateur Radio Club, Inc.
5287 W Belmont Rd
Tucson, AZ 85743
520-664-5070
https://www.laurelvec.com/

MRAC VEC, Inc. (Milwaukee)
2505 S. Calhoun Rd., #203
New Berlin, WI 53151
www.mracvec.org/

MO-KAN VEC Coordinator
228 Tennessee Road
Richmond, KS 66080-9174
785-867-2011

Sandarc-VEC
5511 Maryland Ave
La Mesa, CA 94086-0014
P: 619-843-3747
http://www.sandarc.org/

Sunnyvale VEC Amateur Radio Club, Inc.
P.O. Box 70014
Sunnyvale, CA 94086-0014
408-255-9000
www.amateur-radio.org

W4VEC
1150 Rock Grove Church Rd.
Salisbury, NC 28146-0325
901-691-8810
www.w4vco.com

W5YI-VEC
P.O. Box 260916
Plano, TX 75026-0916
972-412-5221
www.w5yi-vec.org

Western Carolina Amateur Radio Society VEC, Inc.
7 Skylyn Ct.
Asheville, NC 28806-3922
828-253-1192
www.wcars-vec.org/

2024-2028 ELEMENT 4 Q&A CROSS REFERENCE

The following cross reference presents all 603 question numbers in numerical order that are included in the 2024-2028 Element 4 Extra Class question pool, followed by the page number on which the question begins in this book. This will allow you to locate specific questions by their question number.

ELEMENT 4 (EXTRA CLASS) QUESTION POOL SYLLABUS

The syllabus used by the QPC for the development of the question pool is included here as an aid in studying the subelements and topic groups. Review the syllabus before you start your study to gain an understanding of how the question pool is used to develop the Element 4 written examination. Remember, one question will be taken from each topic group within each subelement to create your exam. The QPC topic groups are not the same as those used by Gordon West in this book.

E1 - COMMISSION RULES

[6 Exam Questions - 6 Groups]

E1A Frequency privileges; signal frequency range; automatic message forwarding; stations aboard ships or aircraft; power restriction on 630- and 2200-meter bands

E1B Station restrictions and special operations: restrictions on station location; general operating restrictions; spurious emissions; antenna structure restrictions; RACES operations

E1C Automatic and remote control; band-specific regulations; operating in and communicating with foreign countries; spurious emission standards; HF modulation index limit; band-specific rules

E1D Amateur Space and Earth stations; telemetry and telecommand rules; identification of balloon transmissions; one-way communications

E1E Volunteer examiner program: definitions; qualifications; preparation and administration of exams; reimbursement; accreditation; question pools; documentation requirements

E1F Miscellaneous rules: external RF power amplifiers; prohibited communications; spread spectrum; auxiliary stations; Canadian amateurs operating in the US; special temporary authority

E2 - OPERATING PROCEDURES

[5 Exam Questions - 5 Groups]

E2A Amateur radio in space: amateur satellites; orbital mechanics; frequencies and modes; satellite hardware; satellite operations

E2B Television practices: fast-scan television standards and techniques; slow scan television standards and techniques

E2C Contest and DX operating; remote operation techniques; log data format; contact confirmation; RF network systems

E2D Operating methods: digital modes and procedures for VHF and UHF; APRS; EME procedures; meteor scatter procedures

E2E Operating methods: digital modes and procedures for HF

E3 - RADIO WAVE PROPAGATION

[3 Exam Questions - 3 Groups]

E3A Electromagnetic Waves and Specialized Propagation: Earth-Moon-Earth (EME) communications; meteor scatter; microwave tropospheric and scatter propagation; auroral propagation; daily variation of ionospheric propagation; circular polarization

E3B Transequatorial propagation; long-path propagation; ordinary and extraordinary waves; chordal hop; sporadic-E mechanisms; ground-wave propagation

E3C Propagation prediction and reporting: radio horizon; effects of space-weather phenomena

E4 - AMATEUR PRACTICES

[5 Exam Questions - 5 Groups]

E4A Test equipment: analog and digital instruments; spectrum analyzers; antenna analyzers; oscilloscopes; RF measurements

E4B Measurement technique and limitations: instrument accuracy and performance limitations; probes; techniques to minimize errors; measurement of Q; instrument calibration; S parameters; vector network analyzers; RF signals

E4C Receiver performance: phase noise, noise floor, image rejection, minimum detectable signal (MDS), increasing signal-to-noise ratio and dynamic range, noise figure, reciprocal mixing; selectivity; SDR non-linearity; use of attenuators at low frequencies

E4D Receiver performance characteristics: dynamic range; intermodulation and cross-modulation interference; third-order intercept; desensitization; preselector; sensitivity; link margin

E4E Noise and interference: external RF interference; electrical and computer noise; line noise; DSP filtering and noise reduction; common-mode current; surge protectors; single point ground panel

E5 - ELECTRICAL PRINCIPLES

[4 Exam Questions - 4 Groups]

E5A Resonance and Q: characteristics of resonant circuits; series and parallel resonance; definitions and effects of Q; half-power bandwidth

E5B Time constants and phase relationships: RL and RC time constants; phase angle in reactive circuits and components; admittance and susceptance

E5C Coordinate systems and phasors in electronics: rectangular coordinates; polar coordinates; phasors; logarithmic axes

E5D RF effects in components and circuits: skin effect; real and reactive power; electrical length of conductors

E6 - CIRCUIT COMPONENTS

[6 Exam Questions - 6 Groups]

E6A Semiconductor materials and devices: semiconductor materials; bipolar junction transistors; operation and types of field-effect transistors

E6B Diodes

E6C Digital ICs: families of digital ICs; gates; programmable logic devices

E6D Inductors and piezoelectricity: permeability, core material and configuration; transformers; piezoelectric devices

E6E Semiconductor materials and packages for RF use

E6F Electro-optical technology: photoconductivity; photovoltaic devices; optical sensors and encoders; optically isolated switching

E7 - PRACTICAL CIRCUITS
[8 Exam Questions - 8 Groups]

E7A Digital circuits: digital circuit principles and logic circuits; classes of logic elements; positive and negative logic; frequency dividers; truth tables
E7B Amplifiers: class of operation; vacuum tube and solid-state circuits; distortion and intermodulation; spurious and parasitic suppression; switching-type amplifiers
E7C Filters and matching networks: types of networks; types of filters; filter applications; filter characteristics; impedance matching
E7D Power supplies and voltage regulators; solar array charge controllers
E7E Modulation and demodulation: reactance, phase, and balanced modulators; detectors; mixers
E7F Software defined radio fundamentals: digital signal processing (DSP) filtering, modulation, and demodulation; analog-digital conversion; digital filters
E7G Operational amplifiers: characteristics and applications
E7H Oscillators and signal sources: types of oscillators; synthesizers and phase-locked loops; direct digital synthesizers; stabilizing thermal drift; microphonics; high-accuracy oscillators

E8 - SIGNALS AND EMISSIONS
[4 Exam Questions - 4 Groups]

E8A Fourier analysis; RMS measurements; average RF power and peak envelope power (PEP); analog/digital conversion
E8B Modulation and demodulation: modulation methods; modulation index and deviation ratio; frequency- and time-division multiplexing; orthogonal frequency-division multiplexing (OFDM)
E8C Digital signals: digital communication modes; information rate vs. bandwidth; error correction; constellation diagrams
E8D Keying defects and overmodulation of digital signals; digital codes; spread spectrum

E9 - ANTENNAS AND TRANSMISSION LINES
[8 Exam Questions - 8 Groups]

E9A Basic antenna parameters: radiation resistance, gain, beamwidth, efficiency; effective radiated power (ERP) and effective isotropic radiated power (EIRP)
E9B Antenna patterns and designs: azimuth and elevation patterns; gain as a function of pattern; antenna modeling
E9C Practical wire antennas; folded dipoles; phased arrays; effects of ground near antennas
E9D Yagi antennas; parabolic reflectors; feed point impedance and loading of electrically short antennas; antenna Q; RF grounding
E9E Impedance matching: matching antennas to feed lines; phasing lines; power dividers
E9F Transmission lines: characteristics of open and shorted feed lines; coax versus open wire; velocity factor; electrical length; coaxial cable dielectrics; microstrip
E9G The Smith chart
E9H Receiving antennas: radio direction finding (RDF) techniques; Beverage antennas; single- and multiple-turn loops

E0 - SAFETY
[1 exam question - 1 group]

E0A Safety: RF radiation hazards; hazardous materials; grounding

SCIENTIFIC NOTATION

Prefix	Symbol			Multiplication Factor
exa	E	10^{18}	=	1,000,000,000,000,000,000
peta	P	10^{15}	=	1,000,000,000,000,000
tera	T	10^{12}	=	1,000,000,000,000
giga	G	10^{9}	=	1,000,000,000
mega	M	10^{6}	=	1,000,000
kilo	k	10^{3}	=	1,000
hecto	h	10^{2}	=	100
deca	da	10^{1}	=	10
(unit)		10^{0}	=	1

Prefix	Symbol			Multiplication Factor
deci	d	10^{-1}	=	0.1
centi	c	10^{-2}	=	0.01
milli	m	10^{-3}	=	0.001
micro		10^{-6}	=	0.000001
nano	n	10^{-9}	=	0.000000001
pico	p	10^{-12}	=	0.000000000001
femto	f	10^{-15}	=	0.000000000000001
atto	a	10^{-18}	=	0.000000000000000001

SCHEMATIC SYMBOLS

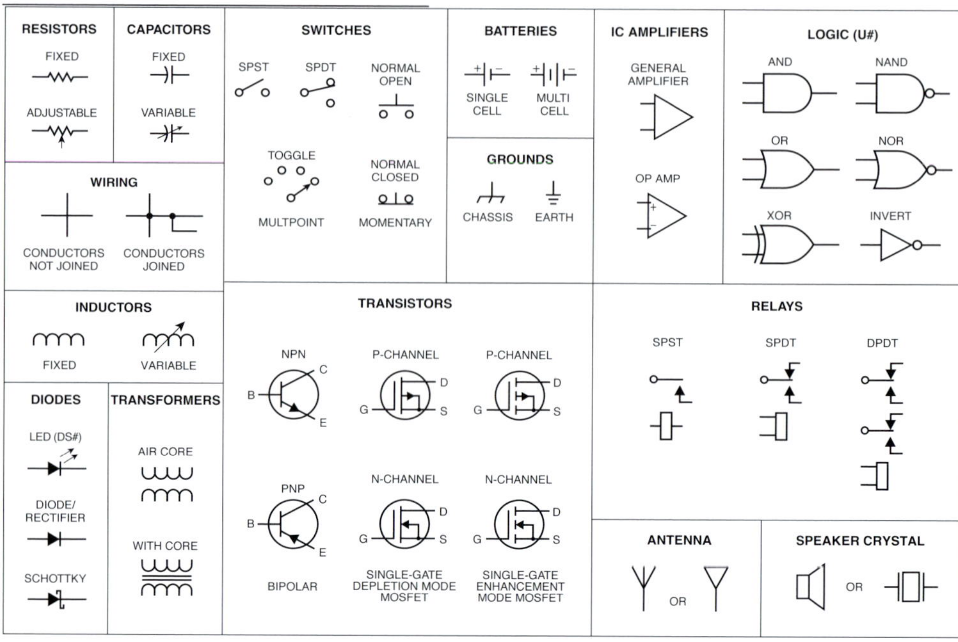

Decibels

It is important to know about decibels because they are used extensively in electronics. Look at the derivation for decibels and note it is a measure of the ratio of two powers, P_1 and P_2. Remember the power changes for different dB values. Also, since 6 dB is a four times change, 16 dB (6 dB + 10 dB) will be a 4 × 10 = 40 times change, and 26 dB (6 dB + 20 dB) will be a 4 × 100 = 400 times change. Thus, any dB value from 10 dB and above can be evaluted by using the power change of 1 dB to 9 dB and multiplying it by the 10 dB, 20 dB, 30 dB, 40 dB, 50 dB, 60 dB, etc. change. Thus, 57 dB (7 dB + 50 dB) will be a 5 × 100,000 = 500,000 times change.

E

dB	Power Change $\frac{P_1}{P_2}$	
1 dB	1 ¼X	Power change
2 dB	1 ½X	Power change
3 dB	2X	Power change
4 dB	2 ½X	Power change
5 dB	3X	Power change
6 dB	4X	Power change
7 dB	5X	Power change

dB	Power Change $\frac{P_1}{P_2}$	
8 dB	6 ¼X	Power change
9 dB	8X	Power change
10 dB	10X	Power change
20 dB	100X	Power change
30 dB	1000X	Power change
40 dB	10,000X	Power change
50 dB	100,000X	Power change
60 dB	1,000,000X	Power change

Derivation:

If $\text{dB} = 10 \log_{10} \frac{P_1}{P_2}$

then what power ratio is 60 dB?

$$60 = 10 \log_{10} \frac{P_1}{P_2}$$

$$\frac{60}{10} = \log_{10} \frac{P_1}{P_2}$$

$$6 = \log_{10} \frac{P_1}{P_2}$$

Remember: logarithm of a number is the exponent to which the base must be raised to get the number.

$$\therefore 10^6 = \frac{P_1}{P_2}$$

$$1{,}000{,}000 = \frac{P_1}{P_2}$$

Or $P_1 = 1{,}000{,}000\ P_2$

60 dB means P_1 is one million times P_2

AUTHORIZED FREQUENCY BANDS – AMATEUR SERVICE (for U.S. Amateur Stations operating from ITU-Region 2–North and South America)

METERS	Novice	Technician	General	Advanced	Extra Class
Current License Class[1]		Technician	General		Extra Class
Grandfathered[2]	Novice			Advanced	
2200			135.7-137.8 kHz/All	135.7-137.8 kHz/All	135.7-137.8 kHz/All
630			472-479 kHz/All	472-479 kHz/All	472-479 kHz/All
160			1800-2000 kHz/All	1800-2000 kHz/All	1800-2000 kHz/All
80 75	3525-3600 kHz/CW	3525-3600 kHz/CW	3525-3600 kHz/CW 3800-4000 kHz/Ph	3525-3600 kHz/CW 3700-4000 kHz/Ph	3500-4000 kHz/CW 3600-4000 kHz/Ph
60			5 channels	5 channels	5 channels
40	7025-7125 kHz/CW	7025-7125 kHz/CW	7025-7125 kHz/CW 7175-7300 kHz/Ph	7025-7125 kHz/CW 7125-7300 kHz/Ph	7000-7300 kHz/CW 7125-7300 kHz/Ph
30			10.1-10.15 MHz/CW	10.1-10.15 MHz/CW	10.1-10.15 MHz/CW
20			14.025-14.15 MHz/CW 14.225-14.35 MHz/Ph	14.025-14.15 MHz/CW 14.175-14.35 MHz/Ph	14.0-14.35 MHz/CW 14.15-14.35 MHz/Ph
17			18.068-18.11 MHz/CW 18.11-18.168 MHz/Ph	18.068-18.11 MHz/CW 18.11-18.168 MHz/Ph	18.068-18.11 MHz/CW 18.11-18.168 MHz/Ph
15	21.025-21.2 MHz/CW	21.025-21.2 MHz/CW	21.025-21.2 MHz/CW 21.275-21.45 MHz/Ph	21.025-21.2 MHz/CW 21.225-21.45 MHz/Ph	21.0-21.45 MHz/CW 21.2-21.45 MHz/Ph
12			24.89-24.99 MHz/CW 24.93-24.99 MHz/Ph	24.89-24.99 MHz/CW 24.93-24.99 MHz/Ph	24.89-24.99 MHz/CW 24.93-24.99 MHz/Ph
10	28.0-28.5 MHz/CW 28.3-28.5 MHz/Ph	28.0-28.5 MHz/CW 28.3-28.5 MHz/Ph	28.0-28.3 MHz/CW 28.3-29.7 MHz/Ph	28.0-28.3 MHz/CW 28.3-29.7 MHz/Ph	28.0-29.7 MHz/CW 28.3-29.7 MHz/Ph
6		50-54 MHz/CW 50.1-54 MHz/Ph	50-54 MHz/CW 50.1-54 MHz/Ph	50-54 MHz/CW 50.1-54 MHz/Ph	50-54 MHz/CW 50.1-54 MHz/Ph
2		144-148 MHz/CW 144.1-148 MHz/All	144-148 MHz/CW 144.1-148 MHz/Ph	144-148 MHz/CW 144.1-148 MHz/All	144-148 MHz/CW 144.1-148 MHz/All
1.25	222-225 MHz/All	222-225 MHz/All	222-225 MHz/All	222-225 MHz/All	222-225 MHz/All
0.70		420-450 MHz/All	420-450 MHz/All	420-450 MHz/All	420-450 MHz/All
0.33		902-928 MHz/All	902-928 MHz/All	902-928 MHz/All	902-928 MHz/All
0.23	1270-1295 MHz/All	1240-1300 MHz/All	1240-1300 MHz/All	1240-1300 MHz/All	1240-1300 MHz/All

GRID-SQUARE MAP FOR UNITED STATES

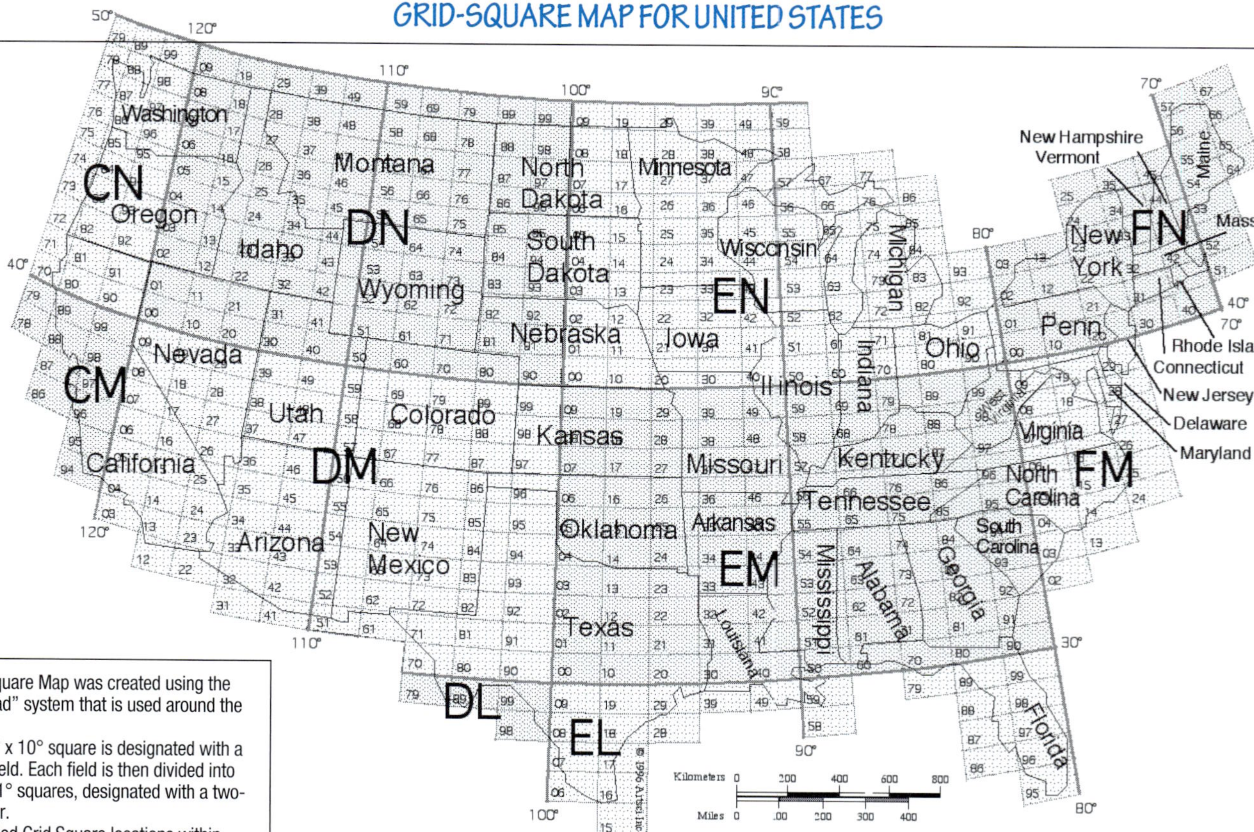

This Grid Square Map was created using the "Maidenhead" system that is used around the world.

Each 20° x 10° square is designated with a two-letter field. Each field is then divided into 100 – 2° x 1° squares, designated with a two-digit number.

For detailed Grid Square locations within each state, refer to the U.S. Repeater Mapbook 1996/97 edition.

http://home.earthlink.net/~artsci

ITU Regions

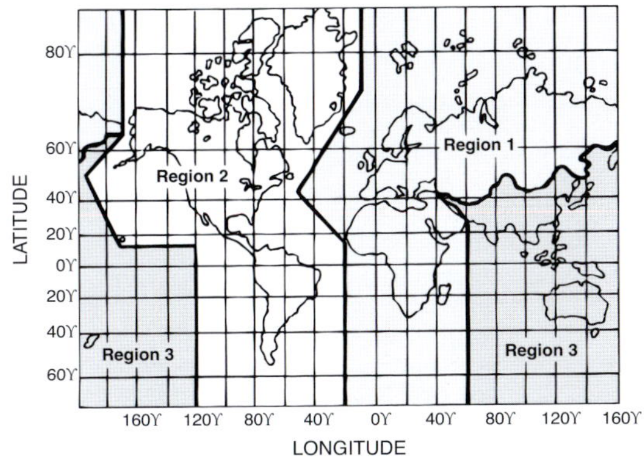

List of Countries Permitting Third-Party Traffic

Country	Call Sign Prefix	Country	Call Sign Prefix	Country	Call Sign Prefix
Antigua/Barbuda	V2	Gambia	C5	Peru	OA-OC
Argentina	LO-LW	Ghana	9G	Philippines	DU-DZ
Australia	VK	Grenada	J3	Pitcairn Island	VR6
Belize	V3	Guatemala	TG	St. Kitts/Nevis	V4
Bolivia	CP	Guyana	8R	St. Lucia	J6
Bosnia-Herzegovina	E7	Haiti	HH	St. Vincent & Grenadines	J8
Brazil	PP-PY	Honduras	HQ-HR	Sierra Leone	9L
Canada	VE, VO, VY	Israel	4X, 4Z	South Africa	ZR-ZU
Chile	CA-CE	Jamaica	6Y	Swaziland	3DA
Colombia	HJ-HK	Jordan	JY	Trinidad/Tobago	9Y-9Z
Comoros	D6	Liberia	EL	Turkey	TA-TC
Costa Rica	TI, TE	Marshall Islands	V7	United Kingdom	GB
Cuba	CM, CO	Mexico	XA-XI	Uruguay	CV-CX
Dominica	J7	Micronesia	V6	Venezuela	YV-YY
Dominican Republic	HI	Nicaragua	YN	ITU – Geneva	4U1ITU
Ecuador	HC-HD	Panama	HO-HP	VIC – Vienna	4U1VIC
El Salvador	YS	Paraguay	ZP		

Countries Holding U.S. Reciprocal Agreements

Antigua, Barbuda
Argentina
Australia
Austria
Bahamas
Barbados
Belgium
Belize
Bolivia
Bosnia-Herzegovina
Botswana
Brazil
Canada[1]
Chile
Colombia
Costa Rica
Croatia
Cyprus
Denmark
Dominica
Dominican Rep.
Ecuador
El Salvador
Fiji
Finland
France[2]
Germany
Greece
Greenland
Grenada
Guatemala
Guyana
Haiti
Honduras
Iceland
India
Indonesia
Ireland
Israel
Italy
Jamaica
Japan
Jordan
Kiribati
Kuwait
Liberia
Luxembourg
Macedonia
Marshall Is.
Mexico
Micronesia
Monaco
Netherlands
Netherlands Ant.
New Zealand
Nicaragua
Norway
Panama
Paraguay
Papua New Guinea
Peru
Philippines
Portugal
Seychelles
Sierra Leone
Solomon Islands
South Africa
Spain
St. Lucia
St. Vincent and Grenadines
Surinam
Sweden
Switzerland
Thailand
Trinidad, Tobago
Turkey
Tuvalu
United Kingdom[3]
Uruguay
Venezuela

1. Do not need reciprocal permit
2. Includes all French Territories
3. Includes all British Territories

POPULAR Q SIGNALS

Given below are a number of Q signals whose meanings most often need to be expressed with brevity and clarity in amateur work. (Q abbreviations take the form of questions only when each is sent followed by a question mark.)

QRG Will you tell me my exact frequency (or that of _____)? Your exact frequency (or that of _____) is _____ kHz.
QRH Does my frequency vary? Your frequency varies.
QRI How is the tone of my transmission? The tone of your transmission is _____ (1. Good; 2. Variable; 3. Bad).
QRJ Are you receiving me badly? I cannot receive you. Your signals are too weak.
QRK What is the intelligibility of my signals (or those of _____)? The intelligibility of your signals (or those of _____) is _____ (1. Bad; 2. Poor; 3. Fair; 4. Good; 5. Excellent).
QRL Are you busy? I am busy (or I am busy with _____). Please do not interfere.
QRM Is my transmission being interfered with? Your transmission is being interfered with _____ (1. Nil; 2. Slightly; 3. Moderately; 4. Severely; 5. Extremely).
QRN Are you troubled by static? I am troubled by static _____ (1-5 as under QRM).
QRO Shall I increase power? Increase power.
QRP Shall I decrease power? Decrease power.
QRQ Shall I send faster? Send faster (_____ WPM).
QRS Shall I send more slowly? Send more slowly (_____ WPM).
QRT Shall I stop sending? Stop sending.
QRU Have you anything for me? I have nothing for you.
QRV Are you ready? I am ready.
QRW Shall I inform _____ that you are calling on _____ kHz? Please inform _____ that I am calling on _____ kHz.
QRX When will you call me again? I will call you again at _____ hours (on _____ kHz).
QRY What is my turn? Your turn is numbered _____.
QRZ Who is calling me? You are being called by _____ (on _____ kHz).
QSA What is the strength of my signals (or those of _____)? The strength of your signals (or those of _____) is _____ (1. Scarcely perceptible; 2. Weak; 3. Fairly good; 4. Good; 5. Very good).
QSB Are my signals fading? Your signals are fading.
QSD Is my keying defective? Your keying is defective.
QSG Shall I send _____ messages at a time? Send _____ messages at a time.
QSK Can you hear me between your signals and if so can I break in on your transmission? I can hear you between my signals; break in on my transmission.
QSL Can you acknowledge receipt? I am acknowledging receipt.
QSM Shall I repeat the last message which I sent you, or some previous message? Repeat the last message which you sent me [or message(s) number(s) _____].
QSN Did you hear me (or _____) on _____ kHz? I heard you (or _____) on _____ kHz.
QSO Can you communicate with _____ direct or by relay? I can communicate with _____ direct (or by relay through _____).
QSP Will you relay to _____? I will relay to _____.
QST General call preceding a message addressed to all amateurs and ARRL members. This is in effect "CQ ARRL."
QSU Shall I send or reply on this frequency (or on _____ kHz)?
QSW Will you send on this frequency (or on _____ kHz)? I am going to send on this frequency (or on _____ kHz).
QSX Will you listen to _____ on _____ kHz? I am listening to _____ on _____ kHz.
QSY Shall I change to transmission on another frequency? Change to transmission on another frequency (or on _____ kHz).
QSZ Shall I send each word or group more than once? Send each word or group twice (or _____ times).
QTA Shall I cancel message number _____? Cancel message number _____.
QTB Do you agree with my counting of words? I do not agree with your counting of words. I will repeat the first letter or digit of each word or group.
QTC How many messages have you to send? I have messages for you (or for _____).
QTH What is your location? My location is _____.
QTR What is the correct time? The time is _____.

Source: ARRL

Glossary

Advanced: An amateur operator who has passed Element 2, 3A, 3B, and 4A written theory examinations and Element 1A and 1B code tests to demonstrate Morse code proficiency to 13 wpm.

Amateur communication: Non-commercial radio communication by or among amateur stations solely with a personal aim and without personal or business interest.

Amateur operator/primary station license: An instrument of authorization issued by the Federal Communications Commission comprised of a station license, and also incorporating an operator license indicating the class of privileges.

Amateur operator: A person holding a valid license to operate an amateur station issued by the FCC. Amateur operators are frequently referred to as ham operators.

Amateur Radio services: The amateur service, the amateur-satellite service and the radio amateur civil emergency service.

Amateur-satellite service: A radiocommunication service using stations on Earth satellites for the same purpose as those of the amateur service.

Amateur service: A radiocommunication service for the purpose of self-training, intercommunication and technical investigations carried out by amateurs; that is, duly authorized persons interested in radio technique solely with a personal aim and without pecuniary interest.

Amateur station: A station licensed in the amateur service embracing necessary apparatus at a particular location used for amateur communication.

AMSAT: Radio Amateur Satellite Corporation, a non-profit scientific organization.
(850 Sligo Avenue, Silver Spring, MD 20910-4703)

Antenna gain: The increase as a result of the physical construction of the antenna which confines the radiation to desired or useful directions. Usually specified in dB, referenced to the gain of a dipole.

ARES: The emergency division of the American Radio Relay League. See RACES.

ARRL: American Radio Relay League, national organization of U.S. Amateur Radio operators. (225 Main Street, Newington, CT 06111)

ATV: Amateur fast-scan television.

Audio Frequency (AF): The range of frequencies that can be heard by the human ear, generally 20 hertz to 20 kilohertz.

Authorized bandwidth: The allowed frequency band, specified in kilohertz, and centered on the carrier frequency.

Automatic control: The use of devices and procedures for station control without the control operator being present at the control point when the station is transmitting.

Automatic Position Reporting System (APRS): Links a remote GPS unit to a ham transceiver which transmits the location to a base station.

Automatic Volume Control (AVC): A circuit that continually maintains a constant audio output volume in spite of deviations in input signal strength.

Beam or Yagi antenna: An antenna array that receives or transmits RF energy in a particular direction. Usually rotatable.

Block diagram: A simplified outline of an electronic system where circuits or components are shown as boxes.

Broadcasting: Information or programming transmitted by means of radio intended for the general public.

Business communications: Any transmission or communication the purpose of which is to facilitate the regular business or commercial affairs of any party. Business communications are prohibited in the amateur service.

Call Book: A published list of all licensed amateur operators available in North American and Foreign editions.

Call sign assignment: The FCC systematically assigns each amateur station their primary call sign.

Carrier frequency: The frequency of an unmodulated electromagnetic wave, usually specified in kilohertz or megahertz.

Certificate of Successful Completion of Examination (CSCE): A certificate verifying successful completion of a written and/or Morse code examination Element. Credit for passing is valid for 365 days.

Coaxial cable, Coax: A concentric, two conductor cable in which one conductor surrounds the other, separated by an insulator.

Control point: Any place from which a transmitter's function may be controlled.

Control station: A station directly associated with the control point of a radio system, commonly used to control repeaters.

Control operator: An amateur operator designated by the licensee of an amateur station to be responsible for the station transmissions.

Coordinated repeater station: An amateur repeater station for which the transmitting and receiving frequencies have been recommended by the recognized repeater coordinator.

Coordinated Universal Time (UTC): Some-times referred to as Greenwich Mean Time, UCT or Zulu time. The time at the zero-degree (0∞) Meridian which passes through Greenwich, England. A universal time among all amateur operators.

Crystal: A quartz or similar material which has been ground to produce natural vibrations of a specific frequency. Quartz crystals produce a high-degree of frequency stability in radio transmitters.

CW: Continuous wave, another term for the International Morse code.

Dipole antenna: The most common wire antenna. Length is equal to one-half of the wavelength.

Direct Digital Frequency Synthesizer: Modern frequency synthesizer employing technology to reduce phase noise to an insignificant level.

Dummy antenna: A device or resistor which serves as a transmitter's antenna without radiating radio waves. Generally used to tune up a radio transmitter.

Duplex: Transmitting on one frequency, and receiving on another, commonly used in mobile telephone use, as well as mobile business radio on UHF frequencies.

Duplexer: A device that allows a single antenna to be simultaneously used for both reception and transmission.

Effective Radiated Power (ERP): The product of the transmitter (peak envelope) power, expressed in watts, delivered to the antenna, and the relative gain of an antenna over that of a half-wave dipole antenna.

Emergency communication: Any amateur communication directly relating to the immediate safety of life of individuals or the immediate protection of property.

Examination Element: The written theory exams and Morse code test required to obtain amateur radio licenses. Technicians are required to pass written Element 2 theory; Generals are required to pass written Element 3 theory plus Element 1 Morse code test at five (5) words-per-minute; and Extras are required to pass written Element 4 theory.

FCC Form 605: The universal application form used to renew or modify an existing license. (See NCVEC Form 605)

Federal Communications Commission (FCC): A board of five Commissioners, appointed by the President, having the power to regulate wire and radio telecommunications in the United States.

Feedline transmission line: A system of conductors that connects an antenna to a receiver or transmitter.

Field Day: Annual activity sponsored by the ARRL to demonstrate emergency preparedness of amateur operators.

Filter: A device used to block or reduce alternating currents or signals at certain frequencies while allowing others to pass unimpeded.

Frequency: The number of cycles of alternating current in one second.

Frequency coordinator: An individual or organization recognized by amateur operators eligible to engage in repeater operation that recommends frequencies and other operating and/or technical parameters for amateur repeater operation in order to avoid or minimize potential interferences.

Frequency Modulation (FM): A method of varying a radio carrier wave by causing its frequency to vary in accordance with the information to be conveyed.

Frequency privileges: The transmitting frequency bands available to the various classes of amateur operators. The privileges are listed in Part 97.7(a) of the FCC rules.

General: An amateur operator who has passed the Element 3 written theory examination and Element 1 code test to demonstrate Morse code proficiency to 5-wpm.

Ground: A connection, accidental or intentional, between a device or circuit and the earth or some common body and the earth or some common body serving as the earth.

Ground wave: A radio wave that is propagated near or at the earth's surface.

Half-Duplex: A method of operation on a duplex (separate transmit, separate receive) frequency pair where the operator only receives when the microphone button is released. Half-duplex does not permit simultaneous talk and listen.

Handi-Ham system: Amateur organization dedicated to assisting handicapped amateur operators. (3915 Golden Valley Road, Golden Valley, MN 55422)

Harmful interference: Interference which seriously degrades, obstructs or repeatedly interrupts the operation of a radio communication service.

Harmonic: A radio wave that is a multiple of the fundamental frequency. The second harmonic is twice the fundamental frequency, the third harmonic, three times, etc.

Hertz: One complete alternating cycle per second, abbreviated Hz. Named after Heinrich R. Hertz, a German physicist. The number of hertz is the frequency of the audio or radio wave.

High Frequency (HF): The band of frequencies that lie between 3 and 30 megahertz. It is from these frequencies that radio waves are returned to earth from the ionosphere.

High-Pass filter: A device that allows passage of high frequency signals but attenuates the lower frequencies.

Hysteresis: a lagging in the values of resulting magnetization in a magnetic material (as iron) due to a changing magnetizing force

Interference: The effect that occurs when two or more radio stations are transmitting at the same time. This includes undesired noise or other radio signals on the same frequency.

International Telecommunication Union (ITU): World organization composed of delegates from member countries, which allocates frequency spectrum for various purposes, including amateur radio.

Ionosphere: Outer limits of atmosphere from which HF amateur communications signals are returned to earth.

Jamming: The intentional, malicious interference with another radio signal.

Key clicks, Chirps: Defective keying of a telegraphy signal sounding like tapping or high varying pitches.

Lid: Amateur slang term for poor radio operator.

Linear amplifier: A device that accurately reproduces a radio wave in magnified form.

Long wire: A horizontal wire antenna that is one wavelength or longer in length.

Low-Pass filter: A device that allows passage of low frequency signals but attenuates the higher frequencies.

Machine: A ham slang word for an automatic repeater station.

Malicious interference: Willful, intentional jamming of radio transmissions.

MARS: The Military Affiliate Radio System. An organization that coordinates the activities of amateur communications with military radio communications.

Maximum authorized transmitting power: Amateur stations must use no more than the maximum transmitter power necessary to carry out the desired communications.

Maximum usable frequency (MUF): The highest frequency that will be returned to earth from the ionosphere.

Medium frequency (MF): The band of frequencies that lies between 300 and 3,000 kHz (3 MHz).

Mobile operation: Radio communications conducted while in motion or during halts at unspecified locations.

Mode: Type of transmission such as voice, teletype, code, television, facsimile.

Modulate: To vary the amplitude, frequency or phase of a radio frequency wave in accordance with the information to be conveyed.

Monolithic Microwave Integrated Circuit (MMIC): A four-legged device designed for RF amplification. Well-suited for VHF, UHF and microwave applications.

Morse code (see CW): The International Morse code, A1A emission. Interrupted continuous wave communications conducted using a dot-dash code for letters, numbers and operating procedure signs.

NCVEC Form 605: The form provided by VECs that is used to apply for an amateur radio license or license upgrade.

Noise level: The strengths of extraneous audible sounds in a given location, usually measured in decibels.

Novice operator: An FCC licensed, entry-level operator in the amateur service. Novices may operate a transmitter in the following meter wavelength bands: 80, 40, 15, 10, 1.25 and 0.23.

Ohm's law: The basic electrical law explaining the relationship between voltage, current and resistance. The current I in a circuit is equal to the voltage E divided by the resistance R, or I = E/R.

OSCAR: Acronym for "Orbiting Satellite Carrying Amateur Radio," the name given to a series of satellites designed and built by amateur operators of several nations.

Oscillator: A device for generating oscillations or vibrations of an audio or radio frequency signal.

Output power: The radio-frequency output power of a transmitter's final radio-frequency stage as measured at the output terminal while connected to a load of the impedance recommended by the manufacturer.

Packet radio: A digital method of communicating computer-to-computer. A terminal-node controller makes up the packet of data and directs it to another packet station.

Peak Envelope Power (PEP): 1. The power during one radio frequency cycle at the crest of the modulation envelope, measured under normal operating conditions. 2. The maximum power that can be obtained from a transmitter.

Phone patch: Interconnection of amateur service to the public switched telephone network, and operated by the control operator of the station.

Power supply: A device or circuit that provides the appropriate voltage and current to another device or circuit.

Propagation: The travel of electromagnetic waves or sound waves through a medium.

Q-signals: International three-letter abbreviations beginning with the letter Q used primarily to convey information using the Morse code.

QSL Bureau: An office that bulk processes QSL (radio confirmation) cards for (or from) foreign amateur operators as a postage saving mechanism.

RACES (radio amateur civil emergency service): A radio service using amateur stations for civil defense communications during periods of local, regional, or national civil emergencies.

Radiation: Electromagnetic energy, such as radio waves, traveling forth into space from a transmitter.

Radio Frequency (RF): The range of frequencies over 20 kilohertz that can be propagated through space.

Radio wave: A combination of electric and magnetic fields varying at a radio frequency and traveling through space at the speed of light.

Repeater operation: Automatic amateur stations that retransmit the signals of other amateur stations.

Repeater station: An intermediate station in a system which is arranged to receive a signal from a station, amplify and retransmit the signal to another station. Usually performs this function in both directions, simultaneously.

RST Report: A telegraphy signal report system of Readability, Strength and Tone.

S-meter: A voltmeter calibrated from 0 to 9 that indicates the relative signal strength of an incoming signal at a radio receiver.

Selectivity: The ability of a circuit (or radio receiver) to separate the desired signal from those not wanted.

Sensitivity: The ability of a circuit (or radio receiver) to detect a specified input signal.

Short circuit: An unintended, low-resistance connection across a voltage source resulting in high current and possible damage.

Shortwave: The high frequencies that lie between 3 and 30 megahertz that are propagated long distances.

Simplex: A method of operation of a communication circuit which can receive or transmit, but not both simultaneously. Thus, system stations are operating on the same transmit and receive frequency.

Single-Sideband (SSB): A method of radio transmission in which the RF carrier and one of the sidebands is suppressed and all of the information is carried in the one remaining sideband.

Skip wave, Skip zone: A radio wave reflected back to earth. The distance between the radio transmitter and the site of a radio wave's return to earth.

Sky wave: A radio wave that is refracted back to earth in much the same way that a stone thrown across water skips out. Sometimes called an ionospheric wave.

Spectrum: A series of radiated energies arranged in order of wavelength. The radio spectrum extends from 20 kilohertz upward.

Sporadic-E: Radio propagation caused by refraction of signals by dense ionization at the E-layer of the ionosphere.

Spread spectrum: Modulation method which employs lightning-fast frequency "hopping" to allow multiple users to share a band.

Spurious Emissions: Unwanted radio frequency signals emitted from a transmitter that sometimes causes interference.

Station license, location: No transmitting station shall be operated in the amateur service without being licensed by the FCC. Each amateur station shall have one land location, the address of which appears on the station license.

Sunspot Cycle: An 11-year cycle of solar disturbances which greatly affects radio wave propagation.

TAPR: Tucson Amateur Packet Radio Corporation, a non-profit research and development organization. (8987-309 E. Tangue Verde Rd. #337, Tucson, AZ 85749-9399)

Technician operator: An amateur radio operator who has successfully passed Element 2. This operator has not passed Element 1, the 5-wpm code test.

Technician with Code Credit: An amateur operator who has passed a 5-wpm code test in addition to Technician Class requirements.

Telecommunications: The electrical conversion, switching, transmission and control of audio signals by wire or radio. Also includes video and data communications.

Telegraphy: Telegraphy is communications transmission and reception using CW International Morse Code.

Telephony: Telephony is communications transmission and reception in the voice mode.

Temporary operating authority: Authority to operate your amateur station while awaiting arrival of an upgraded license.

Terrestrial station location: Any location within the major portion of the Earth's atmosphere, including air, sea and land locations.

Third-party traffic: Amateur communication by or under the supervision of the control operator at an amateur station to another amateur station on behalf of others.

Transceiver: A combination radio transmitter and receiver.

Transmatch: An antenna tuner used to match the impedance of the transmitter output to the transmission line of an antenna.

Transmitter: Equipment used to generate radio waves. Most commonly, this radio carrier signal is amplitude or frequency modulated with information and radiated into space.

Transmitter power: The average peak envelope power (output) present at the antenna terminals of the transmitter. The term "transmitted" includes any external radio frequency power amplifier which may be used.

Ultra High Frequency (UHF): Ultra high frequency radio waves that are in the range of 300 to 3,000 MHz.

Upper Sideband (USB): The operating mode for sideband transmissions where the carrier and lower sideband are suppressed.

Vanity call sign: Custom-chosen call sign instead of "next-in-sequence" assignment from the FCC.

Very High Frequency (VHF): Very high frequency radio waves that are in the range of 30 to 300 MHz.

Volunteer Examiner: An amateur operator of at least a General Class level who administers or prepares amateur operator license examinations. A VE must be at least 18 years old and not related to the applicant.

Volunteer Examiner Coordinator (VEC): A member of an organization which has entered into an agreement with the FCC to coordinate the efforts of volunteer examiners in preparing and administering examinations for amateur operator licenses.

Index

About ARRL

We're the American Radio Relay League, Inc. — better known as ARRL. We're the largest membership association for the amateur radio hobby and service in the US. For over 100 years, we have been the primary source of information about amateur radio, offering a variety of benefits and services to our members, as well as the larger amateur radio community. We publish books on amateur radio, as well as four magazines covering a variety of radio communication interests. In addition, we provide technical advice and assistance to amateur radio enthusiasts, support several education programs, and sponsor a variety of operating events.

One of the primary benefits we offer to the ham radio community is in representing the interests of amateur radio operators before federal regulatory bodies advocating for meaningful access to the radio spectrum. ARRL also serves as the international secretariat of the International Amateur Radio Union, which performs a similar role internationally, advocating for amateur radio interests before the International Telecommunication Union and the World Administrative Radio Conferences.

Today, we proudly serve nearly 150,000 members, both in the US and internationally, through our national headquarters and flagship amateur radio station, W1AW, in Newington, Connecticut. Every year we welcome thousands of new licensees to our membership, and we hope you will join us. Let us be a part of your amateur radio journey. Visit www.arrl.org/join for more information.

ARRL The National Association for Amateur Radio®

225 Main Street
Newington, CT 06111-1400 USA
Tel: 860-594-0200
FAX: 860-594-0259
Email: membership@arrl.org

www.arrl.org

Membership Application

ARRL The National Association for Amateur Radio®

☐ **New** ☐ **Renew** ☐ **Previous Member** ☐ **Unlicensed**

Name ______________________________ Call Sign ______________

Address ____

City______________________ State ________ ZIP ____________

Email ______________________________ Phone ______________

Date of Birth _____ / _____ / _____

My Family Member is Joining or Renewing: ($12 per member)

Name ______________________________ Call Sign ______________

Name ______________________________ Call Sign ______________

☐ Please note my new address ☐ I do not want my name and address made available for non-ARRL related mailings

Your Annual Membership Dues*

Circle Your Choice (rates effective Jan. 1, 2024)

	1 Year	3 Years
Standard membership	$59	$174
Family	$12	$36
Student (must be under age 26)	$30	
Blind (requires one-time statement of legal blindness)	$12	$36

Add-on ARRL Subscriptions

QST, ARRL's membership journal for active radio amateurs.

☐ 1 Year $25* ☐ 3 Years $75*

On the Air, For beginner-to-intermediate-level radio amateurs.

☐ 1 Year $25* ☐ 3 Years $75*

Member Benefits

Your membership supports benefits, services, and programs that keep you active and on the air.

Membership Includes:

- Access to four digital magazines and archives (*QST*, *On the Air*, *QEX*, & *NCJ*)
- Unlimited courses through the ARRL Learning Center (learn.arrl.org)
- Logbook of The World®, contests, and award programs

...and more!

*A print subscription for *QST* and/or *On the Air* requires an ARRL membership. Dues and subscription rates are subject to change without notice and are non-refundable

Payment Information

$______________ Total Charge to: ☐ Visa ☐ MasterCard ☐ AmEx ☐ Discover ☐ Check Enclosed

Card Number ______________________________ Expiration Date ______________

Card Holder's Signature ____

Toll Free (US) 1-888-277-5289 or 860-594-0200 • ARRL, 225 Main St., Newington, CT 06111-1400

membership@arrl.org • www.arrl.org/join

CLUB

form rev 1/24

US Amateur Radio Bands

Operator license classes: **E** = Amateur Extra **A** = Advanced **G** = General **T** = Technician **N** = Novice
CW operation is permitted throughout all amateur bands. Except as noted, all frequencies are in megahertz (MHz).

= RTTY, data, phone, image = USB phone, RTTY, data and CW = RTTY and data = phone and image = SSB phone = CW only

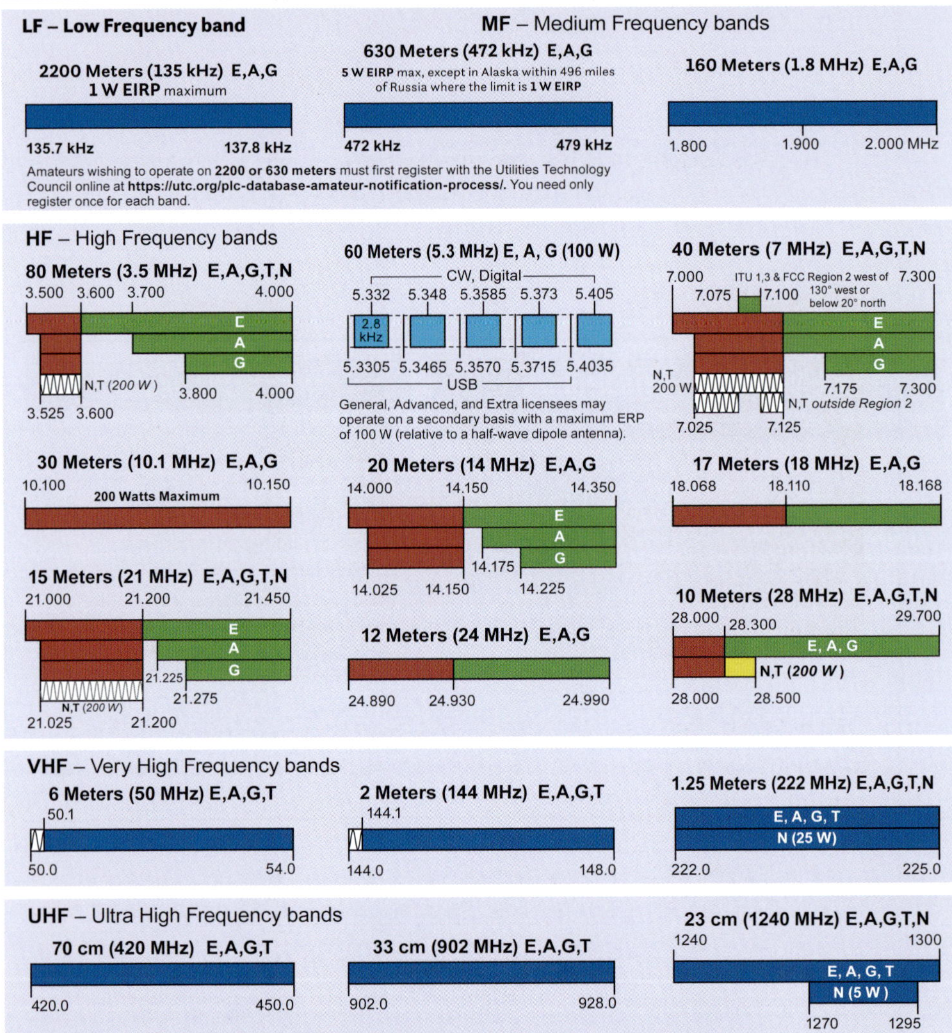

SHF&EHF – Super and Extremely High Frequency bands

All licensees except Novices are authorized all modes on the following frequencies:

2300-2310 MHz	3300-3450 MHz	10.0-10.5 GHz	47.0-47.2 GHz	122.25-123.0 GHz	241-250 GHz
2390-2450 MHz	5650-5925 MHz	24.0-24.25 GHz	76.0-81.0 GHz	134-141 GHz	All above 275 GHz

See www.arrl.org/band-plan for detailed band plans.

OTAbands **rev. 11/16/2023**

Notes

Notes

Notes

Notes